BEST MANAGEMENT PRACTICES

FOR DRIP IRRIGATED CROPS

Research Advances in Sustainable Micro Irrigation

VOLUME 6

BEST MANAGEMENT PRACTICES FOR DRIP IRRIGATED CROPS

Edited by
Kamal Gurmit Singh, PhD
Megh R. Goyal, PhD, PE
Ramesh P. Rudra, PhD, PE

Apple Academic Press Inc. | Apple Academic Press Inc.
3333 Mistwell Crescent | 9 Spinnaker Way
Oakville, ON L6L 0A2 | Waretown, NJ 08758
Canada | USA

© 2016 by Apple Academic Press, Inc.

First issued in paperback 2021

Exclusive worldwide distribution by CRC Press, a member of Taylor & Francis Group

No claim to original U.S. Government works

ISBN-13: 978-1-77463-371-7 (pbk)
ISBN-13: 978-1-77188-095-4 (hbk)

Library and Archives Canada Cataloguing in Publication

Best management practices for drip irrigated crops / edited by Kamal Gurmit Singh, PhD, Megh R. Goyal, PhD, PE, Ramesh P. Rudra, PhD, PE.

(Research advances in sustainable micro irrigation; volume 6)
Includes bibliographical references and index.
ISBN 978-1-77188-095-4 (bound)
1. Microirrigation. 2. Microirrigation--Management. 3. Sustainable agriculture. I. Singh, Kamal Gurmit, author, editor II. Goyal, Megh Raj, editor III. Rudra, R. P. (Ramesh Pall), author, editor IV. Series: Research advances in sustainable micro irrigation; v. 6

S619.T74B48 2015   631.5'87   C2015-903833-2

Library of Congress Cataloging-in-Publication Data

Singh, Kamal Gurmit, author.
Best management practices for drip irrigated crops / authors: Kamal Gurmit Singh, PhD, Megh R. Goyal, PhD, PE, Ramesh P. Rudra, PhD, PE. -- 1st ed.

pages cm. -- (Research advances in sustainable micro irrigation ; volume 6)
Includes bibliographical references and index.
ISBN 978-1-77188-095-4 (alk. paper)
1. Microirrigation--Handbooks, manuals, etc. I. Goyal, Megh Raj, author. II. Rudra, Ramesh P., author. III. Title. IV. Series: Research advances in sustainable micro irrigation ; v. 6.

S619.T74.S562 2015   631.5'87--dc23   2015019670

Apple Academic Press also publishes its books in a variety of electronic formats. Some content that appears in print may not be available in electronic format. For information about Apple Academic Press products, visit our website at **www.appleacademicpress.com** and the CRC Press website at **www.crcpress.com**

CONTENTS

LIST OF CONTRIBUTORS

Pooja Behal
Ex-undergraduate Student, Department of Soil and Water Engineering, Punjab Agricultural University, Ludhiana. India–141004, Phone: +919646111866

Nilesh Biwalkar, PhD
Assistant Professor, Department of Soil and Water Engineering, Punjab Agricultural University, Ludhiana. India–141004, Mobile: +918146977883, E-mail: nileshbiwalkar@gmail.com

Amanpreet Chawla, PhD
Research Associate, Department of Soil and Water Engineering, Punjab Agricultural University, Ludhiana. India–141004. Mobile: +918872109021, E-mail: aman_chawlapau@yahoo.com

Pradeep K. Goel, PhD
Senior Surface Water Scientist, Water Monitoring and Reporting Section, Environmental Monitoring and Reporting Branch of Ontario Ministry of the Environment, 125 Resources Road, West Wing, Etobicoke, ON, M9P 3 V6, Canada, Tel: 001–4162356060, E-mail: pradeep.goel@ontario.ca

Megh R. Goyal, PhD
Retired Professor in Agricultural and Biomedical Engineering, University of Puerto Rico – Mayaguez Campus; and Senior Technical Editor-in-Chief in Agriculture Sciences and Biomedical Engineering, Apple Academic Press Inc., PO Box 86, Rincon – PR – 00677 – USA. E-mail: goyalmegh@gmail.com

H. A. W. S. Gunathilake, PhD
Director, Sri Lanka Institute of Technological Education, NO 320, T.B Jaya Mawatha, Colombo 10, Sri Lanka. E-mail: haws.gunathilaka@gmail.com

Marvin E. Jensen, PhD, PE
Retired Research Leader at USDA – ARS. 1207 Spring Wood Drive, Fort Collins, Colorado 80525. E-mail: mjensen419@aol.com

Pawan Preet Kaur
Ex-undergraduate Student, Department of Soil and Water Engineering, Punjab Agricultural University, Ludhiana. India–141004

Arun Kaushal, PhD
Associate Professor, Department of Soil and Water Engineering, Punjab Agricultural University, Ludhiana. India–141004, Mobile: +919855338437, E-mail: arunkaushalarun@rediffmail.com

Ashwani Kumar, PhD
Director, Directorate of Water Management, ICAR, Opposite Rail Vihar, Chandrasekharpur, Bhubaneswar, 751023, Odisha, India. Tel.: 91-0674-2300060. E-mail: director.dwm@icar.org.in; ashwani_wtcer@yahoo.com

Santosh Kumar, MS
Ex graduate Student, Department of Soil and Water Engineering, Punjab Agricultural University, Ludhiana. India–141004

Satyendra Kumar, PhD
Senior Scientist, CSSRI Zarifa Farm, Kachawa Road, Karnal – 132001, India. Tel.: 91-184-2292004.
E-mail: skumar@cssri.ernet.in

A. S. Lodhi, PhD
Research Scholar (Hydraulics Eng.), Department of Civil Engineering, IIT-Roorkee-247667, Uttarkhand,
India. Mobile: +919548759089. E-mail: ajay0312@gmail.com

Gulshan Mahajan, PhD
Agronomist (Rice), Department of Plant Breeding and Genetics, Punjab Agricultural University, Ludhiana. India–141004. Mobile: +919417352312. E-mail: mahajangulshan@rediffmail.com

V. M. Mayande, PhD
Vice Chancellor, Dr. Panjabrao Deshmukh Krishi Vidyapeeth, Akola, 444104, Maharashtra-India. Phone:
+91 9423174299, E-mail: vmmayande@yahoo.com

Sushant Mehan
Ex Graduate Student, Department of Soil and Water Engineering, College of Agricultural Engineering
and Technology, Punjab Agricultural University, Ludhiana, India–141004. Tel.: +91- 9592772996. E-
mail: sushantmehan@gmail.com

Miguel A. Muñoz-Muñoz, PhD
Ex-President of University of Puerto Rico, University of Puerto Rico, Mayaguez Campus, College of
Agriculture Sciences, Call Box 9000, Mayagüez, PR. 00681–9000. Tel.: 787-265-3871, E-mail: miguel.
munoz3@upr.edu

D. D. Nangare, PhD
Scientist (SWCE), National Institute of Abiotic Stress Management, Malegaon, Baramati, Pune, Maha-
rashtra-413115. Mobile: +9109665204895. E-mail: ddnangare@rediffmail.com

Rahul Patole, MS
Ex-Graduate Student, Department of Soil and Water Engineering, Punjab Agricultural University, Ludhiana. India–141004. Mobile: +91 8968459044; E-mail: rsp1887@gmail.com

Ramesh P. Rudra, PhD, PE
Professor in Water Resources Engineering, University of Guelph, Guelph, Ontario N1G2W1, Canada.
Phone: 001–519824-4120 ext 52110. Fax: 001–5198360227. E-mail: rrudra@uoguelpgh.ca

A. K. Saini, PhD
Ex Research Engineer, Department of Soil and Water Engineering, Punjab Agricultural University, Ludhiana. India–141004. Mobile: 001-416-524-7030. E-mail: aksaini_2000@rediffmail.com

Sharda Rakesh, PhD
Extension Specialist, Department of Soil and Water Engineering, Punjab Agricultural University, Ludhiana, India–141004. Mobile: +919855545189. E-mail: rakeshadra@yahoo.com

Mukesh Siag, PhD
Associate Professor, Department of Soil and Water Engineering, Punjab Agricultural University, Ludhiana. India–141004. Mobile: +919872999640, E-mail: siagmukesh@rediffmail.com

Angrej Singh, PhD
Agronomist, Department of Soil and Water Engineering, Punjab Agricultural University, Ludhiana. India–141004. Mobile: +9181463888111. E-mail: angrejsingh30@yahoo.com

Gajendra Singh, PhD
Former Vice President, Asian Institute of Technology, Thailand. C-86, Millennium Apartments, Plot
E-10A, Sector −61, NOIDA – U.P. – 201301, India, Mobile: (011)-(91) 9971087591, E-mail: <prof.
gsingh@gmail.com>

Kamal Gurmit Singh, PhD
Senior Research Engineer, Department of Soil and Water Engineering, Punjab Agricultural University, Ludhiana. India–141004. E-mail: kamalgurmit@yahoo.com

S. R. Singh, PhD
Retired Professor, House No.561, Sector 2, Udyan II, ELDECO Colony, Rae Bareli Road, Lucknow. Mobile: +919935631990. E-mail: strmsingh@yahoo.com

Chetan, Singla, MS
Assistant Agricultural Engineer, Director (Farm), Punjab Agricultural University, Ludhiana, India 141004, Phone: +91-9872034222. E-mail: chetan_singla@yahoo.com

R. K. Sivanappan, PhD
Former Professor and Dean, College of Agricultural Engineering and Technology, Tamil Nadu Agricultural University (TNAU), Coimbatore. Mailing address: Consultant, 14, Bharathi Park, 4th Cross Road, Coimbatore-641043, India. E-mail: sivanappanrk@hotmail.com

A. K. Tiwari, PhD
Ex Professor, Department of Soil and Water Engineering, Punjab Agricultural University, Ludhiana, India–141004

LIST OF ABBREVIATIONS

ASABE	American Society of Agricultural and Biological Engineers
CU	Coefficient of Uniformity
DIS	Drip Irrigation System
DOY	Day of the Year
EPAN	Pan Evaporation
FAO	Food and Agricultural Organization, Rome
FC	Field Capacity
FUE	Fertilizers Use Efficiency
GPIS	Gated Pipe Irrigation System
gpm	Gallons per Minute
ICAR	Indian Council of Agriculture Research
IR	Water Intake Rate Into the Soil
ISAE	Indian Society of Agricultural Engineers
LAI	Leaf Area Index
lps	Liters per Second
lph	Liters per Hour
MAD	Maximum Allowable Depletion
MSL	Mean Sea Level
MWD	Mean Weight Diameter
PE	Polyethylene
PET	Potential Evapotranspiration
PM	Penman-Monteith
ppm	One Part Per Million
psi	Pounds Per Square Inch
PVC	Polyvinyl Chloride
PWP	Permanent Wilting Point
RA	Extraterrestrial Radiation
RH	Relative Humidity
RMSE	Root Mean Squared Error
RS	Solar Radiation
SAR	Sodium Absorption Rate
SDI	Subsurface Drip Irrigation
SRW	Simulated Rain Water
SW	Saline Water
SWB	Soil Water Balance
TE	Transpiration Efficiency

TEW	Total Evaporable Water
TR	Temperature Range
TSS	Total Soluble Solids
TUE	Transpiration Use Efficiency
USDA	US Department of Agriculture
USDA-SCS	US Department of Agriculture-Soil Conservation Service
WSEE	Weighed Standard Error of Estimate
WUE	Water Use Efficiency

LIST OF SYMBOLS

A	cross sectional flow area (L^2)
AL	average life of wells
AW	available water (Θ_w, %)
C	concentration of chlorine wanted, ppm
Cp	specific heat capacity of air, in $J/(g \cdot °C)$
C_v	coefficient of variation
D	accumulative intake rate (mm/min)
d	depth of effective root zone
D	depth of irrigation water (mm)
E	evapotranspiration rate, in $g/(m^2 \cdot s)$
e	vapor pressure, in kPa
e_a	actual vapor pressure (kPa)
Ecp	cumulative class A pan evaporation
eff	irrigation system efficiency
E_i	irrigation efficiency of drip system
E_p	pan evaporation as measured by Class-A pan evaporimeter (mm/day)
E_{pan}	class A pan evaporation
ER	cumulative effective rainfall (mm)
e_s	saturation vapor pressure (kPa)
Es	saturation vapor pressure, in kPa
$e_s - e_a$	vapor pressure deficit (kPa)
ET	evapotranspiration rate, in mm/year
ETa	reference ET, in the same water evaporation units as Ra
ETc	crop-evapotranspiration (mm/day)
ET_o	the reference evapotranspiration obtained (mm/day)
ET_{pan}	the pan evaporation-derived evapotranspiration
EU	emission uniformity
F	flow rate of the system (GPM)
F.C.	field capacity (v/v, %)
G	soil heat flux at land surface, in W/m^2
H	plant canopy height (m)
h	soil water pressure head (L)
I	infiltration rate at time t (mm/min)
IR	injection rate, GPH
IRR	irrigation
K	unsaturated hydraulic conductivity (LT^{-1})
K_c	crop coefficient

Kc	crop-coefficient for bearing 'Kinnow' plant
Kg	kilograms
K_p	pan coefficient
Kp	pan factor
Ks	hydraulic conductivity
n	number of emitters
P	percentage of chlorine in the solution*
P.W.P.	permanent wilting point ($\ominus_w$%)
Pa	atmospheric pressure, in Pa
pH	acidity/alkalinity measurement scale
Q	flow rate in gallons per minute
q	the mean emitter discharges of each lateral (lh^{-1})
R	rainfall
r_a	aerodynamic resistance (s m^{-1})
Ra	extraterrestrial radiation
R_e	effective rainfall depth (mm)
R_i	individual rain gauge reading in mm
R_{MAX}	maximum relative humidity
R_{MIN}	minimum relative humidity
R_n	net radiation at the crop surface (MJ m$^{-\square}$day^{-1})
RO	surface runoff
Rs	incoming solar radiation on land surface
r_s	bulk surface resistance (s m^{-1})
S	sink term accounting for root water uptake (T^{-1})
Se	effective saturation
S_p	plant-to-plant spacing (m)
S_r	row-to-row spacing (m)
SU	statistical uniformity (%)
S_ψ	water stress integral (MPa day)
t	time that water is on the surface of the soil (min)
T	time in hours
T_{MAX}	maximum temperature
T_{MIN}	minimum temperature
V	volume of water required (liter/day/plant)
V_{id}	irrigation volume applied in each irrigation (liter tree^{-1})
V_{pc}	plant canopy volume (m^3)
W	canopy width
W_p	fractional wetted area
z	vertical coordinate positive downwards (L)

GREEK SYMBOLS

α	inverse of a characteristic pore radius (L^{-1})
Δ	slope of the vapor pressure curve ($kPa°C^{-1}$)
γ	psychometric constant ($kPa°C^{-1}$)
θ	volumetric soil water content (L^3L^{-3})
$\theta(h)$	soil water retention (L^3L^{-3})
θr	residual water content (L^3L^{-3})
θ_s	saturated water content (L^3L^{-3})
θ_{vol}	volumetric moisture content (cm^3/cm^3)
λ	latent heat of vaporization ($MJ\ kg^{-1}$)
λE	latent heat flux, in W/mo
ρa	mean air density at constant pressure ($kg\ m^{-3}$)
$\ominus_w$	dry weight basis

PREFACE

Due to increased agricultural production, irrigated land has increased in the arid and subhumid zones around the world. Agriculture has started to compete for water use with industries, municipalities and other sectors. This increasing demand along with increments in water and energy costs have made it necessary to develop new technologies for the adequate management of water. The intelligent use of water for crops requires understanding of evapotranspiration processes and use of efficient irrigation methods.

An informative article was published on the importance of micro irrigation in India (weblink: http://www.newindianexpress.com/cities/bengaluru/Micro-irrigation-to-be-promoted/2013/08/17/article1738597.ece). Every day, similar news appears all around the world indicating that government agencies at central/state/local levels, research and educational institutions, industry, sellers and others are aware of the urgent need to adopt micro irrigation technology, which can have an irrigation efficiency up to 90% compared to 30–40% for conventional irrigation systems. I stress the urgent need to implement micro irrigation systems in water scarcity regions.

Micro irrigation is sustainable and is one of the best management practices. I attended the 17th Punjab Science Congress on February 14–16, 2014 at Punjab Technical University in Jalandhar. I was shocked to know that the underground water table has lowered to a critical level in Punjab. My father-in-law in Dhuri told me that his family bought the 0.10 acres of land in the city for US $100.00 in 1942 because the water table was at 2 feet depth. In 2012, it was sold for US $200,000 because the water table had dropped to greater than 100 feet depth. This has been due to luxury use of water by wheat-paddy farmers. The water crisis is similar in other countries, including Puerto Rico where I live. We can therefore conclude that the problem of water scarcity is rampant globally, creating the urgent need for water conservation. The use of micro irrigation systems is expected to result in water savings, increased crop yields in terms of volume and quality. The other important benefits of using micro irrigation systems include expansion in the area under irrigation, water conservation, optimum use of fertilizers and chemicals through water, and decreased labor costs, among others. The worldwide population is increasing at a rapid rate, and it is imperative that food supply keeps pace with this increasing population.

Micro irrigation, also known as trickle irrigation or drip irrigation or localized irrigation or high frequency or pressurized irrigation, is an irrigation method that saves water and fertilizer by allowing water to drip slowly to the roots of plants, either onto the soil surface or directly onto the root zone, through a network of valves, pipes, tubing, and emitters. It is done through narrow tubes that deliver water directly to the base of the plant. It supplies controlled delivery of water directly to individual plants and can be installed on the soil surface or subsurface. Micro irrigation systems are often used in farms and large gardens but are equally effective in the home garden or even for houseplants or lawns.

The mission of this compendium is to serve as a reference manual for graduate and undergraduate students of agricultural, biological, and civil engineering; horticulture, soil science, crop science, and agronomy. I hope that it will also be a valuable reference for professionals that work with micro irrigation and water management; for professional training institutes, technical agricultural centers, irrigation centers, agricultural extension services, and other agencies that work with micro irrigation programs.

After my first textbook, *Drip/Trickle or Micro Irrigation Management* by *Apple Academic Press Inc.*, and response from international readers, I was motivated to bring out for the world community this ten-volume series on, *Research Advances in Sustainable Micro Irrigation*. This book series will complement other books on micro irrigation that are currently available on the market, and my intention is not to replace any one of these. This book series is unique because it is complete and simple, a one-stop manual, with worldwide applicability to irrigation management in agriculture. This series is a must for those interested in irrigation planning and management, namely, researchers, scientists, educators and students.

The contributions by the cooperating authors to this book series have been most valuable in the compilation of this volume. Their names are mentioned in each chapter and in the list of contributors. This book would not have been written without the valuable cooperation of these investigators, many of whom are renowned scientists who have worked in the field of micro irrigation throughout their professional careers. I am glad to introduce Dr. Kamal Gurmit Singh, Senior Research Engineer and Professor at Punjab Agricultural University, and Dr. Ramesh P. Rudra, Distinguished Professor in Water Resources Engineering at University of Guelph, Canada. They join as editors for this volume. We all three are alumni of the College of Agricultural Engineering at Punjab Agricultural University, Ludhiana – Punjab. Dr. Ramesh was one year senior to me and graduated in 1970. Kamal graduated in 1983, after me. Without their support and extraordinary work, readers will not have this quality publication. Most of the research studies in this volume were conducted by Dr. Kamal G. Singh, his colleagues, and students.

I would like to thank editorial staff, Sandy Jones Sickels, Vice President, and Ashish Kumar, Publisher and President at Apple Academic Press, Inc., (www.appleacademicpress.com) for making every effort to publish the book when the diminishing water resources are a major issue worldwide. Special thanks are due to the AAP Production staff for the quality production of this book.

We request that readers offer us your constructive suggestions to help to improve the next edition.

I express my deep admiration to my family for understanding and collaboration during the preparation of this ten-volume book series. With my whole heart and best affection, we dedicate this volume to the late (Dr.) C. M. Jacob, who in 1965 founded College of Agricultural Engineering at Punjab Agricultural Engineering, Ludhiana – India; and was Dean. Dr. Jacob held admirable professional and human values as I saw during my college years. He has been my master, councilor, professional father and guru since 1966. He helped me to trickle on to add my drop to the ocean of service to the world of humanity. Without his advice and patience, I would not have been a *"Father of Irrigation Engineering of twentieth century in Puerto Rico"* with zeal for service to others. My salutes to him for his legacy. As an educator, I offer this advice to one and all in the world: *"Permit that our Almighty God, our Creator and excellent Teacher, irrigate the life with His Grace of rain trickle by trickle, because our life must continue trickling on..."*

—Megh R. Goyal, PhD, PE
Senior Editor-in-Chief
August 1, 2014

FOREWORD 1

With only a small portion of cultivated area under irrigation and with the need to expand this area, which can be brought about by irrigation, it is clear that the most critical input for agriculture today is water. It is important that all available supplies of water should be used intelligently to the best possible advantage. Recent research around the world has shown that the yields per unit quantity of water can be increased if the fields are properly leveled, the water requirements of the crops as well as the characteristics of the soil are known, and the correct methods of irrigation are followed. Significant gains can also be made if the cropping patterns are changed so as to minimize storage during the hot summer months when evaporation losses are high, if seepage losses during conveyance are reduced, and if water is applied at critical times when it is most useful for plant growth.

Irrigation is mentioned in the Holy Bible and in the old documents of Syria, Persia, India, China, Java, and Italy. The importance of irrigation in our times has been defined appropriately by N.D. Gulati: "In many countries irrigation is an old art, as much as the civilization, but for humanity it is a science, the one to survive." The need for additional food for the world's population has spurred rapid development of irrigated land throughout the world. Vitally important in arid regions, irrigation is also an important improvement in many circumstances in humid regions. Unfortunately, often less than half the water applied is used by the crop—irrigation water may be lost through runoff, which may also cause damaging soil erosion, deep percolation beyond that required for leaching to maintain a favorable salt balance. New irrigation systems, design and selection techniques are continually being developed and examined in an effort to obtain high practically attainable efficiency of water application.

The main objective of irrigation is to provide plants with sufficient water to prevent stress that may reduce the yield. The frequency and quantity of water depends upon local climatic conditions, crop and stage of growth, and soil-moisture-plant characteristics. The need for irrigation can be determined in several ways that do not require knowledge of evapotranspiration (ET) rates. One way is to observe crop indicators such as change of color or leaf angle, but this information may appear too late to avoid reduction in the crop yield or quality. Other similar methods of scheduling include determination of the plant water stress, soil moisture status, or soil water potential. Methods of estimating crop water requirements using ET and combined with soil characteristics have the advantage of not only being useful in determining when to irrigate, but also enables us to know the quantity of water needed. ET estimates have not been made for the developing countries though basic information on weather data is available. This has contributed to one of the existing problems that the vegetable crops are over irrigated and tree crops are under irrigated.

Water supply in the world is dwindling because of luxury use of sources; competition for domestic, municipal, and industrial demands; declining water quality; and losses through seepage, runoff, and evaporation. Water rather than land is one of the limiting factors in our goal for self-sufficiency in agriculture. Intelligent use of water will avoid problem of sea water seeping into aquifers. Introduction of new irrigation methods has encouraged marginal farmers to adopt these methods without taking into consideration economic benefits of conventional, overhead, and drip irrigation systems. What is important is "net in the pocket" under limited available resources. Irrigation of crops in tropics requires appropriately tailored working principles for the effective use of all resources peculiar to the local conditions. Irrigation methods include border-, furrow-, subsurface-, sprinkler-, sprinkler, micro, and drip/ trickle, and xylem irrigation.

Drip irrigation is an application of water in combination with fertilizers within the vicinity of plant root in predetermined quantities at a specified time interval. The application of water is by means of drippers, which are located at desired spacing on a lateral line. The emitted water moves due to an unsaturated soil. Thus, favorable conditions of soil moisture in the root zone are maintained. This causes an optimum development of the crop. Drip/micro or trickle irrigation is convenient for vineyards, tree orchards, and row crops. The principal limitation is the high initial cost of the system that can be very high for crops with very narrow planting distances. Forage crops may not be irrigated economically with drip irrigation. Drip irrigation is adaptable for almost all soils. In very fine textured soils, the intensity of water application can cause problems of aeration. In heavy soils, the lateral movement of the water is limited, thus more emitters per plant are needed to wet the desired area. With adequate design, use of pressure compensating drippers and pressure regulating valves, drip irrigation can be adapted to almost any topography. In some areas, drip irrigation is used successfully on steep slopes. In subsurface drip irrigation, laterals with drippers are buried at about 45 cm depth, with an objective to avoid the costs of transportation, installation, and dismantling of the system at the end of a crop. When it is located permanently, it does not harm the crop and solve the problem of installation and annual or periodic movement of the laterals. A carefully installed system can last for about 10 years.

The publication of this book series is an indication that things are beginning to change, that we are beginning to realize the importance of water conservation to minimize the hunger. It is hoped that the publisher will produce similar materials in other languages.

In providing this book series on micro irrigation, Megh Raj Goyal, as well as the Apple Academic Press, is rendering an important service to the farmers. Dr. Goyal, *Father of Irrigation Engineering in Puerto Rico*, has done an unselfish job in the presentation of this series that is simple and thorough. I have known Megh Raj since

1973 when we were working together at Haryana Agricultural University on an ICAR research project in "Cotton Mechanization in India."

Dr. Gajendra Singh, PhD,
Former Vice Chancellor, Doon University, Dehradun, India.
Adjunct Professor, Indian Agricultural Research Institute, New Delhi
Ex-President (2010–2012), Indian Society of Agricultural Engineers.
Former Deputy Director General (Engineering), Indian Council of Agricultural Research (ICAR), New Delhi.
Former Vice-President/Dean/Professor and Chairman, Asian Institute of Technology, Thailand.

Dr. Gajendra Singh, PhD
New Delhi
August 1, 2014

FOREWORD 2

Monsoon failure during June of 2014 has created shock waves once again across India. The Indian Meteorological Department has reported a shortage of rains in major parts of India with the country average of 42%, Karnataka 35%, Konkan and Goa 56%, Kerala 24%, Gujarat 88% and Rajasthan 80% during June 2014. India still is 62% agriculture dependent on monsoon rain, and most of the 83% small and marginal farmers are living in these regions. Monsoon failure in June affects food production and livelihood of the majority population of India. The Government of India has taken timely and laudable initiatives to develop a contingency program. India has observed this type of monsoon situation 12 times during the last 113 years, meaning a huge deficit of rain once in 10 years. Although contingency plans provide some relief, there is a need to address fundamental issues of water management in India. India has 1896 km^3 total renewable water resources; in addition only 5% of the total precipitation is harvestable. Improving water productivity is a major challenge. Improving irrigation efficiency, effective rainwater management, and recycling of industrial and sewage water will get enough water available for agriculture in the state. Micro irrigation can mitigate abiotic stress situation by saving over 50% of irrigation water and can be useful in a late monsoon situation for timely sowing.

Agricultural engineers across India have made several specific recommendations on water conservation practices, ground water recharge, improving water productivity, land management practices, tillage/cultivation practices and farm implements for moisture conservation. These technologies have potential to conserve water that will facilitate timely sowing of crops under the delayed monsoon situation that has occurred this year and provide solutions to monsoon worries. Agricultural engineers need to provide leadership opportunities in the water resources and water management sector, which include the departments of Command Area Development, Rural Development, Panchayat Raj, Water Resources, Irrigation, Soil Conservation, Watersheds, Environment and Energy for Stability of Agriculture, and in turn the stable growth of Indian economy.

This book series on micro irrigation addresses the urgent need to adopt this water saving technology not only in India but throughout the world. I would like to see more literature on micro irrigation for use by the irrigation fraternity. I appeal to all irrigation engineering fraternities to bring such issues to the forefront through research publications, organizing symposiums, seminars and discussions with planners and policymakers at the regional, state and national level so that agricultural engineers will get a well-deserved space in the development process of the country.

Dr. V. M. Mayande, PhD

President 2012–15, Indian Society of Agricultural Engineers,

Vice Chancellor,

Dr. Panjabrao Deshmukh Krishi Vidyapeeth

Akola–444104, Maharashtra, India. Tel.: +91 9423174299.

E-mail: vmmayande@yahoo.com

Dr. V. M. Mayande, PhD
August 1, 2014

FOREWORD 3

In the world, water resources are abundant. The available fresh water is sufficient even if the world population is increased by four times the present population, that is about 25 billion. The total water present in the earth is about 1.41 billion Km^3 of which 97.5% is brackish and only about 2.5% is fresh water. Out of 2.5% of fresh water, 87% is in ice caps or glaciers, in the ground, or deep inside the earth. According to Dr. Serageldin, 22 of the world's countries have renewable water supply of less than 1000 cubic meter per person per year. The World Bank estimates that by the year 2025, one person in three, in other words 3.25 billion people in 52 countries, will live in conditions of water shortage. In the last two centuries (1800–2000) the irrigated area in the world has increased from 8 million ha to 260 million ha for producing the required food for the growing population. At the same time, the demand of water for drinking and industries has increased tremendously. The amount of water used for agriculture, drinking, and industries in developed countries are 50% in each and in developing countries it is 90% and 10%, respectively. The average quantity of water is about 69% for agriculture and 31% for other purposes. Water scarcity is now the single largest threat to global food production. To overcome the problem, there is a compulsion to use the water efficiently and at the same time increase the productivity from the unit area. It will involve spreading the whole spectrum of water-thrifty technologies that enable farmers to get more crops per drop of water. This can be achieved only by introducing drip/trickle/micro irrigation in large scale throughout the world.

Micro irrigation is a method of irrigation with high frequency application of water in and around the root zone of plant (crop) and consists of a network of pipes with suitable emitting devices. It is suitable for all crops except rice, and especially for widely spaced horticultural crops. It can be extended to wastelands, hilly areas, coastal sandy belts, water scarcity areas, semi arid zones, and well irrigated lands. By using micro irrigation, the water saving compared to conventional surface irrigation is about 40–60% and the yield can be increased up to 100%. The overall irrigation efficiency is 30–40% for surface irrigation, 60–70% for sprinkler irrigation, and 85–95% for micro irrigation. Apart from this, one has the advantage of saving of costs related to labor and fertilizer and weed control. The studies conducted and information gathered from various farmers in India have revealed that micro irrigation is technically feasible, economically viable, and socially acceptable. Since the allotment of water is going to be reduced for agriculture, there is a compulsion to change the irrigation method to provide more area under irrigation and to increase the required food for the growing population.

Most farmers in the developing countries are poor, and hence it is not possible for them to adopt/install the micro irrigation with fertigation, though it would be economically viable and profitable for them. In Tamil Nadu, India, the number of marginal farmers (holding less than 1.0 hectare) and small farmers (holding 1 to 2 ha) has increased from 5,076,915 in 1967–1968 to 7,184,940 in 1995–1996 and area owned by them has also decreased in the same period from 0.63 ha to 0.55 ha. In addition, the small farmers category is about 89.68% in 1995–1996, of the total farmers in the state. At the same time if micro irrigation is used for all crops, yield can be increased and water savings will be 50%. In the case of sugarcane crop, the yield can be increased to 250 tons/ha from the present average yield of 100 tons/ha, which is the highest at present in India. Therefore, to popularize the micro irrigation system among this group of farmers, more books like this, not only in English but also in the respective national languages, should be published.

Volumes 1 through 6 in this book series contain 72 chapters covering all areas of micro irrigation as well as it potentials, and reviews of the system, presents principles of micro irrigation, discusses the experience of micro irrigation in desert region mainly in Middle East, and presents its application in the field for various crops, especially in water requirements for crops such as citrus, banana, eggplant, papaya, plantations, sugarcane, tanier, blueberries, etc. The book series also includes wetting patterns under various sustainable practices; evaluation of the micro irrigation systems; the software to design the systems; ornamental nursery production. Volume three includes an extensive bibliography in addition the references at the end of each chapter. The chapters are written by experienced scientists from various parts of the world, bringing their findings, which will be useful for all users of the micro irrigation in the world in the coming years.

I must congratulate Dr. Goyal for contacting many experts who are involved in the subject to bring their experience and knowledge about micro irrigation to this book series. He has also given many figures, illustrations, and tables to understand the subject. I congratulate the editor(s) for the volume in this book series.

The editors of these volumes are reputed agricultural engineers in the world and have wide knowledge and experience in soil and water conservation engineering particularly micro irrigation. After the big success first book titled, *Management of Drip/Trickle or Micro Irrigation* by Dr. Goyal, this compendium book series is unique. Dr. Goyal, Senior Editor-in-Chief of ten-volume book series, has taken into account the fate of marginal farmers and thus serves the poor. The information provided in this book series will go a long way in bringing large areas under micro irrigation in the world, especially in water scarcity countries. On behalf of international scientists and agricultural engineers on micro irrigation, I am indebted to Dr. Megh R. Goyal and Apple Academic Press for undertaking this project.

Professor (Dr.) R. K. Sivanappan,

Former Dean-cum-Professor of College of Agricultural Engineering and Founding Director of Water Technology Centre at Tamil Nadu Agricultural University [TAMU], Coimbatore – India. Ex-member of Tamil Nadu State Planning Commission (2005–2006).

Recipient of Honorary PhD degree by Linkoping University – Sweden; and conferment of the honorary D.Sc. degree by the TAMU – India.

E-mail: sivanappan@hotmail.com

Dr. R. K. Sivanappan, PhD
August 1, 2014

FOREWORD 4

The microirrigation system, more commonly known as the drip irrigation system, has been one of the greatest advancements in irrigation system technology developed over the past half century. The system delivers water directly to individual vines or to plant rows as needed for transpiration. The system tubing may be attached to vines, placed on or buried below the soil surface.

This book series, written by experienced system designers/scientists, describes various systems that are being used around the world, the principles of microirrigation, chemigation, filtration systems, water movement in soils, soil-wetting patterns, design principles, use of wastewater, crop water requirements and crop coefficients for a number of crops. The book series also includes chapters on hydraulic design, emitter discharge and variability, and pumping station. Irrigation engineers will find this book series to be a valuable reference.

Dr. Marvin E. Jensen, PhD, PE

Retired Research Program Leader at USDA-ARS; and Irrigation Consultant

1207 Spring Wood Drive, Fort Collins, Colorado 80525, USA.

E-mail: mjensen419@aol.com

Dr. Marvin E. Jensen, PhD

August 1, 2014

BOOK SERIES: RESEARCH ADVANCES IN SUSTAINABLE MICRO IRRIGATION

Volume 1: Sustainable Micro Irrigation: Principles and Practices
Senior Editor-in-Chief: Megh R. Goyal, PhD, PE

Volume 2: Sustainable Practices in Surface and Subsurface Micro Irrigation
Senior Editor-in-Chief: Megh R. Goyal, PhD, PE

Volume 3: Sustainable Micro Irrigation Management for Trees and Vines
Senior Editor-in-Chief: Megh R. Goyal, PhD, PE

Volume 4: Management, Performance, and Applications of Micro Irrigation
Senior Editor-in-Chief: Megh R. Goyal, PhD, PE

Volume 5: Applications of Furrow and Micro Irrigation in Arid and Semi-Arid Regions
Senior Editor-in-Chief: Megh R. Goyal, PhD, PE

Volume 6: Best Management Practices for Drip Irrigated Crops
Editors: Kamal Gurmit Singh, PhD, Megh R. Goyal, PhD, PE, and
Ramesh P. Rudra, PhD, PE

Volume 7: Closed Circuit Micro Irrigation Design: Theory and Applications
Senior Editor-in-Chief: Megh R. Goyal, PhD; Editor: Hani A. A. Mansour, PhD

Volume 8: Wastewater Management for Irrigation: Principles and Practices
Editor-in-Chief: Megh R. Goyal, PhD, PE; Coeditor: Vinod K. Tripathi, PhD

Volume 9: Water and Fertigation Management in Micro Irrigation
Senior Editor-in-Chief: Megh R. Goyal, PhD, PE

Volume 10: Innovations in Micro Irrigation Technology
Senior Editor-in-Chief: Megh R. Goyal, PhD, PE; Coeditors: Vishal K. Chavan, MTech, and Vinod K. Tripathi, PhD

ABOUT THE EDITORS

Three internationally distinguished scientists have joined together to bring this volume on micro irrigation. They are authorities on the area of soil and water conservation engineering. From left to right: Dr. Kamal Gurmit Singh, Dr. Megh R. Goyal, and Dr. Ramesh P. Rudra obtained the Bachelor of Technology degrees in Agricultural Engineering in 1983, 1971 and 1970, respectively, from Punjab Agricultural University, Ludhiana – Punjab – India.

Kamal Gurmit Singh, PhD, is working as Senior Research Engineer in the Department of Soil & Water Engineering at the Punjab Agricultural Ludhiana, Punjab, India. Dr. Singh is an authority in the Punjab State of India on micro irrigation and protected structures with more than 28 years of academic and consulting experience, particularly in environmental engineering/water resources management, micro irrigation, protected cultivation, and computer simulation modeling such as LEACHN, SALTMED, and MODFLOW. He has worked as Principal Investigator on several research projects, including "Plasticulture" and "best management practices" at Punjab Agricultural University. He has extensive experience in designing, installation, and use of protected cultivated structures, i.e. polyhouses, net houses, and micro irrigation systems. He has been awarded a gold medal for recognition in the field of soil and water engineering by the Society of Recent Developments in Agriculture (SRDA), India. In addition, he has attended training on net house designing at AVRDC—The World Vegetable Centre, Taiwan. He has also acted as chairman/ member of several technical committees on micro irrigation and greenhouse technology. A dedicated professor teaching courses such as advanced hydrology, watershed hydrology, system engineering for graduate and postgraduate students, he has also advised five MS and three PhD students. He has published over 87 papers including 38 refereed journal articles. He received a BTech, a MTech and a PhD (Agricultural Engineering) from Punjab Agricultural University, Ludhiana, India.

Megh R. Goyal, PhD, PE, is a Retired Professor in Agricultural and Biomedical Engineering from the General Engineering Department in the College of Engineering at University of Puerto Rico–Mayaguez Campus; and Senior Acquisitions Editor and Senior Technical Editor-in-Chief in Agriculture and Biomedical Engineering for Apple Academic Press Inc. He received his BSc degree in engineering in 1971 from Punjab Agricultural University, Ludhiana, India; his MSc degree in 1977 and PhD degree in 1979 from the Ohio State University, Columbus; and his Master of Divinity degree in 2001 from Puerto Rico Evangelical Seminary, Hato Rey, Puerto Rico, USA. He spent one-year sabbatical leave in 2002–2003 at the Biomedical Engineering Department at Florida International University in Miami, Florida, USA. Since 1971, he has worked as Soil Conservation Inspector (1971); Research Assistant at Haryana Agricultural University (1972–75) and Ohio State University (1975–79); Research Agricultural Engineer/Professor at the Department of Agricultural Engineering of UPRM (1979–1997); and Professor in Agricultural and Biomedical Engineering in the General Engineering Department of UPRM (1997–2012).

He was first agricultural engineer to receive the professional license in Agricultural Engineering in 1986 from College of Engineers and Surveyors of Puerto Rico. On September 16, 2005, he was proclaimed as "Father of Irrigation Engineering in Puerto Rico for the twentieth century" by the ASABE, Puerto Rico Section, for his pioneer work on micro irrigation, evapotranspiration, agroclimatology, and soil and water engineering. During his professional career of 45 years, he has received awards such as Scientist of the Year, Blue Ribbon Extension Award, Research Paper Award, Nolan Mitchell Young Extension Worker Award, Agricultural Engineer of the Year, Citations by Mayors of Juana Diaz and Ponce, Membership Grand Prize for ASAE Campaign, Felix Castro Rodriguez Academic Excellence, Rashtrya Ratan Award and Bharat Excellence Award and Gold Medal, Domingo Marrero Navarro Prize, Adopted Son of Moca, Irrigation Protagonist of UPRM, and Man of Drip Irrigation by Mayor of Municipalities of Mayaguez/Caguas/Ponce and Senate/Secretary of Agriculture of ELA, Puerto Rico.

He has authored more than 200 journal articles and textbooks, including Elements of Agroclimatology (Spanish) by UNISARC, Colombia, and two *Bibliographies on Drip Irrigation*. Apple Academic Press Inc. (AAP) has published his books, namely *Biofluid Dynamics of Human Body, Management of Drip/Trickle or Micro Irrigation, Evapotranspiration: Principles and Applications for Water Management, Sustainable Micro Irrigation Design Systems for Agricultural Crops: Practices and Theory, Biomechanics of Artificial Organs and Prostheses*, and *Sci-*

entific and Technical Terms in Bioengineering and Biotechnology. During 2014–15, AAP is publishing his ten-volume set, Research Advances in Sustainable Micro Irrigation. Readers may contact him at goyalmegh@gmail.com.

Ramesh P. Rudra, PhD, PE, is currently Professor of Water Resources Engineering at the University of Guelph in Guelph, Ontario, Canada. With more than 34 years of experience as a water resource engineer, researcher, and educator in Canada, the United States, and India, he is involved in research related to source water protection, development of procedures for watershed system capacity for water quality (TMDL), and tools for site-specific design of vegetative filter strips to protect and improve stream water quality in rural areas. He is a member of many professional organizations and has received the Canadian Society of Agricultural Engineers' Jim Beamish Award for excellence (research and teaching) in soil and water conservation. Dr. Rudra received his BSc (Engineering) from Punjab Agricultural University, Ludhiana, India; his MSc; and his PhD from the Pennsylvania State University, University Park, Pennsylvania, USA.

Dr. Rudra's work has primarily focused on agricultural water management through investigation of fundamental process of soil erosion, hydrology, irrigation and drainage, irrigation with saline water, source water protection, modeling and management of non-point source pollution, as evident by the publications in Asia, Europe and North America. Several graduate students from Asia, Africa, Far East and Middle East have worked on soil and water management problems related to their countries under his supervision. In addition, he has provided technical assistance to extension agents, government officials in planning, design, development, management and evaluation of source water protection schemes, water quality monitoring projects, irrigation and drainage systems, both surface and sub-surface drainage, soil erosion control practices and structures, and water harvesting. He has been involved in the development and evaluation of models for source water protection and non-point source pollution management, nomographs and computer software for management of water quantity and quality at the farm as well as watershed scale. His present research activities include development of watershed-based tools to help quantify the health status of water bodies in rural watersheds, and also economically feasible, environmentally sustainable and socially acceptable strategies to improve their status by identifying the source of pollution and their relative contribution to the impairment of water quality. This approach provides a linkage between needs for nutrient management and watershed-based source water protection. He is also involved in the development of tools for the site-specific design of vegetative strips to improve stream water quality in rural and urban environment.

WARNING/DISCLAIMER

The goal of this compendium, **Best Management Practices for Drip Irrigated Crops,** is to guide the world community on how to manage efficiently for economical crop production. The reader must be aware that dedication, commitment, honesty, and sincerity are most important factors in a dynamic manner for complete success. This reference is not intended for a one-time reading; we advise you to consult it frequently. To err is human. However, we must do our best. Always, there is a place for learning new experiences.

The editor, the contributing authors, the publisher, and the printer have made every effort to make this book as complete and as accurate as possible. However, there still may be grammatical errors or mistakes in the content or typography. Therefore, the contents in this book should be considered as a general guide and not a complete solution to address any specific situation in irrigation. For example, one size of irrigation pump does not fit all sizes of agricultural land and work for all crops.

The editor, the contributing authors, the publisher and the printer shall have neither liability nor responsibility to any person, organization, or entity with respect to any loss or damage caused, or alleged to have caused, directly or indirectly, by information or advice contained in this book. Therefore, the purchaser/reader must assume full responsibility for the use of the book or the information therein.

The mention of commercial brands and trade names are only for technical purposes and does not imply endorsement. The editor, contributing authors, educational institutions, and the publisher do not have any preference for a particular product.

All web links that are mentioned in this book were active on December 31, 2014. The editors, the contributing authors, the publisher, and the printing company shall have neither liability nor responsibility if any of the web links are inactive at the time of reading of this book.

PART I
IRRIGATION METHODS

CHAPTER 1

LOW TUNNEL TECHNOLOGY FOR VEGETABLE CROPS IN INDIA

A. S. LODHI, ARUN KAUSHAL, and KAMAL G. SINGH

1.1 INTRODUCTION

Agriculture was developed by human beings for their survival against hunger. As the time passed, human beings learnt that maximum crop yield is achieved when the crops are grown during different seasons under favorable climatic conditions. Vegetables are rich source of vitamins, carbohydrates, salts and proteins. There is a year round high demand for fresh vegetables in the country in domestic and export market due to: increased health awareness, high population growth rate, changing dietary patterns of increasingly affluent middle class, and availability of packaged vegetables. But due to unfavorable climatic conditions, there is a flood of vegetables in the season and high priced vegetables in off-season. Vegetables can be cultivated in off-season, with the introduction of green houses, low and high poly tunnel technology, in which temperature and moisture are controlled for specific growth of vegetables. The production of vegetables all around the year enables the growers to fully use the resources and supplement income from vegetable growing as compared to other normal agricultural crops.

Low tunnels are miniature structures producing green house like effect. In these tunnels, plastic sheets are used for roof covering of the tunnel with shaped construction having low height, which is built with steel bars. These tunnels facilitate the entrapment of carbon dioxide, thereby enhancing the photosynthetic activity of the plants and hence the yield. These structures also protect the plants from the high winds, rain, frost and snow. Besides being inexpensive, they are easy to construct and dismantle. Low tunnels are being used for producing high quality, high valued nurseries and crops such as tomatoes, cucumber, radish, beans and capsicum. With this technology, the farmers can capture the market in the early season and may get good return of the produce. Another advantage of such technology is that low tunnels can be easily dismantled and used in the next year.

---

*In this chapter, the currency is expressed in Indian Rupees (1.00 US\$ = Rs. 60.93; 1.00 Rs. = 0.02 US\$).

The use of low tunnels conserves warmth climate, stimulates germination and early growth, protects plants from injury and improves the quality of crop. Other beneficial effects include: maintaining soil texture and protecting crops from the attacks of birds and pest.

This chapter presents the research review in production of vegetables under low tunnel technology (LTT).

1.2   EFFECT OF LOW TUNNEL TECHNOLOGY ON YIELD OF VEGETABLES

Saini and Singh [13] conducted a research study on growth and yield of chilly crop under low tunnel polyhouse, at research farm of Soil and Water Engineering Department at Punjab Agricultural University (PAU), Ludhiana – India. They found that there was no significant effect on the yield of chili due to variation in perforations on polythene cover. Drip irrigation system with IW/CPE ratio of 0.50 and 30 cm low tunnel polythene cover gave the best yield and water saving.

Helbacka [4] conducted a study on row covers for vegetable gardens. It was reported that many cucurbits (squash, cucumber, and melons) respond well under row covers with increased yield of as much as 25%.

Joublan and Vergara [7] conducted a study on vegetative and productive development of strawberry (*Fragaria X ananassa Duch.*), using row cover of spunbonded polyester with different densities. Row covers were placed directly over the plants as a tunnel without any support structure. Treatments were comprised of a control treatment (without row covers), row covers of 20 g/m^2 and row covers of 30 g/m^2. Fruit production started 4.8 and 2.2 days earlier under 20 and 30 g/m^2 row covers, respectively, than under the control treatment. The use of row covers also increased the number of fruit and weight, yield per plant and sugar concentration compared to the control treatment. The best results were obtained with 30 g/m^2 row covers.

Henandez et al. [5] conducted studies on row covers for quality improvement of Chinese cabbage for three years in the area of Granada, Spain, under a Mediterranean continental temperate climate, on 55-day cycles with transplanting in midmarch. The mean commercial yield for the 3-years was 1 1.9 kg/m^2 under row cover compared to only 2.1 kg/m^2 in open air, owing primarily to important number of noncommercial cabbages.

Vishnuvardhana et al. [18] conducted a study on the economics on the propagation of cashew grafts in a mist chamber, naturally ventilated green house, low tunnel and shade net during the summer, monsoon and winter season. The initial investment for the establishment of the propagation structure (100 mi) reached Rs. 8,500 for the shade net, Rs. 300,000 for mist chamber, Rs. 36,400 for naturally ventilated green house and Rs. 21,000 for the low tunnel. The highest net profit was obtained with propagation in low tunnels, followed by propagation in a naturally ventilated green house, mist chamber and shade net.

1.3 EARLY HARVESTING OF VEGETABLE UNDER LOW TUNNEL TECHNOLOGY

Meesters [10] conducted a study on early cultivation of strawberry under tunnels. In a study at Tongeren – Belgium, in 1995, the strawberry cultivar Evita was planted in a plastic tunnel on 6 April at densities of 3, 4 or 5 plants/m². The first friots were ready on 24 May, and 80% of the total harvest was picked on 40–50 days between late July and mid-August. The harvest finished on 19 October. Production of Evita was 20–25 days earlier under the tunnel than in the field, and it was 30–35 days earlier than that of field-grown Selva. It therefore filled the gap between field-grown Elsanta and field-grown Selva. Yields in the tunnel were 2.6–3.2 kg/m².

Arin and Ankara [2] conducted a study to determine the effects of low tunnel, mulching and pruning on yield and earliness of tomato in unheated glass house. It was observed that there was an increase of 643.42% in height (relative to height at the planting time) of the plants grown under low tunnel than those grown without tunnel (602.87%). Stem diameter increase was higher in tunneled plants (265.63%) than plants growing without tunnel (233.83%). The number of days to first harvest was 117.97 for tunneled treatment compared to 119.9 days for plant growing without tunnel.

Amer [1] carried out a study on protection effect of low-temperature on some snap bean (*Phaseolus vulgaris L.*) varieties, green yield and some isozyme levels. It was found that protected plants recorded higher vegetative growth and total, early and exportable yields compared with those of the open field. Plants grown under plastic low tunnels recorded higher vegetative growth and total-green yield compared with agrel-covered plants. All the cultivars recorded higher vegetative growth, total, early and exportable yields under plastic protection than under agrel or open field condition.

Singh et al. [15] conducted a study on effects of plastic tunnel and mulching on growth and yield of strawberry. It was found that use of plastic tunnel along with control (without tunnel) were taken as main factors and mulching materials, via black polyethylene, transparent polyethylene and straw mulch as subfactors and laid out in split-plot design replicated four times. Use of plastic tunnel resulted in significantly higher plant spread, dry matter accumulation and yield attributing characters compared to control. Further, plastic tunnel enhanced earliness by 16 days besides 19% higher yield over control. Among different mulching materials, black polyethylene mulch was most suitable and resulted in 41% higher fruit yield compared to straw mulch.

Slezak et al. [16] conducted a study on enhancing earliness of sweet corn by using transplants and plastic row covers. The technological variations were transplanted plants with floating row cover, transplanted plants with no row cover, direct sown plants with floating row cover, and direct sown plants with no row cover. The application of direct sowing and floating row cover increased the earliness by 3 days for germination and by 4 days for the total growing period, compared to the treatment with no row cover. The 25–26 day transplant growing period reduced the growing period by 15–18 days. Covering the seedling in the early season was beneficial for frost protection. The combination of seedling grown plants and floating row cover resulted in a 22-day earlier harvest, compared to the traditional technology.

1.4   FAVORABLE CLIMATIC CONDITIONS FOR PLANT GROWTH UNDER LOW TUNNELS

Libik and Siwek [9] studied the changes in soil temperature affected by the application of plastic covers in field production of lettuce and watermelon. At 8:00 a.m., the highest soil temperature was recorded under a low tunnel, where it was 3°C higher than in the open ground from 29 March to 14 April and 5.9°C higher from 9 to 14 June. However, at 2:00 p.m., the highest soil temperature was recorded under perforated plastic. By 2:00 p.m., the air temperature under the cover was between 35 and 40°C, which was 10°C higher than the ambient temperature. Marketable yield of lettuce was highest under low tunnel (110.9% higher than in the open field).

Lamarrel et al. [8] conducted a study on influence of nitrogen fertilization, row covers and cultivars on the production of day neutral strawberry. It was found that the use of low tunnel was beneficial during winter, when the crop was protected from frost and low temperature for higher productivity.

Hochmuth et al. [6] conducted a study on row covers for commercial vegetable culture in Florida. It was found that row covers are used to enclose one or more rows of plants to enhance the crop growth and production by increasing both air and soil temperatures and reducing wind damage.

1.5   COVERAGE MATERIAL USED IN LOW TUNNEL TECHNOLOGY

Monteiro et al. [11] carried out a study on perforated plastic film for low tunnels cultivated with lettuce. During the spring, tunnels were tested with and without lettuce, with 0, 5, 10, 15 and 20% of perforated film area and in soil without tunnel. The evaluation of the temperature inside the protected atmosphere was inversely related to the percentage of perforation contrary to the humidity loss for the external atmosphere. The production obtained under protected condition was higher and of better quality.

Fu et al. [3] conducted a study on effects of different ventilation methods on seedling growth of chili pepper. Chili pepper seeds were sown in beds under mini plastic tunnels. When the outside temperature dropped to below 12°C, a normal plastic tunnel was set up to cover all the mini tunnels. Different watering and ventilation methods were tested. Based upon the seedling quality and yield, it was suggested that the bed should be thoroughly irrigated before sowing, with no watering needed until the seedlings had six leaves. During this period, the film of mini plastic channels should be removed when 70% of the seedlings emerged and the film should be replaced to cover the mini tunnels completely when the temperature inside the normal plastic tunnel dropped to below 5°C.

Reghin et al. [12] conducted an experiment on mulching and row cover in lettuce crop. The parameters tested were total lettuce leaf number, frost-damaged leaves, plant height, stem length, plant fresh weight, plant dry matter accumulation and biomass. Weed population and dry weight were also assessed. Row cover with white plastic produced positive results on yield, early harvest and quality, even with the occurrence of frost. Frost damage in uncovered plots reduced the plant weight by 34.62%. Black plastic mulch controlled weeds and resulted in 22.12% increase in plant weight compared to rice straw.

Shiraiwa et al. [14] conducted a study on effects of tunnel covering plastic films and fertilization methods on growth, bolting and yield in Welsh onion (*Allium fistulosum L.*) harvested in early summer. It was found that polyolefin plastic film (PO) induced the highest mean air and soil temperatures, while dripped polyethylene plastic film (DP) showed the lowest temperatures. The coefficient of variance on soil water content was higher in PO and nondripped polyethylene plastic film (NDP). Application of an overall layer of fertilizer produced a lower bolting rate than application of fertilizer in a planting furrow when using DP. However, opposite results were demonstrated with PO and NDP. The effects of covering films and fertilizer application methods on the bolting rate and yield showed significant interaction. Higher thermo keeping films suppressed flower initiation. When higher thermo-keeping films were used to cover tunnels, management to control nitrogen concentration is required to inhibit bolting in this culture.

Streck et al. [17] conducted a study on a system to grow lettuce inside low plastic tunnels. Four different covers were tested to obtain a system to produce vegetables throughout the year under low tunnels without ventilation management. Lettuce cv. Regina was grown from October 1994 to July 1997. The winter treatments comprised low tunnels with transparent polyethylene without perforations and with conventional management according to the meteorological conditions; low tunnel with 3% perforated transparent polyethylene and without ventilation management; umbrella-like low tunnel with transparent polyethylene with open laterals throughout the year and without ventilation management; and cropping without a tunnel throughout the year (control). The treatments in the summer were similar to the winter treatments with conventional and perforated covers replaced by a black plastic screen with a 30% reduction in solar radiation. Results showed that umbrella-like tunnels can be used instead of conventional low tunnels, which need daily management. This technique allowed lettuce to grow throughout the year and only required lateral adjustment according to the season.

1.6   FUTURE THRUSTS

Considerable research work has been conducted on low tunnel technology for vegetable production during the past 20 years. However, still a significant work is needed, such as: (a) Testing of low tunnel technology for different vegetables grown under different regions of India; (b) the optimum tunnel heights for various vegetable crops; (c) the optimum polyethylene sheet thickness and effect of perforation in poly sheets on vegetables crop; (d) adoption of low tunnel technology with drip irrigation; (e) the economic analysis of low tunnel; (t) constraints in the adoption of low tunnel technology for a wide variety of vegetable crops.

1.7   SUMMARY

Growing vegetable by low tunnel technology (row cover technology) has many advantages with regards to increase in yield, early harvesting of vegetables, conserving soil warmth, protecting plant from wind and frost and ultimately increasing the net profit for the farmers. This review chapter provides better understanding and facilitates optimal analysis for rational use of low tunnel technology for vegetable production and will help to identify vegetables for adoption of low tunnel technology by farmers.

KEYWORDS

- **cabbage**
- **chili**
- **corn**
- **frost**
- **green house**
- **lettuce**
- **low tunnel technology**
- **plastic film**
- **polysheet**
- **row cover technology**
- **snap bean**
- **strawberry**
- **tomato**
- **vegetable garden**
- **vegetables**
- **yield**

REFERENCES

1. Amer, A. H. (2004). Protection effect of low-temperature on some snap bean (*Phaseolus vulgaris* L.) varieties green yield and some isozyme levels. *Ann. Agric. Sci., Moshtohor, 42,* 661–678.
2. Arin, L., Ankara, S. (2001). Effect of low tunnel, mulch and pruning on the yield and earliness to tomato in unheated green house. *J. Appl. Hort., Lucknow, 3,* 23–27.
3. Fu, D. M., Huang, K. L., Wang, Y. H. (2004). Effects of different ventilation methods on seedling growth of chili pepper. *China Veg., 3,* 35–36.
4. Helbacka, J. (2002). Row covers for vegetable gardens. King County Coop. Extn. Ser., Washington State Univ., Fact Sheet No. 19, USA.
5. Hemandez, J., Soriano, T., Morales, M. L., Castilla, N. (2004). Row covers for quality improvement of Chinese cabbage (Brassica rapa subsp. Penkinensis). New Zeal, *J. Crop Hort. Sci., 32,* 379–388.
6. Hochmuth, G. J., Kostewicz, S., Stall, W. (2000). Row covers for commercial vegetable culture in Florida. Florida Coop. Extn. Ser., Univ. Florida, Circular 728.
7. Joublan, J. P., Vergara, M. (2003). Vegetative and productive development of strawberry (Fragaria Ananassa Duch.) using row cover of spunbonded polyester with different densities. *Agro. Sur., 3,* 37–47.
8. Lamarrel, M., Larcan, M. J., Payette, S., Fortin, C. (1996). Influence of nitrogen fertilization, row covers and cultivars on the production of day neutral strawberry. *Canadian J. Soil Sci., 76,* 29–36.
9. Libik, A., Siwek, P. (1994). Changes in soil temperature affected by the application of 'plastic covers in field production of lettuce and water melon. *Acta Hort., 371,* 269–273.
10. Meesters, P. (1996). Early cultivation of Evita under tunnels. Fruit Belge, *64,* 129–131.
11. Monteiro, J. E. B. A., Silva, I. J. O., Piedade, S. M. (2002). Perforated plastic film for low tunnels cultivated with lettuce. Revista Brasileira-de-Engenharia Agricola -Ambiental, *6,* 535–538.
12. Reghin, M. Y., Purissimo, C., Feltrim, A. L., Foltran, M. A. (2005). Mulching and row cover in lettuce crop. *Scientia Agraria, 3,* 69–77.
13. Saini, A. K., Singh, K. G. (2001). In: Annual report of All India Coordinated Research, Project on application of plastics in agriculture. pp. 69–74. *Dep. Soil Water Engg.,* Punjab Agricultural University, Ludhiana, India.
14. Shiraiwa, N., Kashima, Y., Itai, A., Tanabe, K. (2007). Effects of tunnel covering plastic films and fertilization methods on growth, bolting and yield in Welsh onion (Allium fistulosum L.) harvested in early summer. *Hort. Res.,* Japan, *6,* 17.
15. Singh, R., Asrey, R., Kumar, S. (2006). Effect of plastic tunnel and mulching on growth and yield of strawberry. *Indian J. Hort., 63,* 18–20.
16. Slezak, K., Orosz, F., Osz, A. (2006). Enhancing earliness of sweet corn by using transplants and plastic row covers. Kertgazdasag Hort., *38,* 14–19.
17. Streck, L., Schneider, F. M., Buriol, G. A., Luzza, J., Sandri, M. A. (2007). A system to grow lettuce inside low plastic tunnels. Ciencia Rural, *37,* 667–675.

18. Vishnuvardhana, Lingaiah, H. B., Khan, M. M., Raju, G. T. (2004). Economics of production of cashew grafts in different propagation structures under eastern dry zone of Karnataka. *Cashew, 18,* 39–44.

APPENDIX I: PHOTOS OF TUNNEL TECHNOLOGY

CHAPTER 2

PERFORMANCE OF SWEET PEPPER UNDER LOW TUNNEL TECHNOLOGY

A. S. LODHI, ARUN KAUSHAL, and KAMAL G. SINGH

2.1 INTRODUCTION

Capsicum (*Capsicum annuum* L. var. *grossum*) or sweet pepper is one of the most popular and high value vegetable crops grown for around the world. High demand of fresh vegetables generate domestic and export market throughout the year. However, due to unfavorable climatic conditions, there is a flood of vegetables in the season and very high priced vegetables in offseason. Vegetables can be cultivated in off-season, with the introduction of green houses, low and high poly tunnel technology, in which temperature and moisture are controlled for specific growth of vegetables. The production of vegetables all around the year enables the growers to fully use the resources and supplement income from vegetable growing as compared to other normal agricultural crops.

For sweet pepper, the optimum night temperature for quality fruit production is 16–18°C. When the temperature falls below 16°C for extended periods, growth and yields usually decrease. It can tolerate day temperature above 30°C. Sweet pepper are generally raised in open fields during main season thus causing glut in the market, which lead to price crash in the season. Punjab has extreme low temperature during winter and high temperature during summer, and therefore availability of these vegetables are for a short span. This situation suggests us to modify microclimate, which will not only increase the availability span of vegetables but also the yield. Low tunnel technology (LTT) can help to supply sweet pepper in offseason during early summer. With LTT, the farmers can capture the market in the early season and may get good return of their produce.

Low tunnels or row covers are plastic film covered shelters with small frames, producing greenhouse like effect. For low tunnels the sheet of the film is placed over the plants in a single or double row with an arch-shaped frame for support. The shape of the frame can vary but the farmer cannot work inside the low tunnels.

---

*In this chapter, the currency is expressed in Indian Rupees (1.00 US$ = Rs. 60.93; 1.00 Rs. = 0.02 US$).

These tunnels facilitate the entrapment of carbon dioxide. Low tunnels are being used for producing high quality high valued nurseries and crops such as tomatoes, cucumber, radish, beans, and capsicum.

Libik and Siwek [6] studied the changes in soil temperature affected by the application of plastic covers in field production of lettuce and watermelon. It was reported that at 8:00 a.m., soil temperature was 3°C higher than in the open ground and by 2:00 p.m., the air temperature under the cover was 10–15°C higher than the ambient temperature. Lamarrel et al. [5] was found that the use of low tunnel has been beneficial during winter when the crop has to be protected from frost and low temperature for higher productivity.

2.2   MATERIALS AND METHODS

Field experiment was conducted at the Research Farm of the Department of Soil and Water Engineering, Punjab Agricultural University (PAU), Ludhiana – India from October, 2008 to June, 2009. Ludhiana is situated at latitude of 30° 54'N and longitude of 75° 48'E and at a mean height of 247 meters above sea level. Average minimum and maximum temperature in the region varies from 3°C to 43°C, respectively.

A field plot measuring approximately 550.8 m^2 (54 × 10.2 m^2) was prepared and the experiment was laid out in split plot design keeping five irrigation treatments in main plots and three different heights of low tunnel in sub plots and replicated three times. The irrigation treatments were taken as main plots as they require bigger plot size. The three treatments of different heights of low tunnel made with tunnel frame height of 45 cm (H1), 60 cm (H2) and 75 cm (H3) were used in the experiment. Low tunnels were made with the 6 mm thick steel (iron) rods. The shape of the low tunnel frame was made parabolic with the given base and desired height. A length of 15 cm at the base was kept for inserting frame into the soil and inside loops was provided on both side of frame for the support in addition to low tunnel height (Fig. 2.1).

Nursery production of sweet pepper of "Bharath" variety was done in poly house on 13th of October 2008 and transplanting was done in the field on 17th of November 2008. In paired sowing 60 cm wide beds were raised, row-to-row spacing between paired rows was 45 cm and row spacing between pairs was 75 cm but plant-to-plant spacing was kept as 30 cm. Irrigation was applied as needed. In the single furrow, the row-to-row spacing was 60 cm and plant-to-plant spacing was 30 cm. As per the recommendations of PAU Ludhiana "Package of Practices for Vegetable Crops" [1], full package of plant protection measures were adopted during the growth period of crop so as to have a disease free and weed free crop.

FIGURE 2.1 Low tunnels of different heights in experimental field.

After transplanting of peppers, the crop was covered with poly sheet of 50-micron thickness with width of 150 cm, 185 cm and 240 cm over the low tunnel frame heights of 45 cm, 60 cm and 75 cm, respectively, to protect crop from frost and other injury (Fig. 2.1). The low tunnel frames were kept at beginning and at end of paired row and distance between successive frames was kept as 2.50 m. The crop was completely covered with low tunnels till 4th February 2009 and after that low tunnels were removed.

To observe the effects of low tunnels on soil temperature, soil temperature thermometers were installed at a depth of 10 cm in the field in 45 subplots under low tunnel and outside in adjoining open field.

Maximum and minimum air temperature was recorded by placing thermometer at middle in each subplot and outside in adjoining open field. The minimum temperature was recorded at 7.30 a.m. in the morning and maximum temperature was recorded at 2:30 p.m. in the afternoon. The observations were also recorded daily at 7.30 a.m. and 2:30 p.m. respectively, till the tunnel cover was removed from the crop.

Relative humidity was recorded by placing hygrometer in the middle in each subplot and outside in adjoining open field. These observations were also recorded daily at 7:30 a.m. and 2:30 p.m., till the tunnel cover was removed from the crop.

Maximum solar radiation inside the tunnel and outside in adjoining field was taken out by using digital Lux meter (TES 1332) daily at 2:30 p.m., till the tunnel cover was removed from the crop.

2.3 RESULTS AND DISCUSSION

To observe the effects of different treatments on microclimate inside the low tunnels: Air temperature, soil temperature, relative humidity and solar radiation were recorded daily from the time when the crop was covered with low tunnels, till the low tunnels were removed (i.e., 27th November 2008 to 4th February 2009). Mean of seven days was calculated and used for analysis.

2.3.1   AIR TEMPERATURE

The measurement of recorded temperature showed that the thermal condition depends on the tunnel height and the type of irrigation applied as presented in Table 2.1. At 7:30 a.m., the air temperature in low tunnel varied from 8.75°C to 13.42°C and was higher by 3.25°C to 3.71°C in comparison with open field where temperature varied from 5.50° C to 9.71°C during the low tunnel coverage period.

Among the tunnel height treatments, the air temperature was highest in H2 which varied from 9.33°C to 13.42°C, and was higher by 0.57°C to 0.58°C in comparison with H1 treatment where temperature was lowest that varied from 8.75°C to 12.85°C. Among the irrigation treatments, the air temperature was highest in I2, and varied from 9.61°C to 13.42°C, and was higher by 0.76°C to 0.86°C in comparison with I5 treatment where temperature was lowest with a variation of 8.75°C to 12.66°C. For the treatment combinations, the air temperature was highest with a variation of 9.75° C to 13.42°C and was higher by 1°C H2I2 in comparison with I5H1 treatment where temperature was lowest with a variation of varied 8.75° C to 12.42°C.

At 2:30 p.m., the air temperature varied from 23.8° C to 36.8°C and was higher by 6.38°C to 9.3°C in low tunnel compared to open field where temperature varied from 17.42° C to 27.5°C during the low tunnel coverage period. Among the tunnel height treatments, the air temperature was highest with a variation of 24.42° C to 36.8°C and was higher by 0.14°C to 0.62°C in H2 compared to H1 treatment where temperature was lowest with a variation of 23.8° C to 36.66°C. Among the irrigation treatments, the air temperature was highest with a variation of 25.94°C to 36.8°C and was higher by 1.1°C to 2.14°C in I2 treatment compared to I5 treatment where temperature was lowest with a variation of 23.8°C to 35.7°C. For the treatment combination, the air temperature was highest with a variation of 26.61°C to 36.8°C and was higher by 2.81°C to 3.23°C in I2H2 treatment compared to I5H1 treatment where temperature was lowest with a variation of 23.8°C to 33.57°C. Differences in temperature between low tunnels treatments may be due to many factors including initial air, soil temperature, soil moisture, air volume in tunnel, convective and conductive heat exchange characteristics of the material and transmissibility to long wave radiation.

These results were in close proximity with those of Wolfe et al. [7], who reported 5°C to 20°C rise in daytime air temperature under row covers as compared with open field.

TABLE 2.1 Variation of Air Temperature In Different Heights of Low Tunnel With Various Irrigation Treatments

Week after covering of crop	Mean Air Temp. (°C) (Open Field)	Mean Air Temperature (°C)							
		Low Tunnel Heights							
		45 cm		60 cm			75 cm		
	7:30 a.m.	2:30 p.m.	7:30 a.m.	2:30 p.m.	7:30 a.m.	2:30 p.m.	7:30 a.m.	2:30 p.m.
				I1 = IW/ CPE= 0.60					
1. (27 Nov–3 Dec)	7.85	24.71	10.90	32.47	11.37	33.42	11.14	33.09
3. (11–17 Dec)	7.85	22.42	11.09	31.37	11.42	32.90	11.18	32.37
4. (18–24 Dec)	9.28	20.85	12.33	27.90	12.61	28.94	12.56	28.66
5. (25–31 Dec)	7.42	17.42	10.71	24.94	11.33	25.80	10.99	25.28
6. (1–7 Jan)	6.16	18.00	9.33	26.28	9.56	27.09	9.51	26.95
7. (8–14 Jan)	5.50	23.42	9.04	30.04	9.52	31.24	9.28	30.61
8. (15–21 Jan)	8.28	22.57	11.42	30.85	11.61	31.71	11.52	31.14
9. (22–28 Jan)	8.66	23.50	11.85	30.18	11.99	31.56	11.95	31.28
10. (29 Jan–4 Feb)	7.00	27.50	10.18	35.56	10.33	36.56	10.23	36.14
				I2 = IW/CPE= 0.75					
1. (27 Nov–3 Dec)	7.85	24.71	11.33	33.38	11.90	33.99	11.47	33.94
2. (4–10 Dec)	9.71	23.14	12.85	31.61	13.42	32.56	13.19	32.04

TABLE 2.1 *(Continued)*

3. (11–17 Dec)	7.85	22.42	11.28	32.37	11.76	33.04	11.61	32.90
4. (18–24 Dec)	9.28	20.85	12.61	29.33	12.90	29.80	12.80	29.61
5. (25–31 Dec)	7.42	17.42	11.23	25.94	11.66	26.61	11.42	26.18
6. (1–7 Jan)	6.16	18.00	9.71	27.28	9.95	27.99	9.85	27.47
8. (15–21 Jan)	8.28	22.57	11.71	31.56	11.95	31.95	11.80	31.80
9. (22–28 Jan)	8.66	23.50	12.09	31.71	12.52	31.99	12.23	31.85
10. (29 Jan–4 Feb)	7.00	27.50	10.37	36.66	10.61	36.80	10.47	36.95
I3 = IW/CPE= 0.90								
1. (27 Nov–3 Dec)	7.85	24.71	11.04	32.90	11.76	33.75	11.33	33.61
2. (4–10 Dec)	9.71	23.14	12.76	30.99	13.23	32.23	13.04	31.04
3. (11–17 Dec)	7.85	22.42	11.18	31.8	11.61	32.76	11.33	32.75
4. (18–24 Dec)	9.28	20.85	12.47	28.75	12.71	29.42	12.66	28.94
5. (25–31 Dec)	7.42	17.42	11.14	25.33	11.52	26.37	11.28	25.90
6. (1–7 Jan)	6.16	18.00	9.56	26.99	9.71	27.47	9.61	27.23
7. (8–14 Jan)	5.50	23.42	9.18	30.23	9.56	31.61	9.47	31.18
8. (15–21 Jan)	8.28	22.57	11.52	31.18	11.85	31.66	11.66	31.33
9. (22–28 Jan)	8.66	23.50	12.04	31.66	12.33	31.71	12.13	31.61
10. (29 Jan–4 Feb)	7.00	27.50	10.23	36.42	10.47	36.75	10.37	36.52
I4 = Paired row planting								
1. (27 Nov–3 Dec)	7.85	24.71	11.04	32.90	11.76	33.75	11.33	33.61

TABLE 2.1   *(Continued)*

2. (4–10 Dec)	9.71	23.14	12.76	30.99	13.23	32.23	13.04	31.04
3. (11–17 Dec)	7.85	22.42	11.18	31.8	11.61	32.76	11.33	32.75
4. (18–24 Dec)	9.28	20.85	12.47	28.75	12.71	29.42	12.66	28.94
5. (25–31 Dec)	7.42	17.42	11.14	25.33	11.52	26.37	11.28	25.90
6. (01–7 Jan)	6.16	18.00	9.56	26.99	9.71	27.47	9.61	27.23
7. (8–14 Jan)	5.50	23.42	9.18	30.23	9.56	31.61	9.47	31.18
8. (15–21 Jan)	8.28	22.57	11.52	31.18	11.85	31.66	11.66	31.33
9. (22–28 Jan)	8.66	23.50	12.04	31.66	12.33	31.71	12.13	31.61
10. (29 Jan–4 Feb)	7.00	27.50	10.23	36.42	10.47	36.75	10.37	36.52

I5 = Single row planting

1. (27 Nov–3 Dec)	7.85	24.71	10.47	31.23	10.95	32.18	10.61	31.33
3. (11–17 Dec)	7.85	22.42	10.85	30.18	11.09	31.90	10.94	31.33
4. (18–24 Dec)	9.28	20.85	11.90	27.32	12.33	27.66	12.14	27.56
5. (25–31 Dec)	7.42	17.42	10.42	23.80	10.99	24.42	10.71	24.18
6. (1–7 Jan)	6.16	18.00	8.85	24.42	9.33	25.04	9.04	24.90
7. (8–14 Jan)	5.50	23.42	8.75	27.90	9.37	29.04	8.90	29.23
8. (15–21 Jan)	8.28	22.57	11.23	30.28	11.47	30.94	11.33	30.47
9. (22–28 Jan)	8.66	23.50	11.56	28.99	11.76	30.42	11.66	29.94
10. (29 Jan–4 Feb)	7.00	27.50	10.04	33.57	10.18	35.70	10.09	34.61

2.3.2 *SOIL TEMPERATURE*

The measurement of recorded soil temperature also depends on the tunnel height and the type of irrigation applied as presented in Table 2.2. At 7:30 a.m., the soil temperature in low tunnel varied from 8.71°C to 16.61°C and was higher by 1.29°C to 4.76°C in comparison with open field where soil temperature varied from 7.42°C to 11.85°C during the low tunnel coverage period. Among the tunnel height treatments, the soil temperature in H2 treatment was highest with a variation of 9.23°C to 16.61°C and was higher by 0.52°C to 1°C in comparison with H1 treatment where soil temperature was lowest with a variation of 8.71°C to 15.61°C.

Among the irrigation treatments, the soil temperature in I2 treatment was highest with a variation of 10.57° C to 16.61°C and was higher by 0.95°C to 1.86°C in comparison with I5 treatment where soil temperature was lowest with a variation of 8.7°C to 15.7°C. For the treatment combinations, the soil temperature in I2H2 treatment was highest with a variation of 11.13°C to 16.61°C and was higher by 2.42°C to 2.62°C in comparison with I5H1 treatment where soil temperature was lowest with a variation of 8.71° C to 13.99°C.

At 2:30 p.m., the soil temperature in low tunnel varied from 20.52° C to 27.95°C and was higher by 3.52°C to 6.38°C in comparison with open field where temperature varied from 17° C to 21.57°C during the low tunnel coverage period. Among the tunnel height treatments, the soil temperature in H2 treatment was highest with a variation of 21.76° C to 27.95°C and was higher by 0.48°C to 1.24°C in comparison with H1 treatment where temperature was lowest with a variation of 20.52°C to 27.47°C. Among the irrigation treatments, the soil temperature in I2 treatment was highest with a variation of 23°C to 27.95°C and was higher by 1.43°C to 2.48°C in comparison with I5 treatment where soil temperature was lowest with a variation of 20.52° C to 26.52°C. For the treatment combinations, the soil temperature in I2H2 treatment was highest with a variation of 23.76°C to 27.95°C and was higher by 1.43°C to 2°C in comparison with I5H1 treatment where temperature was lowest with a variation of 21.76°C to 26.52°C. Differences in soil temperature among low tunnels treatments may be due to many factors including initial air, soil temperature, soil moisture, air volume in tunnel, convective and conductive heat exchange characteristics of the material, surface area of the tunnel to volume ratio, and transmissibility to long wave radiation.

TABLE 2.2 Variations In Soil Temperature In Different Heights of Low Tunnel With Various Irrigation Treatments

Week after covering of crop	Mean Soil Temp. (°C) (Open Field)	Mean Soil Temperature (°C)							
		Low Tunnel Heights							
		45 cm			60 cm			75 cm	
		7:30 a.m.	2:30 p.m.	7:30 a.m.	2:30 p.m.	7:30 a.m.	2:30 p.m.	7:30 a.m.	2:30 p.m.
					I1 = IW/ CPE= 0.60				
1. (27 Nov–3 Dec)	11.00	19.14	13.28	23.14	13.75	24.19	13.61	23.85
2. (4–10 Dec)	11.85	20.57	13.95	25.23	14.66	26.14	14.28	25.71
3. (11–17 Dec)	9.71	17.42	12.37	22.47	12.90	22.95	12.66	22.66
4. (18–24 Dec)	11.14	20.85	13.47	25.38	14.09	26.85	13.76	25.9
5. (25–31 Dec)	11.28	18.14	13.71	22.52	14.04	23.38	13.99	23.00
6. (01–7 Jan)	10.66	19.14	13.28	23.95	13.94	25.04	13.61	24.47
7. (08–14 Jan)	7.42	17.00	10.18	22.33	10.80	23.19	10.42	22.61
8. (15–21 Jan)	10.42	19.57	15.04	23.61	16.00	24.47	15.56	24.14
9. (22–28 Jan)	9.71	19.71	12.38	23.19	12.75	24.04	12.61	23.57
10. (29 Jan–4 Feb)	10.50	21.57	13.85	26.28	14.04	27.28	13.99	26.61

Best Management Practices for Drip Irrigated Crops

I2 = IW/CPE= 0.75

1. (27 Nov −3 Dec)	11.00	19.14	13.80	24.23	14.28	25.18	14.14	24.62
2. (4–10 Dec)	11.85	20.57	14.37	26.38	15.09	26.85	14.71	26.52
3. (11–17 Dec)	9.71	17.42	12.95	23.00	13.37	23.76	13.18	23.28
4. (18–24 Dec)	11.14	20.85	13.94	26.85	14.71	27.52	14.47	27.09
5. (25–31 Dec)	11.28	18.14	14.23	23.42	15.09	24.33	14.47	23.71
6. (1–7 Jan)	10.66	19.14	13.99	25.04	14.37	25.66	14.13	25.23
7. (8–14 Jan)	7.42	17.00	10.57	23.00	11.13	23.76	10.76	23.23
8. (15–21 Jan)	10.42	19.57	15.61	24.71	16.61	26.09	15.85	25.61
9. (22–28 Jan)	9.71	19.71	12.90	24.19	13.23	25.33	13.00	24.95
10. (29 Jan–4 Feb)	10.50	21.57	14.04	27.47	14.61	27.95	14.37	27.66

I3 = IW/CPE= 0.90

1. (27 Nov−3 Dec)	11.00	19.14	13.42	24.09	13.95	24.85	13.71	24.57
2. (4–10 Dec)	11.85	20.57	14.23	26.00	14.80	26.62	14.56	26.43
3. (11–17 Dec)	9.71	17.42	12.75	22.71	13.18	23.38	12.99	23.14
4. (18–24 Dec)	11.14	20.85	13.85	26.14	14.42	27.04	14.18	26.9
5. (25–31 Dec)	11.28	18.14	13.99	23.14	14.61	23.95	14.33	23.57

TABLE 2.2   *(Continued)*

6. (1–7 Jan)	10.66	19.14	13.75	24.76	14.14	25.33	13.95	24.99
7. (8–14 Jan)	7.42	17.00	10.33	22.66	10.94	23.52	10.52	22.95
8. (15–21 Jan)	10.42	19.57	15.37	24.33	16.13	24.80	15.71	24.76
9. (22–28 Jan)	9.71	19.71	12.61	23.99	13.04	24.95	12.90	24.52
10. (29 Jan–4 Feb)	10.50	21.57	13.94	26.99	14.37	27.52	14.14	27.14
I4 = Paired row planting								
1. (27 Nov–3 Dec)	11.00	19.14	13.09	22.57	13.61	23.81	13.47	23.28
2. (4–10 Dec)	11.85	20.57	13.80	24.42	14.47	25.71	14.14	25.09
3. (11–17 Dcc)	9.71	17.42	12.09	21.19	12.70	22.76	12.37	22.33
4. (18–24 Dec)	11.14	20.85	13.28	24.95	13.85	25.95	13.71	25.57
5. (25–31 Dec)	11.28	18.14	13.51	21.71	13.85	23.09	13.71	22.19
6. (1–7 Jan)	10.66	19.14	13.04	23.05	13.71	24.61	13.37	24.19
7. (8–14 Jan)	7.42	17.00	9.94	21.38	10.33	22.23	10.04	21.90
8. (15–21 Jan)	10.42	19.57	14.95	23.43	15.75	24.00	15.37	23.76
9. (22–28 Jan)	9.71	19.71	12.23	22.76	12.52	23.62	12.37	23.04
10. (29 Jan–4 Feb)	10.50	21.57	13.66	25.90	13.85	26.81	13.76	26.52

TABLE 2.2　*(Continued)*

I5 = Single row planting							
1. (27 Nov–3 Dec)	11.00	19.14	12.95	22.09	13.33	23.47	13.18	22.62
2. (4–10 Dec)	11.85	20.57	13.28	24.75	14.13	25.38	13.80	24.19
3. (11–17 Dec)	9.71	17.42	11.89	20.52	12.33	21.76	12.09	21.57
4. (18–24 Dec)	11.14	20.85	13.04	23.76	13.56	25.61	13.33	24.90
5. (25–31 Dec)	11.28	18.14	13.37	21.33	13.71	22.42	13.47	21.90
6. (1–7 Jan)	10.66	19.14	12.90	22.28	13.42	24.38	13.04	23.42
7. (8–14 Jan)	7.42	17.00	8.71	21.09	9.23	22.14	8.95	21.42
8. (15–21 Jan)	10.42	19.57	13.99	22.81	15.66	23.62	14.56	23.19
9. (22–28 Jan)	9.71	19.71	11.61	22.24	12.23	22.80	11.99	22.47
10. (29 Jan–4 Feb)	10.50	21.57	13.04	25.42	13.56	26.52	13.23	26.14

The above mentioned results were in close proximity with those of Hemphill [4] who reported that row covers increased daily mean soil temperature by 1 to 4°C over bare ground. The results are also in line with those of Gaye and Maurer [3] who reported that row cover elevated mean soil temperature 1 to 6°C and promoted plant growth compared with bare soil.

2.3.3　RELATIVE HUMIDITY

The measurement of recorded relative humidity also depends on the tunnel height and the type of irrigation applied as presented in Table 2.3. At 7:30 a.m., the relative humidity in low tunnel varied from 87.59% to 94.13% and was lower by 4.58% to 6.55% in comparison with open field where relative humidity varied from 94.14% to 98.71% during the low tunnel coverage period. Among the tunnel height treatments, the relative humidity in H3 treatment was highest with a variation of 88.4% to 94.00% and was higher by 0.01% to 0.81% in comparison with H1 treatment where relative humidity was lowest with a variation of 87.59% to 93.99%. Among the irrigation treatments, the relative humidity in I3 treatment was highest with a variation of 89.3% to 94.13% and was higher by 1.23% to 1.71% in comparison with I5 treatment where relative humidity was lowest with a variation of 87.59% to 92.9%. For the treatment combinations, the relative humidity in I3H3 treatment was highest with a variation of 89.8% to 93.8% and was higher by 1.19% to 2.23% in comparison with I5H1 treatment where relative humidity was lowest with a variation of 87.57% to 92.61%. The decrease in the relative humidity from the open field may be due to dew effect on the poly sheet in the morning.

At 2:30 p.m., the relative humidity in low tunnel varied from 59.71% to 91.66% and was higher by 11.52% to 17.57% in comparison with open field where relative humidity varied from 42.14% to 80.14% during the low tunnel coverage period. Among the tunnel height treatments, the relative humidity in H2 treatment was highest with a variation of 61.75% to 91.66% and was higher by 1.86% to 2.04% in comparison with H1 treatment where relative humidity was lowest with a variation of 59.71% to 89.8%. Among the irrigation treatments, the relative humidity in I2 treatment was highest with a variation of 63.71% to 91.66% and was higher by 2% to 4% in comparison with I5 treatment where relative humidity was lowest with a variation of 59.71% to 89.66%. For the treatment combinations, the relative humidity in I2H2 treatment was highest with a variation of 67.04% to 91.66% and was higher by 3.57% to 7.28% in comparison with I4H3 treatment where relative humidity was lowest with a variation of 59.76% to 88.09%. Differences in relative humidity between low tunnels treatments may be due to many factors including initial air, soil temperature, soil moisture, air volume in tunnel, convective and conductive heat exchange characteristics of the material, surface area of the tunnel to volume ratio and transmissibility to long wave radiation. The data also reveals that the low tunnel effect decreases the relative humidity in the morning but in the daytime it increases the relative humidity to provide favorable condition for plant growth. These results were in close proximity with those of Chaugule et al. [2].

TABLE 2.3 Variation of Relative Humidity In Different Heights of Low Tunnel With Various Irrigation Treatments

Week After Covering of Crop	Mean Relative Humidity (%) (Open Field)	Mean Relative Humidity (%)						
		Low Tunnel Heights						
		45 cm		60 cm		75 cm		
	7:30 a.m.	2:30 p.m.	7:30 a.m.	2:30 p.m.	7:30 a.m.	2:30 p.m.	7:30 a.m.	2:30 p.m.
I1 = IW/ CPE= 0.60								
1. (27 Nov–3 Dec)	98.71	42.14	93.76	63.75	93.52	65.33	94.00	64.8
2. (4–10 Dec)	98.14	48.85	91.8	66.33	91.9	67.04	92.09	66.99
3. (11–17 Dec)	97.71	51.42	90.47	67.47	90.61	69.37	90.87	68.09
4. (18–24 Dec)	94.14	65.42	88.78	77.99	88.87	80.47	89.21	78.76
5. (25–31 Dec)	98.14	80.14	91.9	88.04	92.47	90.8	93.18	89.09
6. (1–7 Jan)	96.85	62	89.95	75.85	90.42	76.8	90.8	76.37
7. (8–14 Jan)	97.42	58.85	92.13	74.09	92.61	75.56	92.76	74.85
8. (15–21 Jan)	96.57	68.71	90.33	82.18	90.66	83.09	91.13	82.66
9. (22–28 Jan)	94.14	66	88.71	78.94	88.85	79.56	89.23	79.18
10. (29 Jan–4 Feb)	97.57	57.28	89.66	71.71	90.28	74.23	90.9	73.9
I2 = IW/CPE= 0.75								
1. (27 Nov–3 Dec)	98.71	42.14	93.42	63.71	93.9	67.04	93.61	65.56
2. (4–10 Dec)	98.14	48.85	91.85	66.28	91.9	68.28	92.47	67.52
3. (11–17 Dec)	97.71	51.42	90.4	68.23	90.83	69.9	91.54	68.85
4. (18–24 Dec)	94.14	65.42	88.26	79.18	89.16	82.04	89.73	79.8
5. (25–31 Dec)	98.14	80.14	92.37	89.8	92.47	91.66	92.56	90.57
6. (1–7 Jan)	96.85	62	90.28	75.99	90.61	78.28	90.99	77.71

TABLE 2.2 *(Continued)*

7. (8–14 Jan)	97.42	58.85	92.56	74.75	92.8	77.28	93.04	76.18
8. (15–21 Jan)	96.57	68.71	90.66	82.18	91.18	83.47	91.33	81.85
9. (22–28 Jan)	94.14	66	89.18	81.09	89.23	83.04	89.57	81.61
10. (29 Jan–4 Feb)	97.57	57.28	90.18	73.8	90.37	75.18	91.37	74.04

I3 = IW/CPE= 0.90

1. (27 Nov–3 Dec)	98.71	42.14	93.99	63.61	94.13	66.23	93.8	64.09
2. (4–10 Dec)	98.14	48.85	92.09	66.04	92.13	67.8	92.66	66.09
3. (11–17 Dec)	97.71	51.42	91.21	66.47	91.73	69.61	91.83	68.42
4. (18–24 Dec)	94.14	65.42	89.3	78.75	89.73	80.47	90.11	79.94
5. (25–31 Dec)	98.14	80.14	92.85	88.94	93.33	90.75	93.47	89.39
6. (1–7 Jan)	96.85	62	90.94	75.94	91.42	76.85	91.99	76.71
7. (8–14 Jan)	97.42	58.85	92.61	73.99	92.99	75.85	93.71	74.37
8. (15–21 Jan)	96.57	68.71	91.14	80.23	91.33	83.04	91.9	82.28
9. (22–28 Jan)	94.14	66	89.37	80.51	89.61	81.13	89.8	81.09
10. (29 Jan–4 Feb)	97.57	57.28	91.66	72.23	92.8	74.66	93.13	73.8

I4 = Paired row planting

1. (27 Nov–3 Dec)	98.71	42.14	92.85	61.23	93.23	62.99	93.33	59.76
2. (4–10 Dec)	98.14	48.85	91.04	65.18	91.23	65.42	91.52	63.42
3. (11–17 Dec)	97.71	51.42	90.59	66.14	90.78	67.18	91.44	64.33
4. (18–24 Dec)	94.14	65.42	88.54	77.23	88.73	78.33	89.16	76.66
5. (25–31 Dec)	98.14	80.14	91.94	88.37	91.99	88.8	92.13	88.09
6. (1–7 Jan)	96.85	62	89.47	74.09	90.18	75.42	90.23	74.13
7. (8–14 Jan)	97.42	58.85	91.71	73.61	91.94	72.32	92.32	71.18
8. (15–21 Jan)	96.57	68.71	90.04	82.09	90.56	81.18	90.9	80.37
9. (22–28 Jan)	94.14	66	88.37	75.71	88.42	76.71	88.47	75.56
10. (29 Jan–4 Feb)	97.57	57.28	89.8	72.14	90.09	71.71	90.8	69.75

I5 = Single row planting

TABLE 2.2   *(Continued)*

1. (27 Nov–3 Dec)	98.71	42.14	92.61	59.71	92.85	61.75	92.9	60.04
2. (4–10 Dec)	98.14	48.85	91.04	63.13	91.18	64.42	91.37	63.28
3. (11–17 Dec)	97.71	51.42	89.49	64.13	90.3	65.14	90.78	64.61
4. (18–24 Dec)	94.14	65.42	87.59	75.85	88.11	77.57	88.4	76.85
5. (25–31 Dec)	98.14	80.14	91.13	88.9	91.51	89.66	91.7	89.18
6. (1–7 Jan)	96.85	62	89.37	73.04	89.61	74.09	88.99	73.42
7. (8–14 Jan)	97.42	58.85	91.33	72.56	91.56	72.28	91.66	71.47
8. (15–21 Jan)	96.57	68.71	89.71	81.8	89.76	81.75	89.9	81.56
9. (22–28 Jan)	94.14	66	88.42	76.28	88.47	77.37	87.9	76.75
10. (29 Jan–4 Feb)	97.57	57.28	88.76	71.56	89.71	70.95	89.76	69.85

2.3.4   SOLAR RADIATION

The measurement of solar radiation depends on the tunnel height and the type of irrigation applied as presented in Table 2.4. At 2:30 p.m., the solar radiation in low tunnel, which varied from 19,614 lux to 43,242.66 lux, was lower by 16.65% to 37.45% in comparison with open field whose solar radiation varied from 31,360 lux to 51,885.71 lux during the low tunnel coverage period. Among the tunnel height treatments, in H3 treatment the solar radiation was highest which varied from 19,861.66 lux to 43,242.66 lux and was higher by 1.26% to 9.01% in comparison with H1 treatment whose solar radiation was lowest which varied from 19614 lux to 39,666.33 lux. Among the irrigation treatments, in I5 treatment the solar radiation was highest which varied from 19,718.66 lux to 43,242.66 lux and was higher by 0.53% to 6.97% in comparison with I1 treatment whose solar radiation was lowest which varied from 19,614 lux to 40,423.33 lux. For the treatment combination in I5H3 treatment the solar radiation was highest which varied from 19,994.66 lux to 43,242.66 lux and was higher by 1.94% to 10.97% in comparison with I1H1 treatment whose solar radiation was lowest which varied from 19,614 lux to 38,966.33 lux. Differences in solar radiation between low tunnels treatments may be due to different surface area of the tunnel and soil moisture distribution. The above mentioned results were in close proximity with those of Siwek et al. (1994) who reported that on average 70.3% of radiation reached to the plants inside the tunnel.

TABLE 2.4 Variation of Solar Radiation In Different Heights of Low Tunnel Under Various Irrigation Treatments

Week after covering of crop	Mean Solar Radiation (Lux) (Open Field)	Mean Solar Radiation (Lux)		
		Low Tunnel Heights		
		45 cm	60 cm	75 cm
		I1 = IW/ CPE= 0.60		
1. (27 Nov−3 Dec)	40,328.57	26,361.66	27,647.00	28,399.66
2. (4–10 Dec)	50,471.42	38,304.33	40,166.33	40,423.33
3. (11–17 Dec)	45,514.28	30,399.66	31,409.00	33,032.66
4. (18–24 Dec)	48,614.28	35,270.66	37,866.33	38,147.00
5. (25–31 Dec)	31,360.00	19,614.00	19,747.33	19,866.00
6. (1−7 Jan)	33,342.85	23,380.33	23,557.00	23,904.33
7. (8−14 Jan)	44,128.57	29,647.00	29,871.00	30,285.00
8. (15−21 Jan)	51,885.71	38,966.33	39,728.33	39,842.33
9. (22−28 Jan)	46,342.85	33,914.00	34,499.66	34,694.66
10. (29 Jan–4 Feb)	51,442.85	36,556.66	37,004.33	37,104.66
		I2 = IW/CPE= 0.75		
1. (27 Nov−3 Dec)	40,328.57	27,218.66	27,518.66	28,385.33

TABLE 2.3   *(Continued)*

2. (4–10 Dec)	50,471.42	39,176.00	40,361.66	41,737.66
3. (11–17 Dec)	45,514.28	31,118.66	31,823.33	32,261.33
4. (18–24 Dec)	48,614.28	36,399.66	37,899.66	39,690.00
5. (25–31 Dec)	31,360.00	19,695.00	19,733.00	19,914.00
6. (1–7 Jan)	33,342.85	23,328.33	23,666.33	23,899.66
7. (8–14 Jan)	44,128.57	29,380.66	29,875.66	30,161.66
8. (15–21 Jan)	51,885.71	38,909.00	39,666.00	39,885.33
9. (22–28 Jan)	46,342.85	34,480.33	34,628.33	34,842.66
10. (29 Jan–4 Feb)	51,442.85	36,894.66	37,066.33	37,290.33
I3 = IW/CPE= 0.90				
1. (27 Nov–3 Dec)	40,328.57	27,037.66	27,499.33	27,818.66
2. (4–10 Dec)	50,471.42	39,533.00	40,076.00	41,885.00
3. (11–17 Dec)	45,514.28	30,652.00	31,542.66	32,299.66
4. (18–24 Dec)	48,614.28	37,647.33	37,704.33	37,847.33
5. (25–31 Dec)	31,360.00	19,485.33	19,752.00	19,899.66
6. (1–7 Jan)	33,342.85	23,213.66	23,490.00	23,756.66
7. (8–14 Jan)	44,128.57	29,294.66	29,842.33	30,242.33

TABLE 2.3 *(Continued)*

8. (15–21 Jan)	51,885.71	39,299.66	39,475.66	39,875.66
9. (22–28 Jan)	46,342.85	33,833.00	34,633.00	34,880.66
10. (29 Jan–4 Feb)	51,442.85	36,918.66	37,014.00	37,175.66

I4 = Paired row planting

1. (27 Nov–3 Dec)	40,328.57	26,814.00	27,347.00	27,609.00
2. (4–10 Dec)	50,471.42	39,666.33	40,433.00	41,304.33
3. (11–17 Dec)	45,514.28	30,318.66	31,856.66	32,807.66
4. (18–24 Dec)	48,614.28	35,409.33	37,928.00	38,047.00
5. (25–31 Dec)	31,360.00	19,633.00	19,761.33	19,861.66
6. (1–7 Jan)	33,342.85	23,109.00	23,433.00	23,785.33
7. (8–14 Jan)	44,128.57	29,704.33	29,928.33	30,190.00
8. (15–21 Jan)	51,885.71	39,375.66	39,599.66	39,795.00
9. (22–28 Jan)	46,342.85	33,380.33	34,490.00	34,652.00
10. (29 Jan–4 Feb)	51,442.85	36,809.00	37,128.33	37,290.33

I5 = Single row planting

1. (27 Nov–3 Dec)	40,328.57	28,123.33	28,795.00	29,999.66

TABLE 2.3   *(Continued)*

2. (4–10 Dec)	50,471.42	39,661.66	40,609.00	43,242.66
3. (11–17 Dec)	45,514.28	32,566.00	33,209.00	33,752.00
4. (18–24 Dec)	48,614.28	38,356.66	38,766.33	39,485.33
5. (25–31 Dec)	31,360.00	19,718.66	19,832.66	19,994.66
6. (1–7 Jan)	33,342.85	23,499.33	23,604.33	23,956.66
7. (8–14 Jan)	44,128.57	29,723.33	29,990.00	30,528.33
8. (15–21 Jan)	51,885.71	39,370.66	39,933.00	39,932.66
9. (22–28 Jan)	46,342.85	34,447.00	34,618.66	34,780.66
10. (29 Jan–4 Feb)	51,442.85	36,999.66	37,185.33	37,352.00

2.4   CONCLUSIONS

The mean air temperature was higher in low tunnel than open field by 3.25°C to 3.71°C at 7:30 a.m. and 6.38°C to 9.30°C at 2:30 p.m., respectively. The mean soil temperature was higher in low tunnel than open field by 1.29°C to 4.76°C at 7:30 a.m. and 3.52°C to 6.30°C at 2:30 p.m., respectively. The highest air and soil temperatures were observed in H2 and I2 treatments among the tunnel height and irrigation treatments, respectively. For the combinations, I2H2 treatment gave higher air and soil temperature than other treatment combinations.

The highest relative humidity was observed at 7:30 a.m. and lowest at 2:30 p.m. At 7:30 a.m., mean relative humidity in the low tunnels was lowered by 4.58% to 6.55% than open field, however at 2:30 p.m. it was higher by 11.52% to 17.57%. The highest relative humidity was observed in H3 and I2 treatments among the tunnel height and irrigation treatments, respectively. For the combinations, I3H3 gave higher mean relative humidity at 7:30 a.m., but at 2:30 p.m. combination I2H2 treatment gave higher relative humidity than other treatment combinations.

The mean solar radiation under low tunnels was lowered by 16.65% to 37.45% as compared to open field at 2:30 p.m. The highest solar radiation was observed in H3 and I5 treatments in the tunnel height and irrigation treatments, respectively. For the combinations, higher mean solar radiation in I5H3 treatment was observed than other treatments combinations.

2.5 SUMMARY

Field experiment was conducted in the Department of Soil and Water Engineering, PAU, Ludhiana in 2008–2009 to study the effects of low tunnel environment on sweet pepper (*Capsicum annuum L. var. grossum*). The experiment was laid out in split plot design keeping five irrigation treatments (drip irrigation with IW/CPE ratio of 0.60 (I1), 0.75 (I2), 0.90 (I3), furrow irrigation with paired row planting (I4) and single row planting (I5)), in main plots and three different low tunnel heights (45 cm (H1), 60 cm (H2) and 75 cm (H3)) in sub plots and replicated three times. The air temperature, soil temperature, relative humidity, solar radiations were observed. For the combinations, I2H2 treatment gave higher air and soil temperatures than other treatments combinations; and I3H3 gave higher mean relative humidity at 7:30 a.m. but at 2:30 p.m. combination I2I12 treatment gave higher relative humidity than other treatment combinations.

KEYWORDS

- air temperature
- irrigation
- low tunnel technology
- microclimate
- relative humidity
- soil temperature
- sweet pepper
- tunnel height

REFERENCES

1. Anonymous, (2008). *Package of Practices for Vegetable Crops.* pp. 20–25. Punjab Agricultural University, Ludhiana, India.
2. Chaugule, A. A., Gutal, G. B., Kulkarni, P. V. (1990). The feasibility of plastic polyhouse for capsicum crop. *Proc. International Agricultural Engineering Conference and Exhibition*, Bangkok, Thailand. pp. 1485–1489.

3. Gaye, M. M., Maurer, A. R. (1991). Modified transplant production techniques to increase yield and improve earliness of Brussels sprouts. *J. Amer. Soc. Hort. Sci., 116,* 210–214.
4. Hemphill, D. D. (1986). Response of muskmelon to three floating row covers. *J. Amer. Soc. Hort. Sci., 111,* 513–517.
5. Lamarrel, M., Larean, M. J., Payette, S., Fortin, C. (1996). Influence of nitrogen fertilization, row covers and cultivars on the production of day neutral strawberry. *Canadian J. Soil Sci., 76,* 29–36.
6. Libik, A., Siwek, P. (1994). Changes in soil temperature affected by the application of plastic covers in field production of lettuce and water melon. *Acta Hort., 371,* 269–273.
7. Wolfe, D. W., Albright, L. D., Wyland, J. (1989). Modeling row cover effects on microclimate and yield, I: Growth response of tomato and cucumber. *J. Amer. Soc. Hort. Sci., 114,* 562–568.

CHAPTER 3

ECONOMICS OF GROWING SWEET PEPPER IN LOW TUNNELS

ARUN KAUSHAL, A. S. LODHI, and KAMAL G. SINGH

3.1   INTRODUCTION

The demand for fresh vegetables, which are rich source of vitamins, carbo-hydrates, proteins and salts, is increasing day by day in India, as people have become more health conscious. Population is increasing at an alarming rate and there is change in dietary patterns. Due to unfavorable climatic conditions, there is a flood of vegetables in the season and very high priced vegetables in off-season. Vegetables can be cultivated in off-season with the use of green houses, low and high poly tunnels.

In the low tunnel technology (LTT), the plants in a single or double row are covered with a sheet/film (preferably polythene), w h i c h is placed over an arch-shaped frame for support. These structures protect the plants from the high winds, low temperature, rain, frost and snow. The increase in marketable yield of vegetable under low tunnels as comparison to open field has been well documented [1, 2].

Sweet pepper also called bell pepper (*Capsicum annuum* L. var. *grossum*) is a high value vegetable crop grown in many states of India including Punjab. Sweet pepper is more sensitive to environment (especially soil moisture and tempera-ture), it can be grown under low tunnels. Soil moisture is one of the predominant factors influencing sweet pepper productivity. There is an increasing demand for maximum irrigation efficiency, due to increasing demand for water for agriculture, the limited available supplies and economic considerations. To achieve this goal, drip irrigation is one of the alternatives. Higher water use efficiency of drip irriga-tion system over the conventional irrigation system has been reported by [3, 4]. Higher profit using LTT was reported by Saini [5].

This chapter discusses economic viability of sweet pepper production under low tunnels.

---

*In this chapter, the currency is expressed in Indian Rupees (1.00 US$ = Rs. 60.93; 1.00 Rs. = 0.02 US$).

3.2   MATERIALS AND METHODS

Field experiment was conducted at the Research Farm of the Department of Soil and Water Engineering at Punjab Agricultural University (PAU), Ludhiana – India from October 2008 to June 2009. Ludhiana is situated at latitude of 30°54'N and longitude of 75°48'E and at a mean height of 247 meters above sea level. Average minimum and maximum temperature in the region varies from 3°C to 43°C, respectively.

A field plot measuring approximately 440.64 m² (43.2 × 10.2 m²) was prepared and the experiment was laid out in split plot design keeping four irrigation treatments as main effects and three different heights of low tunnel as sub main effects. All following treatments were replicated three times:

I1   Drip irrigation with IW/CPE ratio of 0.60,
I2   Drip irrigation with IW/CPE ratio of 0.75,
I3   Drip irrigation with IW/CPE ratio of 0.90, and
I4   Furrow irrigation with paired row planting.
H1   Low tunnel height of 45 cm,
H2   Low tunnel height of 60 cm, and
H3   Low tunnel height of 75 cm.

The soil at the experimental site was sandy loam having pH of 8.9, low in organic carbon and available nitrogen, medium in phosphorous and high in potash. Farm yard manure @ 55 tons/ha was added to the field, one month before the field preparation so that it was thoroughly mixed in the soil and got decomposed by the time of sowing of crop [6].

Nursery production of sweet pepper of "Bharath" variety was done in polyhouse on 13th of October 2008 and transplanting was done in the field on 17th of November 2008. In paired sowing, 60 cm wide beds were raised, row-to-row spacing between paired rows was 45 cm and row spacing between pairs was 75 cm, but plant-to-plant spacing was 30 cm.

The low tunnel frame was constructed with 6 mm thick steel (iron) rods and its shape was parabolic with 60 cm base and desired height of tunnel height. A length of 15 cm at the base was kept for inserting frame into the soil and inside loops were provided on both sides of a frame for the support. After transplanting of crop, it was covered with polyethylene sheet of 50-micron thickness over the low tunnel frame. The low tunnel frames were kept at beginning and at end of paired row and distance between successive frames was 2.5 m. The crop was completely covered with low tunnel till February 4, 2009 and after that low tunnel was removed. Full package of plant protection measures were adopted during the crop growth period, according to package practices for vegetable crops by PAU, Ludhiana.

The fruits from each subplot were picked at green mature stage and weighed at each picking. The weight of all pickings was added and yield per plant was used to find yield per hectare.

Economic analysis was carried out by calculating the amount of material needed for one hectare and seasonal cost of production by considering present market rate, depreciation cost, life of material, annual interest. Cost of cultivation for growing furrow irrigated sweet pepper included expenses occurred on cost of seedlings, cost of manures and fertilizers, cost of hoeing, cost of irrigation, cost for labor charges, cost of herbicides and pesticides, cost of marketing and transportation charges. On these charges, R s . 1,000 per month was taken as additional maintenance charges for drip irrigated sweet pepper. Yield of crop was taken from experimental data generated from the study and first three harvestings were considered as early yield, which had market rate R s . 20 per k g, and remaining yield had seasonal rate of R s . 10 per k g.

Table 3.1 gives the material requirement for constructing low tunnel frame for different tunnel height treatments. For the low tunnel frame, life of the material was taken 15 years, annual interest rate of 10% annual, material cost at Rs. 35 per kg, and labor cost at Rs. 5 per tunnel frame construction. Table 3.2 gives the poly sheet requirements for covering the low tunnels for different treatments. For the poly sheet, life of the poly sheet was considered as two years for one crop only and rate of poly sheet Rs. 60 per kg was considered. In the drip irrigation system, for the drip irrigation components (main line, sub main, fertilizer tank, venturi assembly, filters and pumping unit) 15 years life was considered, and for laterals with inline drippers 5 years life was considered for annual cost calculations. It was assumed that this drip irrigation can be used for two crops (i.e., sweet pepper and eggplant). Prices of these drip irrigation components were taken from the market. The annual cost of drip irrigation was calculated by considering no subsidy and maximum subsidy for drip irrigation (Rs 49,680 per hectare) given by Punjab State Government in 2008–2009. The details of material required for one hectare of cultivation of sweet pepper with drip irrigation system are presented in Table 3.3. With data in Tables 3.1–3.3, annual cost was calculated and benefit-cost ratio for each treatment was evaluated.

The yield data were subjected to statistical analysis using split plot experimental design and using analysis of variance (ANOVA) techniques. For the split plot design, irrigation treatments were considered as main plots and different low tunnel heights as subplots. The significance of differences was tested at 5% level.

TABLE 3.1   Total Weight of Tunnel Frame (kg/ha) For Three Tunnel Height Treatments

Irrigation treatments	Tunnel height	Length	Average weight of each frame	No. of low tunnel frames	Total weight of frames
	cm	cm	g/each frame	No. per ha	Kg/ha
I1 to I4	H1 = 45	180	392	3486	1366.51
	H2 = 60	210	446	3486	1554.76
	H3 = 75	235	498	3486	1736.03

TABLE 3.2   Total Weight of Poly Sheet (kg/ha) For Three Tunnel Heights

Tunnel height	Width	Average weight	Total weight, for 83 frames
cm	cm	g per 10 m length	kg/ha
H1 = 45	150	661.62	571.1
H2 = 60	185	816.0	704.37
H3 = 75	240	1058.6	913.78

Note: The length of each
      frame was 104 m.

TABLE 3.3   List of Materials For One Hectare of Sweet Pepper Production

List of materials	Units	Quantity
PVC pipe, 90 mm × 4 kg/cm^2 (for main)	m	60
PVC pipe, 63 mm × 4 kg/cm^2	m	162
PVC pipe, 40 mm × 4 kg/cm^2 (for submain)		60
LDPE lateral inline (16 mm × 2 kg/cm^2) with emitters of 2.6 lph at 30 cm spacing.	m	8400
Fertilizer tank, 30 L	each	1
Venturi assembly for chemigation	each	1
Sand filter of 40 m^3 capacity	each	1
Screen filter of 40 m^3 capacity	each	1
Pumping unit of 5 horsepower	each	1

TABLE 3.4   Commercial Yield (100 *kg*/ha) of Sweet Pepper For Four Irrigation Treatments and For Three Tunnel Heights

Irrigation treatments, I	Yield (100 kg/ha) of sweet pepper			Mean
	Tunnel height, H, cm			
	45	60	75	
I1 Drip irrigation, IW/CPE = 0.60	211.14	229.09	227.17	222.46
I2 Drip irrigation, IW/CPE = 0.75	275.55	298.86	289.92	288.11
I3 Drip irrigation, IW/CPE = 0.90	253.95	297.95	278.95	276.95

TABLE 3.4 *(Continued)*

I4 Furrow irrigation (paired row)	209.39	229.45	222.59	**220.48**
Mean	**237.51**	**263.84**	**254.65**	

CD at 5%, I = 5.06; CD at 5%, H = 5.09; and CD at 5%, IH = 10.19.

TABLE 3.5 Early Yield (100 kg/ha) of Sweet Pepper For Four Irrigation and Three Tunnel Height Treatments

Irrigation treatments, I	Yield (100 kg/ha) of sweet pepper			Mean
	Tunnel height, H, cm			
	45	**60**	**75**	
I1 Drip irrigation, IW/CPE = 0.60	44.63	50.39	49.29	**48.10**
I2 Drip irrigation, IW/CPE = 0.75	48.27	61.11	57.37	**55.58**
I3 Drip irrigation, IW/CPE = 0.90	42.91	55.98	51.77	**50.22**
I4 Furrow irrigation (paired row)	34.04	38.24	35.21	**35.83**
Mean	**42.46**	**51.43**	**48.41**	—

CD at 5%, I = 3.76; CD at 5%, H = 2.22; and CD at 5%, IH = 4.44.

TABLE 3.6 Benefit–Cost Ratio (BCR) of Sweet Pepper Production For Four Irrigation Treatments and Three Tunnel Heights

Irrigation treatment, I	Details	Low tunnel height, H, cm		
		45	**60**	**75**
		Values in Indian Rupees, Rs./ha		
I1 Drip irrigation, IW/CPE = 0.60	Gross returns (Rs./ha)	255,776	279,480	276,460
	Total expenditure (Rs./ha)	117,346	122,512	130,163
	Net seasonal income (Rs./ha)	138,430	156,968	146,297
	BCR	**2.17**	**2.28**	**2.12**
	Delete this row			

TABLE 3.6   *(Continued)*

I2 Drip irrigation, IW/CPE = 0.75	Gross returns (Rs./ha)	323,823	359,976	347,293
	Total expenditure (Rs./ha)	117,346	122,512	130,163
	Net seasonal income (Rs./ha)	206,477	237,464	217,130
	BCR	**2.75**	**2.93**	**2.66**
I3 Drip irrigation, IW/CPE = 0.90	Gross returns (Rs./ha)	296,863	353,933	330,720
	Total expenditure (Rs./ha)	117,346	122,512	130,163
	Net seasonal income (Rs./ha)	179,517	231,421	200,557
	BCR	**2.52**	**2.88**	**2.54**
I4 Furrow irrigation (paired row)	Gross returns (Rs./ha)	243,436	267,690	257,803
	Total expenditure (Rs./ha)	95243	100,409	108,060
	Net seasonal income (Rs./ha)	148,193	167,281	149,743
	BCR	**2.55**	**2.66**	**2.38**

Note: With No Drip Irrigation Subsidy.

TABLE 3.7   Benefit–Cost Ratio With Maximum Subsidy (Rs 49,680) On Drip Irrigation

Irrigation treatment	Details	Low tunnel height, cm		
		45	60	75
Drip irrigation, IW/CPE = 0.60	Gross returns (Rs./ha)	255,776	279,480	276,460
	Total expenditure (Rs./ha)	112,678	117,844	125,495
	Net seasonal income (Rs./ha)	143,098	161,636	150,965
	B/C ratio	**2.26**	**2.37**	**2.20**

TABLE 3.6   *(Continued)*

Drip irriga-tion, IW/CPE = 0.75	Gross returns (Rs./ha)	323,823	359,976	347,293
	Total expenditure (Rs./ha)	112,678	117,844	125,495
	Net seasonal income (Rs./ha)	211,145	242,132	221,798
	B/C ratio	**2.87**	**3.05**	**2.76**
Drip irriga-tion, IW/CPE = 0.90	Gross returns (Rs./ha)	296,863	353,933	330,720
	Total expenditure (Rs./ha)	112,678	117,844	125,495
	Net seasonal income (Rs./ha)	184,185	236,089	205,225
	B/C ratio	**2.63**	**3.00**	**2.63**
Furrow irriga-tion (paired row)	Gross returns (Rs./ha)	243,436	267,690	257,803
	Total expenditure (Rs./ha)	95243	100,409	108,060
	Net seasonal income (Rs./ha)	148,193	167,281	149,743
	B/C ratio	**2.55**	**2.66**	**2.38**

3.3   RESULTS AND DISCUSSION

3.3.1   PERFORMANCE OF SWEET PEPPER

The results obtained for sweet pepper yield per hectare under different irrigation and low tunnel height treatments are presented in Table 3.4. The data revealed that the mean sweet pepper yield in 60 cm tunnel height was maximum followed by 75 cm and 45 cm tunnel height treatments. This may be due to the fact that 60 cm tunnel height provided the best microclimatic conditions to the crop as the volume of entrapped air differed with the tunnel heights. Among tunnel height treatments, 60 cm tunnel height gave an increase of 11.08% yield over 45 cm tunnel height and an increase of 3.60% yield over 75 cm tunnel height. Among the irrigation treatments, drip irrigation with 0.75 IW/CPE ratio gave the maximum mean sweet pepper yield followed by drip irrigation with 0.90 IW/CPE ratio, drip irriga-tion with 0.60 IW/CPE ratio and furrow irrigation paired row planting. Among the irrigation treatments, drip irrigated treatments gave better yield as compared to furrow irrigated treatments. Best drip irrigated treatment (i.e., drip irrigation, IW/CPE = 0.75) gave an increase of 30.67% over the furrow irrigated paired row planting. For treatment combinations, mean sweet pepper yield was maximum (29,886 kg/ha) in the I_2H_2 treatment and minimum (20,939 kg/ha) in I4H1 treat-ment. This may be due to the fact that moisture in optimum level enhances the cell metabolism resulting in better yield. The results are in agreement with Sharma et al. [7], who reported that drip irrigation necessarily enhanced capsicum yield as compared to furrow-irrigated production. Singh [8] reported that tunneling increases the yield of pepper.

Statistical analysis for different irrigation treatments and different tunnel height is given in Table 3.4. There was significant effect of irrigation and tunnel heights on sweet pepper yield. The interaction of irrigation treatments and tunnel heights was also significant.

Out of the total yield, the first three harvests gave early yield till 3 April 2009, which was priced at double the normal rates. The results obtained for early sweet pepper yield under different irrigation and low tunnel height treatments are presented in Table 3.5. The data revealed that the mean sweet pepper early yield in H2 was highest followed by H3 and H1 treatments, due to the fact that H2 provided the best microclimatic conditions to the crop. Among the irrigation treatments, I2 gave the highest mean sweet pepper early yield followed by I3, I1 and I4. For treatment combinations, mean sweet pepper early yield was maximum (61.11 per 100 kg/ha) in the I2H2 treatment and minimum in I4H1 treatment (34.04 per 100 kg/ha). This may be due to early flowering and fruit initiation in the respective treatments. The results are in agreement with Singh et al. [9].

3.3.2 ECONOMIC ANALYSIS

Tables 3.6 and 3.7 summarize the results obtained for benefit–cost ratio (BCR) of growing sweet pepper under different irrigation and low tunnel height treatments with no subsidy and maximum subsidy for drip irrigation (Rs. 49,680 per hectare given by Punjab State Government in 2008–2009). The data clearly revealed that the mean BCR in H2 was highest followed by H1 and H3 tunnel height treatments. Among the irrigation treatments, I2 gave the highest mean BCR followed by drip irrigation I3, I4 and I1. This is due the fact that yields in I2 and I3 were higher as compared to furrow irrigation method (I4). For treatment combinations, BCR was highest (2.93 without subsidy) and (3.05 with maximum subsidy) in the I_2H_2 treatment. BCR was lowest in I_1H_3 treatment (2.12 without subsidy) and (2.20 with maximum subsidy) on drip irrigation. The results are in line with those of (10) who reported that cultivation under low tunnels in the season is highly profitable with a BCR of 3.86.

3.4 CONCLUSIONS

The maximum sweet pepper yield (100 kg/ha) was observed in H2 and I2 treatments among the tunnel heights and irrigation treatments. Among the tunnel height treatments, 60 cm tunnel height gave an increase of 11.08% yield over 45 cm tunnel height and an increase of 3.60% yield over 75 cm tunnel height. Among the irrigation treatments, drip irrigated treatments gave better yield compared to furrow irrigated treatments. Best drip irrigated treatment (i.e., drip irrigation, IW/CPE= 0.75) gave an increase of 30.67% over the furrow irrigated paired row planting. For the treatment combinations, in I2H2 treatment maximum sweet pepper yield (29886 kg/ha) was observed and it was minimum in I4H1 treatment (20,939 kg/ha). There was significant effect of irrigation and tunnel height treatment on sweet pepper yield. Also the effect of their interaction on sweet pepper yield was significant. The maximum BCR was observed in H2 and I2 treatments among the tunnel heights and irrigation treatments. For the treatment combinations, in I2H2 treatment BCR was maximum and I1H3 it was minimum. This study concludes that it is economically feasible to grow sweet pepper in low tunnel with 60 cm tunnel height, which is drip irrigated with IW/CPE ratio of 0.75 as it gives maximum yield and maximum BCR.

3.5 SUMMARY

To study the economics of growing sweet pepper under low tunnels, an experiment was laid out in split plot design with four irrigation treatments (drip irrigation with IW/CPE ratio of 0.60 (I1), 0.75 (I2), 0.90 (I3) and furrow irrigation with paired row planting (I4)), in main plots and three different low tunnel heights (45 cm (H1), 60 cm (H2) and 75 cm (H3)) in sub plots. All treatments were replicated three times.

Total yield was highest in H2 and I2 treatments. There was significant effect of irrigation, tunnel height on sweet pepper yield and their interaction was also significant. Among the tunnel height treatments, H2 gave an increase of 11.08% yield over lowest H1 treatment. Best drip irrigated treatment (I2) gave an increase of 30.67% over the furrow irrigated paired row planting (I4). For treatment combinations, in the I2H2 treatment mean sweet pepper yield was maximum (29,886 kg/ha) and minimum in I4H1 treatment (20,939 kg/ha). The early yield varying from 16–20% was obtained under low tunnels. The treatment combination of I2H2 treatment gave highest BCR (2.93 without subsidy) and (3.05 with maximum subsidy) on drip irrigation.

KEYWORDS

- bell pepper
- benefit–cost ratio, BCR
- drip irrigation
- economics
- low tunnel technology, LTT
- row cover
- Sweet pepper
- tomato
- tunnel height
- vegetable crop
- yield

REFERENCES

1. Anonymous, (2008). *Package of Practices for Vegetable and Horticultural Crops*. Pp 35–37. Punjab Agricultural University, Ludhiana.
2. Arin, L., Ankara, S. (2001). Effect of low tunnel, mulch and pruning on the yield and earliness to tomato in unheated green house. *Journal of Applied Hortculture, 3,* 23–27.
3. Gerber, J., Mohd-Khir, M. I., and Splittoesser, W. E. (1988). Row tunnel effect on growth, yield and fruit quality of bell pepper. *Hort. Science, 26(3–4),* 191–197.
4. Helbacka, J. (2002). Row covers for vegetable gardens. *King County Cooperative Extension Service.* Fact Sheet No. 19, Washington State University, Washington, USA.
5. Saini, A. K., Singh, K. G. (2001). *Annual Report of AICRP on Application of Plastics in Agriculture.* pp. 69–74. Punjab Agricultural University, Ludhiana, India.
6. Shara, E. A., George, H. C. (1998). Spun bonded row cover and capsicum fertilization improve quality and yield in bell pepper. *HortScience, 33,* 1150–1152.
7. Sharma, P. K., Sharma, H. G., Singh, P. N. (2004). Effects of methods/levels and colored plastic mulches on weeds incidence in capsicum (*Capsicum annuum* var. *grossum* L.) crop. *Agricultural Science Digest, 24,* 42–44.
8. Singh, A. (2008). Economic feasibility to growing capsicum under drip irrigation in West Bengal, India. *Irrigation & Drainage System 22,* 179–188.
9. Singh, Sirohi, B. N. P. S., Neubauer, E., Chin, A. (2001). Off-season production of muskmelon under plastic low tunnels. *Indian Horticulture, 46,* 15–17.
10. Vishuvardhana, L. H. B., Khan, M. M., Raju, G. T. T. (2004). Economics of production of cashew grafts in different propagation structures under eastern dry zone of Karnataka. *Cashew, 18,* 39–44.

CHAPTER 4

PERFORMANCE OF GREENHOUSE SWEET PEPPER

KAMAL G. SINGH, ANGREJ SINGH, and G. MAHAJAN

4.1 INTRODUCTION

Maximization of crop yields requires combination of an optimum genotype with an optimum environment. The productivity of a particular crop in an area depends on the overall environment [3]. Polyhouse farming, also known as protected cultivation, is one of the farming systems widely used to provide and maintain a controlled environment suitable for optimum crop production. This technology is a breakthrough in agricultural production technology that integrates market driven quality parameters with production system profits [1]. In addition to this, polyhouse technology can contribute to solve global issues such as the shortage of artificial energy, water, environmental problems and instability of ecological system in various ways. In north India, Capsicum is very popular crop for production in polyhouse, because in open field, fruit yield and quality is poor due to very low temperature during the winter season, when it is grown [5, 13]. This crop is chilling sensitive and cannot tolerate extended periods of temperature below 10°C, the mean growing temperature of the crop is range from 18–30°C. Irena [7] reported that under plastic covered greenhouse the night temperature was favorable for the development of capsicum fruit. The night temperature of 18–20°C before anthesis and 8–10°C after anthesis recorded the maximum fruit set. Water is another important input for poly house Capsicum crop because irrigation is the only source for application of water to the Capsicum plants in polyhouse. Several efforts have been made to use irrigation water as efficient as possible under protected cultivation system. The use of drip irrigation and fertigation saves water, fertilizer and gives better plant yield and quality as it reduces the humidity build up inside the polyhouse after irrigation due to precise application of irrigation water to the root zone of the crop [12]. In polyhouse, Capsicum due to indeterminate nature of crop, vegetative and reproductive stage overlaps and the plant needs nutrients even up to fruit maturity, so, fertilizer application method such as fertigation may be very effective in polyhouse Capsicum. Further,

---

*In this chapter, the currency is expressed in Indian Rupees (1.00 US$ = Rs. 60.93; 1.00 Rs. = 0.02 US$).

fertilizer costs and nitrate losses can be reduced and nutrient applications can be better timed to plant's needs. A limited research has been done on drip irrigation and fertigation on Capsicum grown in low cost naturally ventilated polyhouse [12].

The present experiment was conducted to study water and nitrogen requirement of drip-irrigated Capsicum grown in naturally ventilated polyhouse.

4.2 MATERIALS AND METHODS

The experiment was conducted in a naturally ventilated polyhouse at the research farm of the Department of Soil and Water Engineering at Punjab Agricultural University (PAU), Ludhiana during 2005–2007. Ludhiana is located 30°54″N latitude and 75°48″E longitude with altitude of 247 m above mean sea level. It represents semi arid and sub tropical climate with typical monsoon conditions.

A semicircular shaped polyhouse covering a floor area 6.25 × 16 m (100 m²) with sidewall ventilation was used for the study. The orientation of the polyhouse was east-west direction. The polyhouse was covered with an ultra violet (UV) stabilized low-density polyethylene film having 200-micron thickness. The Capsicum cultivar's Bharath (F_1 hybrid), widely grown by the farmers in polyhouse was used. The soil of the experimental plot was loamy sand in texture having pH 8.2 and electrical conductivity (EC) of 0.14 mmhos/cm. The soil was low in organic carbon (0.36%). The soil was low in available nitrogen (244.8 Kg.ha^{-1}) and very high in phosphorus (65.0 Kg.ha^{-1}) and potassium (240 Kg.ha^{-1}). The experiment consisted of 6 treatments. One N level (100% of recommended dose) was tested against check basin methods of irrigation at recommended dose of N, when the crops was sown at recommended spacing of 60 × 30 cm² both inside and outside the polyhouse in randomized block deign design. In check-basin method (surface flooding) method the irrigations were provided on the basis of 1.0 cumulative pan evaporation (Epan). The details of the treatments are given below:

   T_1 – Polyhouse crop + drip irrigation (0.5 × E pan) + 100% N
   T_2 – Polyhouse crop + drip irrigation (0.75 × E pan) + 100% N
   T_3 – Polyhouse crop + drip irrigation (1.0 × E pan) + 100% N
   T_4 – Non-Polyhouse crop + drip irrigation (0.5 × E pan) + 100% N
   T_5 – Polyhouse crop + Surface irrigation (1.0 × E pan) + 100% N broadcast
   T_6 – Non- Polyhouse crop + Surface irrigation (1.0 × Epan) + 100% N broadcast.

The irrigation with drip system was done on alternate day on the basis of pan evaporation value of the previous day, while in surface irrigation method; the water depth was 50 mm. In normal sowing, the distance between the rows was 60 cm and plant-to-plant spacing was 30 cm. However, in paired sowing, the row-to-row space among paired rows was 45 cm and row-to-row spacing between pairs was 75 cm but plant-to-plant spacing was 30 cm. So in paired sowing, total number of rows and as well as plants were same. In all the treatments, a basal dose of FYM @ 50 t/ha was also applied before sowing of crop. As indicated by the soil test, soil at the experimental plot was rich in phosphorous and potassium hence these fertilizers were not applied. However, the recommended nitrogen fertilizers in Capsicum @ 125 kg N/ha was applied in three equal split doses. The capsicum seed was sown in the last week of September for raising nursery on a raised bed inside the polyhouse. The nursery was transplanted in the first week of November during both the years. Other cultural operations were same in all the treatments and were attended regularly.

The drip irrigation system consisted of polyethylene laterals of 12 mm in diameter, laid parallel to crop rows (each lateral served 2 rows of crop). The laterals were provided with inline emitters with a discharge of 2 L/h capacity at 0.3 m apart along the length of the lateral. The different levels of water supply were maintained by managing the irrigation time of the system. In drip irrigation system, N was applied at 15 days interval in ten equal doses of N starting from 30 days after transplanting. In the polyhouse, first picking of Capsicum fruit was started in the first fortnight of February and in all 16 pickings of polyhouse Capsicum were done and the last picking ended in the second fortnight of May. The fruits were picked at 45 days after anthesis, at green mature stage for quality testing. Ten fruits were picked at random from each plot. Fruit weight, Capsaicin content, ascorbic acid and chlorophyll content were determined using standard procedure.

4.3 RESULTS AND DISCUSSION

The uniformity coefficients of drip system were 89.6% for the first year and 88.4% for the second year. The high values of uniformity coefficient indicate an excellent performance of irrigation system in polyhouse in supplying water uniformly throughout the lateral lines. Malik et al. [8, 9] observed uniformity coefficient for the same system to the tune of 90 ± 3 and $89 \pm 3\%$ at the beginning and termination of the experiment, respectively.

4.3.1 FRUIT YIELD

In general polyhouse crop resulted an increase in yield at all the levels of irrigation and fertigation, compared to the surface irrigated crop that was grown outside the polyhouse, though the response was comparatively high when irrigation was applied through drip at $0.75 \times E_{pan}$. Higher yield under polyhouse may be due to favorable environment at the early stages of Capsicum (especially in the month of December and January, when the day and night temperature are very low). Climatic data in the polyhouse (Table 4.1) revealed that the soil temperature and air temperature in the polyhouse remained at an average 3.9°C and 3.2°C higher, respectively in polyhouse than the outdoor environment. The relative humidity inside the polyhouse was 10.84 units higher inside the polyhouse but the light intensity reduced by an average of 12,990 lux (Table 4.1), which caused a favorable microclimate for polyhouse Capsicum. These results are in conformity with Chandra et al. [4], who also observed that green house crop of Capsicum gave better growth especially in the early stage and resulted in more early and total yield.

TABLE 4.1 Soil and Air Temperature, Relative Humidity and Light Intensity: Inside/Outside the Polyhouse

Period	Soil temperature		Air temperature		Relative humidity		Light intensity	
	Inside	Outside	Inside	Outside	Inside	Outside	Inside	Outside
	°C				%		lux	
Nov 1–15	30.5	29.3	32.0	28.5	82.6	66.4	12,935	24,370
Nov 16–30	27.0	23.9	28.0	24.9	80.0	65.3	15,642	25,767
Dec 1–15	24.5	19.8	24.9	20.7	76.2	70.3	15,836	25,803
Dec 16–31	23.2	21.2	24.0	19.6	78.7	73.8	11,861	19,677
Jan 1–15	25.4	19.1	23.1	17.9	80.0	71.2	10,120	18,256
Jan 16–31	25.8	21.8	23.0	20.1	79.2	72.1	11,355	22,478
Feb 1–15	24.6	20.7	25.7	22.7	90.2	77.3	15,525	27,492
Feb 16–28	25.8	20.6	27.8	23.9	85.7	75.2	14,658	24,675
Mar 1–15	29.5	25.9	28.5	24.3	82.9	70.3	17,840	33,462
Mar 15–31	33.6	29.8	32.5	28.7	78.7	64.4	18,929	31,489
Apr 1–15	36.0	34.1	36.6	35.0	56.8	45.7	20,787	34,349
Apr 16–30	38.5	36.4	39.8	37.8	47.6	35.7	23,189	37,600
May 1–15	43.0	36.9	40.1	38.3	60.0	48.2	35,967	57,547

TABLE 4.1 *(Continued)*

Period	Soil temperature		Air temperature		Relative humidity		Light intensity	
	Inside	Outside	Inside	Outside	Inside	Outside	Inside	Outside
	°C				%		lux	
May 16–31	44.8	38.7	40.0	38.6	68.5	57.5	36,957	60,503

TABLE 4.2 Effects of Treatments On Different Parameters of Polyhouse Capsicum Pooled Data (2005–2006 and 2006–2007)

Treatments	Fruit yield	Fruit length	Fruit weight	Pooled water applied	Pooled WUE	Capsaicin content	Ascorbic acid content	Chlorophyll content	Dry matter content
	t.ha^{-1}	cm	g	cm	t/ha-cm	%	mg/100 mL	mg/g	%
T$_1$ P.H.C+d	56.9	6.6	50.6	31.0	1.84	0.26	106.0	0.014	7.3
T$_2$ P.H.C+d	63.2	6.8	50.8	44.0	1.44	0.25	106.4	0.013	7.2
T$_3$ P.H.C+d	62.0	8.4	48.3	57.0	1.09	0.24	104.1	0.014	7.1
T$_4$ N.P.H.C+d	43.5	5.2	43.1	51.0	0.85	0.23	98.1	0.012	7.8
T$_5$ P.H.C	45.6	6.2	46.2	68.0	0.67	0.23	102.2	0.012	7.0
T$_6$ N.P.H.C	25.9	4.3	40.3	62.0	0.42	0.22	96.1	0.011	8.40
LSD (0.05)	3.1	2.5	3.17	-	-	0.02	4.27	0.01	0.70

P.H.C = Polyhouse crop; N.P.H.C. = Non-polyhouse crop.

With same quantity of water ($1.0 \times E_{pan}$) and nitrogen was applied through drip irrigation, the data in Table 4.2 revealed that in polyhouse Capsicum a significantly higher yield (62.0 t ha^{-1}) was obtained as compared to 45.6 t ha^{-1} and 25.9 t ha^{-1} in check basin method of irrigation when the crop was sown both inside and outside the polyhouse, respectively. Nimje and Shyam [10] also reported that polyhouse capsicum gave 2.5–3.0 times higher yield as compared to open field cultivation. When the quantity of water applied through drip irrigation was reduced to 0.75xEpan,

Capsicum yield increased further (63.2 t ha^{-1}) significantly as compared to check basin method of irrigation at the same level of nitrogen. The yield was significantly reduced when irrigation was applied at $0.5 \times E_{pan}$ along with fertigation of 100% recommended nitrogen (56.9 t ha^{-1}). The best treatment of drip irrigation at $0.75 \times E_{pan}$ along with fertigation of 100% nitrogen resulted increase in fruit yield by 38.6% over control of recommended practices inside the polyhouse and by 144% over control of recommended practices outside the polyhouse respectively. The increased yield under drip irrigation system might have resulted due to better water utilization [11], higher uptake of nutrients and excellent soil-water-air relationship with higher oxygen concentration in the root zone [2]. Under control treatments, both inside and outside the polyhouse, surface irrigation not only resulted in wastage of water due to deep percolation below root zone, but also sets a chain of undesirable reaction such as leaching of available plant nutrients and consequently development of soil problems and poor aeration resulting in reduced yield.

4.3.2   WATER USE EFFICIENCY (WUE)

Drip irrigation both at $0.5 \times Epan$ and $0.75 \times Epan$ registered much higher WUE as compared to control practices both inside and outside the Polyhouse. For all levels of fertigation, drip irrigation at $0.5 \times Epan$ gave maximum WUE of 1.84 t/ha-cm, followed by $0.75 \times Epan$ with 1.44 t t/ha-cm. At $1.0 \times Epan$, the WUE was 1.08 t/ha-cm. The drip irrigation at $0.75 \times Epan$ with fertigation treatments giving a saving of 35.3% of irrigation water resulted in 38.6% higher fruit yield as compared to recommended practices inside the Polyhouse. Since, the rate of water loss through evaporation from soil surface was much lower under drip irrigation; hence, WUE was higher as compared to surface irrigation. Malik and Kumar [8] also made similar observations on WUE of drip irrigation in pea.

4.3.3   QUALITY CHARACTERISTICS

The investigated factors significantly influenced some quality parameters of sweet pepper (Table 2). Polyhouse environment significantly improved the ascorbic acid content (102.2 mg/100 ml juice) of crop as compared to the outside field crop (96.4 mg/100 mL of juice). The results confirm the findings of Howard et al. [6] who reported that greenhouse environment has favorable influence on the ascorbic acid content of the capsicum fruit. Further drip irrigation enhanced the ascorbic acid in the fruit to 108.4 mg/100 ml juice as compared to surface method of irrigation (102.2 mg/100 mL juice) inside the polyhouse. Howard et al. [6] also reported that the trickle irrigated crop of sweet/bell pepper at green mature stage recorded total ascorbic acid content of 97.5 to 108.7 mg/100 g of the fruit in different varieties of the crop. The chlorophyll content in the fruit did not differ significantly among the different drip irrigation treatments inside the polyhouse, though it was significantly superior to the surface irrigated crop. Results revealed that polyhouse drip irrigated crop was significantly better than the surface irrigated field crop. The dry matter content was comparatively less in the drip-irrigated crop as compared to the surface irrigated crop. Thus the fruits with less dry matter eventually contain more moisture percent and remain fresh for longer duration. Gornat et al. [5] also reported similar results where moisture content in trickle-irrigated crop was 93% at green mature stage of the fruit.

4.4 SUMMARY

A two-year study was conducted during 2005–2007 in Department of Soil and Water Engineering, at Punjab Agricultural University, Ludhiana to investigate the effects of irrigation and fertigation on Polyhouse Capsicum. Drip irrigation at 0.75 E_{pan} along with fertigation of 100% recommended nitrogen resulted an increase in fruit yield by 38.6% over control (recommended practices) inside the polyhouse and by 144% over control (recommended practices) outside the polyhouse, respectively. The drip irrigation at 0.75 E_{pan} irrespective of fertigation treatments gave an irrigation water saving of 35.3% and resulted in 38.6% higher fruit yield as compared to recommended practices inside the polyhouse. Polyhouse Capsicum fruits were found to be superior to fruits of open field crop in view of fruit weight, Capsaicin content, ascorbic acid content and chlorophyll content. Further, drip irrigation and fertigation in polyhouse crop caused significantly improvement in all the quality characteristics.

KEYWORDS

- anthesis
- ascorbic acid content
- basal dose
- Capsaicin
- Capsicum
- check basin method
- chlorophyll content
- drip irrigation
- electrical conductivity
- farm yard manure, FYM
- fertigation
- fertilizer
- genotype

- **green house**
- **loamy sand**
- **naturally ventilated polyhouse**
- **paired sowing**
- **pan evaporation**
- **polyhouse farming**
- **polyhouse technology**
- **randomized block deign**
- **sidewall ventilation**
- **tropical climate**
- **water use efficiency**

REFERENCES

1. Aldrich, R. A., Bartok, J. W. (1989). *Polyhouse Engineering*. North-east Regional Agricultural Engineering Service, Cooperative Extension, Ithaca, NY.
2. Bafna, A. M., Daftardar, S. Y., Khade, K. K., Patel, V. V., Dhotre, R. S. (1993). Utilization of nitrogen and water by Capsicum under drip irrigation system. *Journal Water Management, 1(1)*, 1–5.
3. Buttery, B. R., Buzzel, R. I. (1984). Maximizing crop productivity in crop physiology-advancing frontiers. Oxford and IBH Publishing Company New Delhi, pages 227–280.
4. Chandra, Pritam, Sirohi P. S., Behera, T. K., Singh, A. K. (2000). Cultivating vegetables in polyhouse. *Indian Horticulture, 45(3)*, 17, 32.
5. Gornat, B., Goldberg, D., Rimon, D., Asher, Ben. J. (1973). The physiological effect of water quality and method of application on Capsicum, cucumber and pepper. *Journal of American Society for Horticultural Science, 98(2)*, 202–205.
6. Howard, L. R., Smith, R. T., Wagner, A. B., Villalon, B., Burns, E., E, (1994). Provitamin A and ascorbic acid content of fresh pepper and processed Jalpenos. *Journal of Food Science, 59(2)*, 362–65.
7. Irena, Rylski, (1973). Effect of night temperature on shape and size of Capsicum annum L. *Journal of American Society for Horticultural Science, 98(2)*, 149–152.
8. Malik, R. S., Kumar, K., Bhandari, A. R. (1994). Effect of urea application through drip irrigation system on nitrate distribution in loamy sand soils and pea yield. *J. Indian Soc. Soil Sci., 42(2)*, 6–10.
9. Malik, R. S., Kumar, K. (1996). Effect of drip irrigation levels on yield and water use efficiency of pea. *J. Indian Soc. Soil Sci., 44(3)*, 508–509.
10. Nimje, P. M., Shyam, M. (1991). Greenhouse as an alternative technology for commercial vegetable production. *Indian J. Agricultural Sciences, 61(3)*, 185–189.
11. Manfrinato, H. A. (1971). Effect of drip irrigation on soil water plant relationship. *Second International Drip Irrigation Congress*, 446–451.
12. Papadouplous, K. (1992). Fertigation of vegetables in plastic house: present situation and future aspects. *Acta Horticulture*, No. 323.
13. Steve, S., Linn, M., Hubbel, J. N., Tsou, C. S. (1983). Drip irrigation and Capsicum yield under tropical conditions. *Hort. Science, 18(4)*, 460–461.

CHAPTER 5

NITROGEN FERTIGATION IN DRIP IRRIGATED CAULIFLOWER

CHETAN SINGLA and KAMAL G. SINGH

5.1 INTRODUCTION

Cauliflower (*Brassica Oleracea var. Botrys Linn.*) is very popular vegetable crop and grown widely in northern India. Assured irrigation and nitrogen are the prerequisites for proper growth and development of cauliflower plant [1]. In India, the mid and late season cultivars are more popular and cultivated on a large scale. Due to glut in the market, these cultivars do not fetch lucrative prices to the growers. The cauliflower crop can be raised four times in a year. However, the production of early cauliflower is highly profitable due to nonavailability of the produce from other parts of the country. Inspite of its great importance in bringing good returns to the farmers, very scanty information is available on the agronomical aspects of this crop in greenhouses.

Therefore, this study is designed to evaluate suitable irrigation and nitrogen strategies for getting optimum curd yield of early cauliflower using drip irrigation system in a fan pad cooled greenhouse and naturally ventilated greenhouse.

5.2 MATERIALS AND METHODS

A field study on fertigation of cauliflower (*Brassica Oleracea var. Botrys Linn.*) using drip irrigation system was undertaken during summer season of 2002 at the Irrigation Research Farm of the Department of Soil and Water Engineering, Punjab Agricultural University (PAU), Ludhiana in fan pad cooled greenhouse and naturally ventilated greenhouse, respectively.

There were three irrigation levels (W_1 = 0.5 W/CPE, W_2 = 0.75 W/CPE and W_3 = 1.00 W/CPE) with three rates of nitrogen application 50% (N_1), 75% (N_2) and 100% (N_3) of the recommended dose (125 kg/ha). Nitrogen was applied in three splits, that is, 50% of the dose at planting time, 25% at 30 days after planting (DAP) and remain-

---

*In this chapter, the currency is expressed in Indian Rupees (1.00 US$ = Rs. 60.93; 1.00 Rs. = 0.02 US$).

ing 25% at 60 DAP. Nine treatment combinations were replicated thrice [2]. The measured quantity of irrigation water was applied through drip irrigation system.

The soil at the experimental field was sandy loam, having pH 8.5 and available N, P, K 51.6, 80.6 and 148.9 kg/ha, respectively. The water content (volume basis, cm^3/cm^3) was 0.1739 at − 0.03 MPa (field capacity) and 0.0523 at − 1.5 MPa (permanent wilting point) in 0–30 cm soil depth. The bulk density was 1.58 g/cm^3. The transplanting was done on 17 July of 2002. The harvesting of marketable curds commenced at 80 days after transplanting (DAT) and was completed within four weeks.

5.3 RESULTS AND DISCUSSION

5.3.1 CURD YIELD

Cauliflower yield as influenced by the application of different doses of nitrogen (N) and irrigation water is presented in Table 1. The total cumulative pan evaporation was 534 mm. The amount of water applied at 0.5, 0.75 and 1.0 IW/CPE was 26.70, 40.05 and 53.40 cm, respectively. The amount of fertilizer applied at 50, 75 and 100% of recommended dose was 62.50, 93.75 and 125 kg/ha. Tables 5.1 and 5.2 reveal that cauliflower yield was highest under irrigation schedule based on IW/CPE = 0.5 and 100% of recommended dose of N, which was at par with other drip irrigation treatments. This treatment (IW/CPE = 0.5 and 100% of recommended dose of nitrogen, that is, 125 kg/ha) established its superiority by yielding 120.7 per 100 kg/ha and 105.5 per 100 kg/ha as compared to other eight treatments in fan pad cooled greenhouse and naturally ventilated greenhouse, respectively. An irrigation schedule based on IW/CPE = 0.5 secured yields compared to values for IW/CPE = 0.75 and IW/CPE = 1.0. These results depicted that low water and high fertilizer treatment by drip irrigation were superior to other treatments. It might be due to the fact that optimum requirement of water was met by just IW/CPE = 0.5 (less amount of water) rather than ratio of 1.0 (high amount of water) and 0.75 (medium level of water).

5.3.2 EFFECTS OF NITROGEN ON YIELD

As the quantity of applied nitrogen is increased from 50% to 100% of the recommended dose, the yield goes on increasing (Tables 5.1 and 5.2). In fan pad cooled greenhouse, the yield increased from 93.95 per 100 kg/ha to 106.6 and 120.7 per 100 kg/ha indicating 13.46 and 28.47% increase at N levels of 93.75 and 125 kg/ha, respectively on irrigation with IW/CPE=0.5. While irrigating with IW/CPE=0.75 and 1.0. The relative yield increase was 23.53, 27.52 and 13.13, 20.20% over the nitrogen level of 62.5 kg/ha at the respective nitrogen levels. Similarly in naturally ventilated greenhouse, the yield increase was 13.50, 18.75 and 8.17, 9.77 over the nitrogen level of 62.5 kg/ha with the increased levels of nitrogen, while irrigating with IW/CPE = 0.5 and 0.75.

5.3.3 EFFECTS OF IRRIGATION ON YIELD

The yield of cauliflower curd was more with IW/CPE = 0.5 as compared to IW/CPE = 0.75 and 1.0 at all levels of nitrogen fertilizers (Tables 5.1 and 5.2). Irrigation schedule based on IW/CPE = 0.5 established its superiority by yielding 93.95, 116.6 and 120.7 per 100 kg/ha in fan pad cooled greenhouse and 88.88, 100.88 and 105.5 q/ha in naturally ventilated greenhouse.

TABLE 5.1 Effects of Different Treatments on Cauliflower Yield in Greenhouse Cooled by Fan Pad

Irrigation levels, IW/CPE ratio	N fertilizer levels, %	Crop Yield, 100 kg/ha
	F1 = 50	94.0
T1 = 0.5	F2 = 75	116.6
	F3 = 100	120.7
	F1 = 50	93.9
T2 = 0.75	F2 = 75	116.0
	F3 = 100	119.7
	F1 = 50	91.7
T3 = 1.0	F2 = 75	103.7
	F3 = 100	110.2
LSD, P = 0.05	—	N.S.

TABLE 5.2   Effects of Different Treatments on Cauliflower Yield in Naturally Ventilated Greenhouse

Irrigation levels, IW/CPE ratio	N fertilizer levels, %	Crop Yield, 100 kg/ha
	F1 = 50	88.9
T1 = 0.5	F2 = 75	100.9
	F3 = 100	105.6
	F1 = 50	88.0
T2 = 0.75	F2 = 75	95.19
	F3 = 100	105.4
	F1 = 50	79.6
T3 = 1.0	F2 = 75	76.1
	F3 = 100	92.6
LSD, P = 0.05	—	N.S.

TABLE 5.3   Irrigation Water Used and Water-Use Efficiency (WUE) of Cauliflower in Fan-Pad Cooled Greenhouse

Treatments	Water used	Yield	WUE
	cm	100 kg/ha	kg/(ha-cm)
W_1N_1	26.7	94.0	351.8
W_1N_2	26.7	116.6	436.7
W_1N_3	26.7	120.7	452.6
W_2N_1	40.0	93.9	234.5
W_2N_2	40.0	116.0	289.6
W_2N_3	40.0	119.7	299.0
W_3N_1	53.4	91.7	171.6
W_3N_2	53.4	103.7	194.2
W_3N_3	53.4	110.2	206.3

TABLE 5.4 Irrigation Water Used and Water-Use Efficiency of Cauliflower In Naturally Ventilated Greenhouse

Treatments	Water used	Yield, 100 kg/ha	WUE, kg/ha-cm
	cm	100 kg/ha	kg/(ha-cm)
W_1N_1	26.7	88.9	332.9
W_1N_2	26.7	100.9	377.8
W_1N_3	26.7	105.6	395.3
W_2N_1	40.1	88.0	219.7
W_2N_2	40.1	95.19	237.7
W_2N_3	40.1	105.4	263.2
W_3N_1	53.4	79.6	149.0
W_3N_2	53.4	76.1	142.4
W_3N_3	53.4	92.6	173.3

5.3.4 WATER USE EFFICIENCY (WUE) OF CAULIFLOWER

Total water used during the crop period and the WUE were determined for different treatments and presented in Tables 5.3 and 5.4. The treatment having the maximum WUE was considered superior to other treatments, though treatments were statistically insignificant. The WUE with respect to yield indicated that it was low when irrigations were scheduled at 0.75 and 1.0 IW/CPE. As WUE is a function of crop yield and water used, the decrease in crop yield with increase in water application reduced the WUE under IW/CPE= 0.75 and 1.0. However, at IW/CPE = 0.5, the yield increased linearly with water application and WUE. The WUE at this treatment (W_1N_3) was 452.05 kg/(ha-cm) in fan pad cooled greenhouse and 395.31 kg/(ha-cm) in case of naturally ventilated greenhouse with water saving of 52.43% as compared to conventionally grown cauliflower.

5.4 SUMMARY

Since greenhouse technology has recently been introduced, very limited work has been done for the standardization of different techniques of growing early vegetable crops in greenhouse. Early cultivars of cauliflower are tropical in nature, requiring comparatively higher temperature for reproductive onset and, therefore, these are grown during July–August allowing sufficient vegetative growth. But high monsoon rains coinciding with the period of cropping quite frequently disturb the growth and development of the crop in open field.

A field study on fertigation of cauliflower (*Brassica Oleracea var. Botrys Linn.*) using drip irrigation system was undertaken with three rates of nitrogen application (100%, 75% and 50%) of recommended dose applied in three splits, that is, 50% of the dose at planting time, 25% at 30 days after planting and remaining 25% at 60 days after planting. Three levels of irrigation at IW/CPE (irrigation water to cumulative pan evaporation ratio) were 0.5, 0.75 and 1.0 in fan pad cooled greenhouse and naturally ventilated greenhouse. The maximum yield of 120.7 per 100 kg/ha and 105.5 per 100 kg/ha of early cauliflower was obtained at IW/CPE = 0.5 and 100% of recommended nitrogen dose in fan pad cooled greenhouse and naturally ventilated greenhouse respectively.

KEYWORDS

- cauliflower
- crop yield
- cumulative pan evaporation, CPE
- curd yield
- days after planting, DAP
- days after transplanting, DAT
- drip irrigation
- fan pad cooled greenhouse
- field capacity, FC
- greenhouse
- irrigation schedule
- irrigation water, IW
- IW/CPE ratio
- marketable curd
- naturally ventilated greenhouse
- permanent wilting point, PWP
- vegetable crop
- water use efficiency, WUE

REFERENCES

1. Sharma, R. P., Arora, P. N. (1987). Response of mid-season cauliflower to irrigation, nitrogen and age of seedlings. *J. Vegetable Sci.*, *14(1),* 1–6.
2. Singla, Chetan, (2003). Studies on crop water requirements and fertigation options for off-season drip irrigated cauliflower in a green house. Unpublished M. Tech Thesis, Punjab Agricultural University, Ludhiana, India.

CHAPTER 6

EVALUATION OF IRRIGATION STRATEGIES FOR WHEAT

KAMAL G. SINGH and A. K. TIWARI

6.1 INTRODUCTION

Economically, optimum irrigation practice under limited supplies involves reducing water use from maximizing level of yield at which marginal cost equals the value of the marginal product [1]. This requires an appropriate schedule of water application satisfying the question of how much to apply and when to apply. Therefore, this study aims at: (a) Maximum water use efficiency (WUE), (b) Maximum yields of the crop, and (c) Maximum net return to the farmer.

6.2 MATERIALS AND METHODS

Different types of response functions were developed to predict yield and irrigation water input. A quadratic response function was selected based on the F and r values to fix the level of irrigation application for optimum yield. Keeping this amount of water (34.2 cm) in view, a replicated field experiment was conducted and water was applied at 3 levels of irrigation by 4 irrigation-scheduling techniques namely: (i) Irrigation based on growth stages, (ii) irrigation based on pan evaporation ratio, (iii) stress day index method, and (iv) difference between canopy and air temperature (Table 6.1).

The evaluation of physical characteristics of soil included field capacity, permanent wilting point and bulk density. From these 12 treatments the one which was statistically significant and had maximum water use efficiency (T8) was selected. Based on this treatment, an irrigation-scheduling model was proposed after calculating the sensitivity factor. Both parametric and nonparametric statistical tests satisfied the model predictions. The developed model was used to apportion known quantities of water in a season for getting the optimum yield.

---

*In this chapter, the currency is expressed in Indian Rupees (1.00 US$ = Rs. 60.93; 1.00 Rs. = 0.02 US$).

6.3 RESULTS AND DISCUSSION

6.3.1 DEVELOPMENT OF THE RESPONSE FUNCTION FOR W_D

With the available data, least square regression technique was used to develop: linear, quadratic, square root and exponential type response functions (Anonymous, 1972–1983). Considering F values, all the response functions were significant at $P = 1.00\%$. However, R^2 value for the quadratic response function was maximum. Therefore, the quadratic response function was selected.

$$Y = 0.9922 + 1.721\ W_D - 0.017498\ (W_D)^2, R^2 = 0.55 \tag{1}$$

where: Y is yield in 100 kg per ha; and W_D is the total depth of water applied in cm. From Eq. (1), $W_D = 49.19$ cm, which is the total amount of water needed for getting maximum yield in a growing season including the moisture profile. Taking the average value of moisture profile as 15 cm, the seasonal irrigation depth was:

$$W_D = 49.19 - 15 = 34.19\ \text{cm} \tag{2}$$

Keeping this irrigation depth in consideration, water application was planned to evaluate different irrigation scheduling techniques. Statistical analysis was done for grain yield to test the variability among the treatments. Table 6.1 shows that the treatments T2, T3, T5, T6, T8, T10, T11 and T12 are statistically significant at 5% level of significance. The maximum WUE was obtained in T8. Therefore, T8 was selected as the best. Based on this treatment, the model was used to apportion the seasonal water supply for optimum yield.

TABLE 6.1 Water application, Crop Yield and Water Use Efficiency in Different Treatments, Wheat Crop

	Technique	Irrigation depth, W_D	Profile moisture use	Total water use	Grain yield	Harvest index	WUE
		cm	cm	cm	100 kg/ ha	—	Kg/(ha-cm)
T1	Growth stage: Initial	6,5,5,5,5	10.72	36.72	31.06	33.98	84.5
T2	Growth stage: CRI, tillering,	6,5,6,7,7	9.83	40.83	41.5*	41.06	101.64

TABLE 6.1 *(Continued)*

	Technique	Irrigation depth, W_D	Profile moisture use	Total water use	Grain yield	Harvest index	WUE
		cm	cm	cm	100 kg/ ha	—	Kg/(ha-cm)
T3	Growth stage: Joining, booting & grain filling	6,7,7,7,7	9.52	43.62	43.86*	42.40	100.78
T4	IW/PE = 0.9	6,4,4,4,4,4	11.37	37.37	34.82	36.93	93.17
T5	IW/PE	6,6,6,6	13.26	37.26	42.70*	40.09	113.08
T6	IW/PE	6,8,8	13.89	35.89	38.54	40.28	107.38
T7	Stress Day Index: 35%	6,4,4.4,4.4,4	9.67	39.67	38.09	38.50	96.01
T8	**Stress Day Index: 50%**	**6,6,6,6**	**10.75**	**34.75**	**41.18**	**41.39**	**118.50****
T9	Stress Day Index: 65%	6,8,8,8	9.39	39.39	34.00	36.77	86.46
T10	Tc-Ta > 0	6,4,4,4,4,4	12.75	36.75	39.74*	41.8	107.23
T11	Tc-Ta	6,6,6,6	12.83	36.83	39.40	42.28	106.97
T12	Tc-Ta	6,8,8,8	11.40	41.40	43.17*	41.18	104.27

T_c = Canopy temp.; T_a = Air temp.; IW = Irrigation water; PE = Potential ET.

* Significant at 5%level; and ** Maximum water use efficiency.

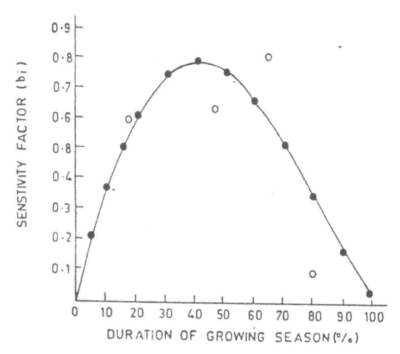

FIGURE 6.1   Sensitivity factor at different growth stages for wheat crop.

6.3.2   DEVELOPMENT OF THE MODEL FOR ESTIMATING YIELD

For the development of the model, sensitivity factor (bi) was calculated [2] as given in Table 6.2. The column 4 in Table 6.2 represents time corresponding to various stages of the crop for a 15-day growing period. A graph was plotted between the percentage of the growing season and the sensitivity factor (Fig. 6.1). The Eq. (3) was developed using least square technique:

$$bi = 0.0051311089 + 0.04320788\ G - 0.0006948206\ G^2 + 0.000002596119\ G^3, R^2 = 0.87$$

$$= 0.51311089 \times 10^{-2} + 0.43207884 \times 10^{-1}G - 0.69482066 \times 10^{-3}\ G^2 + 0.25961199 \times 10^{-5}G^3 \tag{3}$$

where: G = percentage of the growing season; and bi = sensitivity factor (Table 6.2).

TABLE 6.2 Sensitivity Factor For Wheat

	Crop growth stage	No. of days from sowing	% of length of growing season, G	bi Eq. (3)
1	Sowing	0	–	–
2	Crown root initiation	28	18.50	0.610
3	Jointing	72	47.60	0.640
4	Booting	97	64.20	0.816
5	Dough	121	80.17	0.207
6	Finish	151	100	—

6.3.3 APPORTIONING OF AVAILABLE SEASONAL IRRIGATION SUPPLY FOR OPTIMUM YIELD

The following response function was developed to allocate the given quantity of water in a season:

$$[Y/Y_o] = [1-(1-x_i)^2 b_i] \qquad (4)$$

where: Y = Actual yield per 100 kg/ha; Y_o = Maximum yield, per 100 kg/ha; x_i = Ratios of water applied (W) to cumulative pan evaporation (P), W/P; and bi = Crop stage sensitivity factor given by Eq. (3). We want to maximize Y in Eq. (4), where:

$$Y = Y_o[1 - (1 - x_1)^2]^{0.59}[1 - (1 - x_1)^2]^{0.786}[1 - (1 - x_1)^2]^{0.64}[1 - (1 - x_1)^2]^{0.61} \qquad (4)$$

Subject to the condition so that:

$$W_1 + W_2 + W_3 + W_4 = W \qquad (5)$$

where: W_1, W_2, W_3, and W_4 are amounts of water applied in four stages of growth. Water must be applied in such a manner that it is not wasted at all. This implies that depth of water applied does not exceed the field capacity in the active crop root zone.

Since $x = W/P$, Eq. (5) can be rewritten as:

$$P_1x_1 + P_2x_2 + P_3x_3 + P_4x_4 = W \qquad (6)$$

where: P_1, P_2, P_3 and P_4 are cumulative pan evaporation depths, from the date of previous irrigation during the corresponding (first, second, third and fourth) irrigation periods, respectively. Values of P1 for four stages are $P_1 = 6.3$, $P_2 = 6.8$, $P_3 = 7.4$, and $P_4 = 8.0$.

Yield will be maximized, when the water is allocated so that the marginal products (i.e., first derivative of yield with respect to water applied in each period) are equal, as described below:

$$\frac{dY}{dW1} = \frac{dY}{dW2} = \frac{dY}{dW3} = \frac{dY}{dW4} = \frac{dY}{dW}$$

Or $(1/P1)\frac{dY}{dW1} = (1/P2)\frac{dY}{dW2} = (1/P3)\frac{dY}{dW3} = (1/P4)\frac{dY}{dW4} = (1/P)\frac{dY}{dW}$ (7)

To solve the objective function (Eq. (4)), exhaustive search method was employed using the constraints in Eqs. (6) and (7). Results are discussed for the three zones, as shown in Fig. 6.2.

6.3.3.1   ZONE I

If one decides to produce, he must produce up to the level of input, where average product portion (APP) is the highest. In other words, if the product has a value (input use), it should be continued until Zone II is reached. Since the efficiency of the variable inputs keep increasing throughout the Zone I, it is not reasonable to stop using an input when its efficiency on all the units used is increasing.

6.3.3.2   ZONE II

In this zone, the total product is increasing. The marginal product keeps decreasing but remains positive and less than the APP. The average product also decreases.

6.3.3.3   ZONE III

In this zone of response function, the total product is decreasing and the marginal product portion (MPP) is negative. Since additional quantities of input reduced the total output in this zone, it is not profitable even if additional amount of water is available, free of cost. It is, therefore, called as irrational zone of production, for example, if one operates in this region, it will be irrational because of the double loss that may incur (e.g., reduced production and unnecessary additional cost of input).

The above discussion leads to the conclusion that the farmer, who wants to maximize the profit, must operate in Zone II of the physical response function (Fig. 6.2).

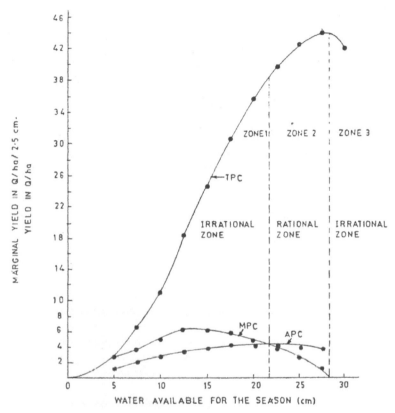

FIGURE 6.2 Total average and marginal productivity of water versus water available for the season (cm) (Note: 1.00 Q = 1.00 quintal = 100 kg).

TABLE 6.3 Proposed Schedule For Different Quantities of Seasonal Water Supplied to Achieve Optimum Yield

Water applied	X1	X2	X3	X4	W1	W2	W3	W4	Yield	Avg. yield	Marg-inal yield
cm						cm			100 Kg/ha	100 Kg/ ha	100 Kg/ [(ha)(2.5 cm)]
5.0	0.15	0.20	0.20	0.15	0.945	1.36	1.48	1.20	2.87	1.43	3.88
7.5	0.20	0.35	0.30	020	1.26	2.38	2.22	1.60	6.75	2.25	5 24
10.0	0.34	0.40	0.40	030	1.89	2.72	2.96	2.40	11.99	92.99	6.42

TABLE 6.3 *(Continued)*

12.5	0.55	0.55	0.45	0.25	3.46	3.74	3.33	2.00	18.41	3.68	6.37
15.0	0.65	0.60	0.55	0.35	4.09	4.08	4.07	2.08	24.78	4.13	6.04
17.5	0.70	0.75	0.60	0.45	4.41	5.10	4.44	3.60	30.82	4.40	—
20.0	0.80	0.80	0.75	0.50	5.04	5.44	5.55	4.00	35.82	4.47	5.00
22.5	0.90	0.90	0.75	0.65	5.67	6.12	5.55	5.20	39.86	4.42	4.04
25.0	0.85	0.90	0.85	0.90	5.35	6.12	6.29	7.20	42.69	4.26	2.83
27.5	0.95	0.95	0.95	1.00	5.98	6.46	7.03	8.00	44.21	4.01	1.52
28.5	0.85	0.95	0.90	1.25	5.35	6.46	6.66	10.00	42.34	3.52	-1.87

6.3.4 USE OF MODEL FOR PREDICTING IRRIGATION LEVELS OF ECONOMIC RELEVANCE

The model developed has been used to apportion hypothetical limited quantities of water in a growing season. The results are given in Table 6.3. From Fig. 6.2, it is clear that the region of 21.66 cm to 28.33 cm of irrigation water input is the area of economic relevance. Thus the producers, who want to maximize their productivity of water, must operate in this economic region (Zone II) with the maximum productivity at 27.5 cm level. However, the irrigation level for maximum profit may lie beyond this level of maximum productivity or water.

6.3.5. USE OF MODEL FOR YIELD PREDICTION

Since the model needs irrigation water to pan evaporation ratio data, all the 12 treatments were transformed to corresponding depth of irrigation water to evaporation ratios, assuming the variations in these values is minimum, year after year. The sensitivity factor was calculated using Eq. (4) for each of the treatments. The predicted yields for all 12 treatments have been compared in Table 6.4.

TABLE 6.4   Comparison of Observed and Predicted Values of Yield

Treatment	Observed	Predicted	Difference
T1	31.06	38.66	−7.60
T2	41.50	43.66	−1.96
T3	43.86	43.72	0.14

TABLE 6.4   *(Continued)*

T4	34.82	43.28	–8.46
T5	42.70	43.68	–0.98
T6	38.54	43.90	–5.36
T7	38.09	43.99	–5.90
T8	41.18	42.61	–1.43
T9	34.06	44.20	–10.14
T10	39.74	37.54	2.20
T11	39.40	40.35	0.95
T12	43.17	42.12	1.05

6.3.6   VALIDATION OF THE MODEL

The validity of the response function developed was tested by the t-test (Parametric test) and Wilconox signed rank test (Non-parametric test).

The observed and the predicted values are given in Table 6.4. From the data in Table 6.4, it is clear that the predicted and observed yields compare reasonably all other treatments, except for the treatment T4 and T9 (where the predicted yields are much higher than the observed yields). Wide variation in T4 and T9 may be attributed to experimental errors. Therefore, further tests were applied for neglecting these two treatments:

a. **t-test:** t-calculated = 2.06; t – tabulated at 5% = 2.26. The difference between the two values is negligible.

b. **Wilconox signed rank test:** t-calculated = 12; t-tabulated at 5% = 8. Hence, the model is good enough for predicting the yield.

Following two equations were used for a test of degree of variation between observed and predicted yields:

$$RES = \sum[Yobserved - Ypredicted] = -20.79 \qquad (8)$$

$$RES_{absolute} = \sum[Yobserved - Ypredicted] = 27.57 \qquad (9)$$

where: RES = Sum of residuals; and $RES_{absolute}$ = Sum of absolute residuals. Comparing the sum of residuals in Eqs. (8) and (9), we observe a large negative value of

RES, indicating that the model is slightly over predicted, in large number of cases. The value of stress day factor is obtained by dividing the average stress day index for 50% depletion of available soil moisture (which is equal to 0.09139) by the crop susceptibility factor for a day.

6.4   CONCLUSIONS

For better planning of water use, irrigation can be applied, when the value of stress day factor is greater than the observed value of the soil moisture status in the field on the same day. Otherwise irrigation is postponed till the value of stress day factor (calculated) becomes greater than the observed soil moisture status in the field.

To apply a known quantity of water in a season for getting the optimum yield, the proposed irrigation scheduling model can be employed which apportions different amounts of available seasonal water supplies to get the optimum yield.

The present irrigation scheduling, model can be used for other crops by replacing the water yield production function of wheat crop by the crop under consideration, provided the crop susceptibility factor is known. Stress day factor can be taken as standard from past experience or published literature.

6.5   SUMMARY

An alternative schedule for optimum yield from seasonal water supply level was worked out using IW/PE ratio (which proved to be the next best), directly using the technique in this study. Four irrigation-scheduling techniques (Crop Growth Stages, Irrigation Water to Pan Evaporation Ratio, Stress Day Index, and Canopy-Air Temperature Difference) each with three irrigation levels have been evaluated using a replicated field experiment in randomized block design. From the 12 treatments, the 'Stress Day Index' method with 50% depletion of available soil moisture was statistically superior, and had maximum water use efficiency. A water input yield response model was developed for this irrigation schedule. Both parametric and nonparametric statistical tests satisfied the model predictions when applied to the data of remaining 11 irrigation treatments. The model was also used for apportioning known quantities of seasonal irrigation supplies for obtaining the optimum yield.

KEYWORDS

- air temperature
- canopy temperature
- crop growth stage
- irrigation scheduling
- irrigation water
- marginal yield
- optimum yield
- pan evaporation
- response function
- seasonal irrigation
- sensitivity factor
- stress day factor
- stress day index
- sum of residuals, RSS
- water use efficiency, WUE
- Wilcoxon signed rank test

REFERENCES

1. Anonymous, (1972–1983). *Annual Progress Report of ICAR Integrated Project for Research on Water Management and Soil Salinity.* Punjab Agricultural University, Ludhiana.
2. Nairizi, S., Rydzewski, J. R. (1977). Effects of Soil Moisture Stress On Crop Yield. *Exp. Agric. 13*, 51–59.

APPENDIX I: PHOTOS OF FIELD EXPERIMENT ON DRIP IRRIGATED WHEAT

CHAPTER 7

EVAPOTRANSPIRATION ESTIMATIONS USING CLIMATOLOGICAL APPROACHES

KAMAL GURMIT SINGH and PAWANPREET KAUR

7.1 INTRODUCTION

Punjab is called the land of irrigated agriculture with 85% of its cultivable area under irrigation. This high percentage of area under irrigation reveals that Punjab has exploited a considerable part of its water resources for irrigation purposes. But irrigation practices adopted by farmers of Punjab are generally arbitrary and not necessarily based on actual water needs of a crop. Therefore, proper management of water resources had become a matter of vital concern for increasing crop production further, on one hand, and to prevent the exhaustion of underground water resources, on the other hand. Evapotranspiration (ET) is the best criterion which needs to be estimated on a scientific basis so that the required amount of water can be applied to the crop at the proper time, taking into account the effective rainfall and irrigation application efficiency. One of the methods of ET estimation is ly-simetery. However, its immobility and high cost restrict its utility. The evaluation of ET by an empirical method is simple and easy to use. The empirical formulae use standard climatological data as the inputs. However, all the proposed empirical formulae hold good in the respective localities where these methods were developed, and cannot be extended beyond these locations. Therefore, a suitable method for determining ET for Punjab is needed to have a reliable estimate with available meteorological data.

Therefore, this study was planned: To evaluate the existing modified empirical approaches for estimation of ET; and to compare and suggest the suitable methods under the climatic conditions of Punjab – India.

---

*In this chapter, the currency is expressed in Indian Rupees (1.00 US$ = Rs. 60.93; 1.00 Rs. = 0.02 US$).

7.2 MATERIALS AND METHODS

Following empirical formulae [4] were selected for estimation of ET: (i) Blaney-Criddle method; (ii) Radiation method; (iii) Modified Penman method; and (iv) Pan evaporation method.

7.2.1 BLANEY–CRIDDLE METHOD

This method represents a mean value over the given month as below:

$$PET = C [P (0.46T + 8)], mm/day \qquad (1)$$

where: PET = reference crop evapotranspiration in mm/day for the month considered; T = mean daily temperature in° C for the month considered; P = mean daily percentage of total annual day time hours for a given month and latitude; and C = adjustment factor which depends relative humidity, sunshine hours and daytime wind speed.

7.2.2 RADIATION METHOD

$$PET = C (W.Rs) \qquad (2)$$

where: PET = reference crop evapotranspiration in mm/day for the periods considered; Rs. = solar radiation in equivalent evaporation in mm/day; W = weighing factor which depends on temperature and altitude; and C = adjustment factor which depends on mean humidity and daytime wind conditions

7.2.3 MODIFIED PENMAN METHOD

$$PET = C [W.Rn + (1–W).f(u) (ea–ed)] \qquad (3)$$

where: PET = reference crop evapotranspiration in mm/day; W = temperature related weighing factor; Rn = net radiation in equivalent evaporation in mm/day; f(u) – wind related function; (ea-ed) = difference between the saturation vapor pressure at the mean air temperature and the mean actual vapor pressure of air in mbar.

7.2.4 PAN EVAPORATION METHOD

The following curvilinear relationship relating the ratio of potential evapotranspiration to pan evaporation (PET/EP) and time t (days after seeding) was used [1]:

$$[PET/EP] = 0.56 + 0.021\ t - 0.000125t^2, \quad R^2 = 0.98. \tag{4}$$

where: t = time following seeding; and EP = Pan evaporation in mm/day.

7.2.5   ESTIMATION OF ACTUAL EVAPOTRANSPIRATION

$$ETcrop = Kc \times PET \tag{5}$$

To find the final value of Kc, the crop season is divided into four stages. The four stages in crop development are initial stage, mid-season stage, crop development stage, and late season stage. Crop coefficient value is determined from the crop coefficient curve [2] for whole growing season.

7.2.6   DATA COLLECTION

The input data (temperature, relative humidity, rainfall, wind speed, and sunshine hours) were those recorded in meteorological observatory at Punjab Agricultural University, Ludhiana. Crop evapotranspiration for the Lysimeter study for the period November 1980 to April 1981 was taken from Singh [3].

7.2.7   COMPUTER PROGRAM

Computer program in FORTRAN-77 was developed for the calculation of PET values for period November 1980 through April 1981 for the four methods (Blaney Griddle method, Radiation method, Modified Penman method, Pan evaporation method). The estimations were made for the year 1980–1981, because lysimeter data was available for this period only.

7.3   RESULTS AND DISCUSSION

7.3.1   COMPARISON OF DIFFERENT ET METHODS

The 15 days average PET computed with various empirical methods are given in Table 7.1. The Table 7.1 reveals that in the initial stages of crop growth ET values were quite high except in case of Pan evaporation method. During the peak growth period, radiation and Modified Penman method showed a sharp rise in ET values. During first 15 days of February, there was a fall in ET values. This may be due to rainfall or cloudy conditions during those days. In general, radiation method showed highest values while Pan evaporation showed lowest value during the whole season. Values of Modified Penman method and Blaney-Criddle method are comparable but for the whole season. Blaney-Criddle method underestimates the ET values as compared to Modified Penman method.

TABLE 7.1   Estimated Potential Evapotranspiration (mm/day) With Different Methods During Wheat Growing Period (Crop was Sown on November 16)

Date	Days	Blaney-Criddle method	Radiation method	Modified Penman method	Pan evaporation method
Nov. 16–30	15	2.853	3.554	3.096	1.986
Dec. 1–15	30	2.406	3.685	2.919	2.154
Dec. 16–30	45	1.704	3.139	2.225	1.805
Dec. 31–Jan. 14	60	2.117	2.090	1.835	1.817
Jan. 15–29	75	1.710	2.758	2.034	1.655
Jan. 30–Feb. 13	90	1.710	2.304	2.031	1.756
Feb. 14–28	105	2.290	2.999	2.57	2.285
Mar. 1–15	120	2.270	3.101	3.07	2.097
Mar. 16–30	135	3.553	4.022	3.685	2.947

7.3.2   COMPARISON OF ET ESTIMATIONS WITH LYSIMETER ET

Lysimeter values are accurate representation of water use by the crop. Therefore, the estimations with climatic approaches were compared with the lysimeter data. The values of ET calculated by four different methods and actual lysimeter ET are given in Table 7.2. The ET estimated by Blaney-Criddle method ranges from 0.998 mm/day to 3.197 mm/day. The values underestimate the lysimeter values in the initial stages, but at the end of season it exceeds lysimeter values. ET estimated by Radiation method ranges from 1.289 mm/day to 3.619 mm/day. The values nearly match the lysimeter values, and only at the end of wheat season the ET value exceeds that may be due to high solar radiation in the month of March–April. ET estimated by Modified Penman methods ranges from 1.084 mm/day to 3.316 mm/day. As the temperature increases in March–April, temperature related factors also show increase in these values. Hence the ET value exceeds lysimeter value. ET estimated by Pan evaporation method ranges from 0.695 mm/day to 3.252 mm/day. The values underestimated the lysimeter observations for the whole season except in month of March–April.

TABLE 7.2 Actual Evapotranspiration and Calculated Evapotranspiration (mm/day) During Wheat Growing Period (Crop was Sown on November 16)

Date	Kc	Blaney-Criddle method	Radiation method	Modified Penman method	Pan evaporation method	Actual ET lysimeter
Nov. 16–30	0.35	0.99	1.294	1.084	0.695	1.01
Dec. 1–15	0.35	0.842	1.289	1.022	0.754	1.71
Dec. 16–30	0.55	0.937	1.726	1.224	0.993	1.94
Dec. 31–Jan. 14	0.82	1.736	1.714	1.505	1.490	1.75
Jan. 15–29	1.05	1.795	2.896	2.136	1.737	2.31
Jan. 30–Feb. 13	1.05	1.795	2.419	2.132	1.844	3.37
Feb. 14 28	1.05	2.404	3.149	2.70	2.399	3.67
Mar. 1–15	1.05	2.383	3.256	3.224	3.252	1.20
Mar. 16–30	0.9 3	.197	3.619	3.316	2.652	1.65
Mar. 31–Apr. 9	0.62	2.851	2.805	2.330	1.173	1.63

TABLE 7.3 Cumulative Evapotranspiration (mm/day) Values by Different Methods For Entire Crop Season

Days	Blaney-Criddle method	Radiation method	Modified Penman method	Pan evaporation method	Lysimeter
15	14.97	19.41	16.26	10.43	15.15
30	27.60	38.75	31.59	21.74	40.80
45	41.65	64.64	49.95	36.30	69.90
60	67.70	90.35	72.53	58.98	96.15
75	94.62	133.79	104.57	85.04	130.80
90	121.55	170.07	136.55	112.70	181.00
105	157.61	217.31	177.04	148.68	206.40
120	193.35	266.15	225.39	197.46	254.40
135	241.31	320.46	275.14	237.24	279.15
145	284.07	362.50	310.10	254.84	303.60

7.3.3   COMPARISON OF DIFFERENT METHODS FOR ESTIMATING CUMULATIVE EVAPOTRANSPIRATION (CET)

The CET values for the entire crop-growing season are given in Table 7.3. Radiation method and lysimeter gave nearly equal values during initial stage of crop growth. CET values calculated by Modified Penman method are nearly equal to lysimeter during initial and then final stage of crop growth. For the rest of season, Modified Penman values underestimate the lysimeter values. Pan evaporation method gave the lowest CET values. From the data in Tables 7.1 to 7.3, it is also clear that lysimeter and Modified Penman gave nearly equal values. Therefore, Modified Penman method may be a reliable method to calculate CET. Under the situations of availability of minimum data, Radiation method may be more suitable.

7.3.4   STATISTICAL ANALYSIS

The t-test at 5% level of significance was used to establish the validity of the ET values with climatological approaches versus lysimetric data. The calculated t-values were 0.91, 0.80, 0.19 and 2.35 for Blaney-Criddle, Radiation, Modified Penman and Pan evaporation method, respectively. Whereas, t-value at 5% level of significance and seven degree of freedom was 1.73. Therefore, Pan evaporation method is discarded as its t-value is more than t-value; and Blaney-Criddle, Radiation and Modified Penman methods are acceptable.

7.4   CONCLUSIONS

Modified Penman and radiation methods to estimate ET showed close resemblance with lysimeter data. But Radiation method appears to be better as if enough data is not available. It is concluded that modified penman method is best for calculating cumulative evapotranspiration, whereas radiation method is best for calculating daily/fortnightly evapotranspiration (for insufficient data).

7.5   SUMMARY

Four modified empirical methods (Blaney-Criddle method, Radiation method, Modified Penman method and Pan evaporation methods of determining potential evapotranspiration) were compared. The potential evapotranspiration values calculated by these methods were multiplied by crop coefficient to find the actual evapotranspiration and these actual ET values were compared with lysimeter data. The study suggested that Modified Penman method is best for CET and radiation method is best for calculating the daily/fortnightly evapotranspiration.

KEYWORDS

- cabbage
- Blaney-Criddle method
- climatological approaches
- effective rainfall
- evapotranspiration, ET
- irrigation application efficiency
- irrigation practices
- lysimeter
- mean actual vapor pressure
- mean air temperature
- mean daily temperature
- Modified Penman method
- net radiation
- Pan evaporation, Epan
- Radiation method
- relative humidity
- saturation vapor pressure
- sunshine hours
- water resources

REFERENCES

1. Arora, V. K., Prihar, S. S., Gajri, P. R. (1987). Synthesis of a simplified water use simulation model for predicting wheat yields. *Water Resource Rev., 23(5),* 903–910.
2. Kumar, Anchal (1983). Simulation of soil moisture profiles under cropped conditions. M. Tech. Thesis, Soil and Water Engineering, College of Agricultural Engineering, PAU, Ludhiana.
3. Singh, Awadh Bihari (1982). To find an empirical method of computing potential evapotranspiration under climatic conditions of Punjab. M. Sc. Thesis, Department of Agricultural Meteorology, PAU, Ludhiana.
4. Singh, Kamal Gurmit, Pawanpreet Kaur (2002). Evaluation of modified climatological approaches for the estimation of evapotranspiration. *J. Res. Punjab Agric. Univ., 39(1),* 97–106.

CHAPTER 8

ADVANCES IN MICRO IRRIGATION FOR ENHANCING RESOURCE USE EFFICIENCY

KAMAL G. SINGH

8.1 INTRODUCTION

It is estimated that world population will grow from 7.0 billion in 2011 to 7.5 billion in 2020. While there is an ever-increasing demand for food, total agricultural land area on earth will not grow from current level of 5.0 billion hectares. In order to ensure food security and nutrition demand, productivity of land has to be increased in the rate of 40–50% in next 10 years by using appropriate technologies of which micro irrigation is very important. Water happens to be the far most crucial input in agriculture sector, which consumes about 90% of the total water resources. In India, including Punjab, the traditional practices of irrigation are still followed in which the overall efficiency is only 30–50%. World Water Council 2000 observed "the present water crisis is the result of the poor management of available water resources that has created an artificial gap between demand and supply of water." Agriculture, being the major consumer of water is responsible to a great extent for the current mismanagement of water resources and consequently a sufferer also. But to meet the increasing food grains need of growing population from the same land area, the need of efficient utilization of all inputs to agriculture especially water is imperative [1–6].

8.1.1 MICRO IRRIGATION IN THE WORLD

Of total 300 million ha of area under irrigation globally, only 45 million ha, that is, 15% is currently under micro irrigation. Of this 80% is under sprinkler irrigation, including Pivot irrigation and 20% is under drip irrigation. Though India has the largest irrigated area at 69 million ha yet it has only 4–5 million ha, that is, less than 7% under micro irrigation.

---

*In this chapter, the currency is expressed in Indian Rupees (1.00 US$ = Rs. 60.93; 1.00 Rs. = 0.02 US$).

8.1.2   PRESENT AGRICULTURE SCENARIO OF INDIA

In India, the current agriculture scenarios followed are: (i) shrinking per capita water availability; (ii) lowering sectoral availability of water; (iii) lower irrigated area; (iv) lower productivity; (v) small land holding. Here, the problem is not water availability but inefficient management of available water resources. Precision farming techniques with optimum use of input resources is one of the solutions. In view of the urgent need to maximize use of the available resources, it is imperative to reduce the use of resources by adopting efficient /advanced methods of irrigation like *drip, sprinkler,* and *micro sprinklers.* By introducing the advanced irrigation methods, in addition to maximizing nutrient use efficiency, more and more cultivated areas can be brought under irrigation resulting in good quality increased agriculture production.

8.1.3.   WHY MODERN IRRIGATION TECHNOLOGIES?

1.   Water is vital input with 80% its consumption in agriculture sector.
2.   The productivity of irrigated land is low compared to its potential.
3.   The productivity per unit water is very low.
4.   Water available for irrigation is becoming scare.
5.   Decline in water table – 0.55 m/year in central Punjab.
6.   The predominance of soils with low water retention capacities and very low hydraulic conductivities makes an ideal case for light and frequent irrigations, that is, Micro-irrigation
7.   Micro irrigation will increase the irrigation cover using existing available water.
8.   Micro irrigation with fertigation will enhance production per unit input in nutrient poor low dense soils.

8.2   MICRO IRRIGATION

Drip/Trickle irrigation represents one of the fastest expanding technologies in modern irrigated agriculture with a great potential of achieving high effectiveness of water use. Crop coverage under drip irrigation in India is shown in Fig. 8.1. Drip irrigation basically involves the slow application of water in the form of discrete drops, continuous drops, tiny streams or miniature sprays through mechanical devices called emitters or applicators located at selected points along water delivery lines where the water infiltrates the soils. The efficiencies of different irrigation systems are given in Table 8.1. Total area under micro irrigation in India is 7.2% where as world average is 15% of the irrigated area. The Department of Soil and Water Engineering PAU, Ludhiana is working on micro irrigation since many years and has developed irrigation and fertigation schedules for pepper, Potatoes, Tomato, egg-plant, early cauliflower, cotton and sugarcane. Results are discussed in this chapter.

TABLE 8.1   Irrigation Efficiencies For Different Methods of Irrigation

Type	Irrigation efficiency, %		
	Surface	**Sprinkler**	**Drip**
Conveyance efficiency	40–50 (Canal) 60–70 (Well)	100	100
Application efficiency	60–70	70–80	90
Surface water moisture evaporation	30 40	30 40	20–25
Overall efficiency	30–35	50–60	80–90

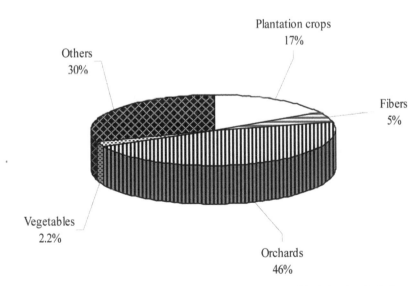

FIGURE 8.1   Percentage distribution of crop coverage under drip irrigation in India.

8.3   AUTOMATION OF DRIP IRRIGATION SYSTEM

Manual operation of the routine practices in agriculture requires lot of attention and care. Also it is difficult to perform desired jobs efficiently and precisely. Ultimately this may result in lower crop production, nonuniform growth and poor quality. Automation provides a faster, precise and reliable operation.

Automation of micro irrigation system refers to operation of the system with no or minimum manual intervention (Fig. 8.2). The introduction of automation in irrigation system has increased application efficiency and drastically reduced labor requirement. The automation is adopted due to:

1. Automation eliminates manual operations
2. Possibility to change frequency of irrigation and fertigation process and also to optimize these processes
3. Adoption of advanced crop systems and new technologies, especially new crops system that are complex and difficult to operate manually
4. Use of water from different sources and increased water and fertilizer use efficiency
5. Smooth and efficient system operation
6. Optimizing energy requirements and increase yield

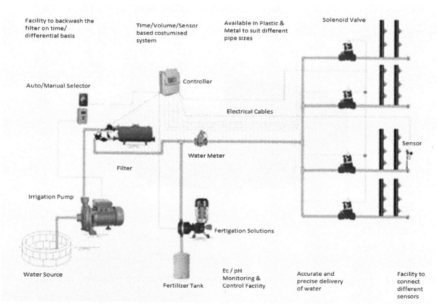

FIGURE 8.2   Automation of drip irrigation system.

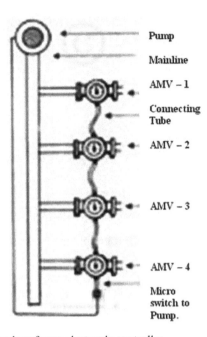

Pump

Mainline

AMV – 1

Connecting Tube

AMV – 2

AMV – 3

AMV – 4

Micro switch to Pump.

FIGURE 8.3   Solenoid valves for an electronic controller.

8.3.1   TYPES OF AUTOMATION SYSTEMS

8.3.1.1   TIME BASED SYSTEM

In this system, time is the basis for operation. The basic objective is to prepare a schedule based on crop water requirements. The operation sequence will be set by user as desired.

8.3.1.2   VOLUME BASED SYSTEM

In this type, every section will receive the preset volume of water. This is possible with the help of two methods: (i) **Electronic Control, and (ii) Mechanical Control (AMV).**

8.3.1.2.1   ELECTRONIC CONTROL

Water meter is an essential component in this method, which gives the feedback to controller, after the preset water volume is delivered. One after the other, every section receives the preset water quantity (Fig. 3).

8.3.1.2.2   MECHANICAL CONTROL (AMV)

This is known as sequential irrigation system. Here controller is not required. Automatic metering valves are interconnected in series. The required quantity has to be set on each AMV. One after the other, all AMVs will operate and deliver the reqd. water quantity. After the last AMV is closed, the pump also shuts off.

8.3.1.3   SENSOR-BASED SYSTEM

In this system, sensors give feedback to controller, depending on which the controller initiates various actions as required.

8.3.1.3   FEATURES OF TIME-BASED IRRIGATION CONTROLLERS

1. Cyclic programming/interval programming.
2. Water window management.
3. Cycle and soak by station.
4. Easily upgradeable for remote control.
5. Global or program level seasonal adjustment.
6. Compatibility with evapotranspiration system, central control and sensors.

8.3.1.2   FEATURES OF VOLUME OR SENSOR BASED CONTROLLERS (FIG. 8.4)

1. Flexibility of start times and cycles.
2. Sequential programming.
3. Powerful software to monitor and control the complete operation of the system.
4. Fully automatic, or manual override operation options.
5. Powerful battery backup.
6. Event logbook to record the last 100 system events (complete/partial logging options).
7. Large, easy to read, wide-angle LCD display.
8. Compatibility with wireless and central control systems

FIGURE 8.4   Sensor based controllers for automation.

8.3.2   BASIC PRINCIPLES OF OPERATION

8.3.2.1   SYSTEM COMPONENTS OF AN AUTOMATIC IRRIGATION SYSTEM (FIG. 8.5)

1. **Controller** controls whole automation system. According to the signals received from sensor, open or close solenoid valve through solenoid coil as shown in Fig. 8.6, we generally use (i) electromechanical, and (ii) electronic controls.

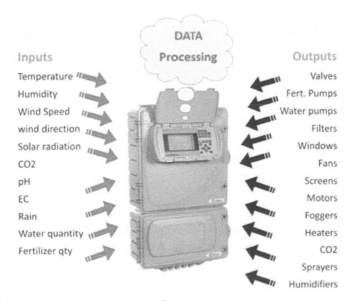

FIGURE 8.5   Components of an automatic system.

Agricultural Controllers (AC/DC)

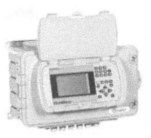

Gal-pro: 8   **Galileo16W/32W**   **Galileo: 40,64,88,112**

FIGURE 8.6   Irrigation controllers.

2. **Sensors** produce electrical signals directly related to parameter to be measured. Available sensors: (i) tensiometer; (ii) dielectric sensor; (iii) thermal soil matric potential sensor; (iv) Gypsum block soil water sensor; and (v) TDR based sensor (Fig. 8.7).

Weather Station   **Tensiometer**   **Ec & pH unit**

FIGURE 8.7   Soil moisture sensors.

3. **Control valves** are solenoid or hydraulic valves open or close according to command by solenoid coil, which converts electrical pulses in to hydraulic pulses to operate control valves (Fig. 8.8).

FIGURE 8.8   Solenoid valves and hydraulic valves.

4. **Water meter**: Measure the rate of flow of water.
5. **Pump:** Lift the water from source and circulate through system.

8.4   SOCIO-ECONOMIC IMPACT OF MI

By converting 0.10 million ha land area under micro irrigation:
1. total water saving is 347 million Cu. meters/annum.
2. fertilizer saving of Rs. 1.05 billions.
3. saving of 27.1 million KWH of energy per year worth Rs. 70 million by pumping less water.
4. employment generation of 0.125 million person.
5. saving on infra-structural investment on Major irrigation projects (for saved) – 2.65 billions.

8.5   FUTURE STRATEGIES FOR PROMOTING DRIP IRRIGATION

In view of the fact that India has a potential of bringing 69 m-ha under drip irrigation especially for horticultural crops, following strategies have to be adopted to conserve precious water resources with simultaneous increase in horticultural production.
1. Initial cost needs to be lowered and loans at concessional rates be provided instead of popular scheme of free power and water to the farmers.
2. The procedure of processing of subsidy cases must be simplified so that these are finalized within one month.
3. More efficient filter(s) need to be designed to prevent clogging of the system.
4. Rural unemployed educated youth may be provided training in operation and maintenance of the system.
5. After sales service and supply of spares need to be strengthened at the block/ tehsil level

6. Special incentives in the form of priority release of connection to the farmers desirous of installing this system for orchards by the State Electricity Board should be given.

8.6 SUMMARY

World's water crisis is a fact. According to the International Water management Institute (IWMI), one-third of the world's population will face absolute water scarcity by the year 2025. Policy makers, researchers, NGOs, and farmers are pursuing various technical, institutional and policy interventions to meet this crucial challenge in the twenty-first century. Micro irrigation technology has now become a time tested tool for making efficient use of water and fertilizers as well as for improving the farm productivity and should be promoted for addressing the needs for Food Security, Water Security and Environmental Security. Micro irrigation technologies constitute one such intervention with the ability to use water efficiently in irrigated agriculture in order to produce 'more crop per drop.' Further, it can raise incomes through improved crop yields and outputs; enhance food security of households and ensure environmental protection. There is an increasing trend to shift from manual to automatic operations. Irrigation can be almost totally automated by using micro irrigation system. Accurate and precise application rates, increased yields, saving in water and energy, reduced labor cost, minimized return flow, improved pest and disease control are among the major advantages that are offered by adopting automation techniques in micro irrigation.

KEYWORDS

- **drip automation**
- **drip irrigation**
- **environment protection**
- **FAO**
- **food security**
- **IARI**
- **ICID**
- **IWMI**

- **labor cost**
- **micro irrigation**
- **mulching**
- **plastics**
- **precision farming**
- **return flow**
- **sensor based irrigation**
- **water crisis**
- **water technology center, WTC**
- **World Water Council, WWC**

REFERENCES

1. Anonymous, (2006–2012). Annual progress reports of AICRP on "Application of Plastics in Agriculture." Department of Soil and Water Eng., P.A.U. Ludhiana – India.
2. Anonymous, (2011). Proceedings of the National seminar on Advances in Micro Irrigation. Organized by NCPAH, Ministry of Agriculture, Govt. of India, New Delhi, and PFDC, WTC, IARI, New Delhi on Feb 15–16.
3. FAO, Water Report/Aquastat.
4. Kulkarni, S. A., Reinders, F. B., Ligetvari, F. Global scenario of sprinkler & micro irrigated areas. <ICID.ORG>.
5. MoWR, (2002). National water policy. Government of India, New Delhi.
6. Rajput, T. B. S., Neelam Patel, (2001). *Drip Irrigation Manual.* Water Technology Center, IARI, New Delhi.

CHAPTER 9

MICRO IRRIGATED SUGARCANE IN INDIA: A REVIEW

ARUN KAUSHAL, RAHUL PATOLE, and KAMAL G. SINGH

9.1 INTRODUCTION

Sugarcane (*Saccharum officinarum L.*) has a unique role in sustaining agro industrial economic growth in India. In the world, India is second largest producer of sugarcane with estimated production of 348.2 million tons per year [38]. This production accounts for approximately one fifth of total production 1743.1 million tons in the world. Sugarcane is an important raw material for production of sugar, industrial alcohol, filter cake (a fertilizer), bagasse (fuel), cattle feed and paper industry. Sugar industry adds Rs. 30,000 crore to the Indian economy and there are about 3.5 crore farmers engaged in sugarcane cultivation.

In India, sugarcane is cultivated in about 4.1 million ha, however the average yield is only about 69 tons per ha [11]. One of the main reasons of low yield is scarcity of water. Sugarcane, being a long duration crop, produces huge amount of biomass, and requires large quantity of water and is mostly grown as an irrigated crop. It is important to judiciously use the already existing water resources by adopting appropriate irrigation technology that not only increases sugarcane production per unit area but also per unit of water used. Thus a scientific and efficient management of water is needed especially in hot dry months of premonsoon period, to enhance water use efficiency (WUE) and cane yield, which is possible by micro irrigation.

In this chapter, sugarcane cultivation is reviewed under micro/drip or trickle irrigation.

9.1.1 WATER REQUIREMENTS OF SUGARCANE

The water requirement of sugarcane is high; and against the background of the rapid decline in irrigation water potential and low WUE of flood (conventional) method of irrigation, micro irrigation is known to save a substantial amount of water and helps to increase the productivity as reported by several workers [24, 27, 31–33].

---

*In this chapter, the currency is expressed in Indian Rupees (1.00 US$ = Rs. 60.93; 1.00 Rs. = 0.02 US$).

The water requirement of sugarcane in Akola – India has been estimated by Zade [38]. The average weekly values of reference evapotranspiration (ET_o) were calculated by modified Penman formula and distribution was fitted for the same. The 70% probability value of ET for sugarcane was 2065.3 mm while the total water requirement was 1633.41 mm.

Gulati and Nayak [9] studied effects of irrigation and planting dates on growth, cane yield and WUE of sugarcane in Orissa – India, during 1994–1996. The study reported four irrigation levels (IW/CPE ratio of 0.6, 0.8, 1.0 and 1.2) and six dates of planting (October, November, December, January, February and March). The highest cane yield of 156.65 t ha^{-1} was recorded from planting sugarcane during the third week of October at 1.2 = IW/CPE, with the corresponding irrigation and water requirements of 132 and 218 cm. The crop extracted more moisture from the 0–30 cm soil depth in all treatments.

Arulkar et al. [2] reported the water requirement estimation of sugarcane crop from climatological data by probability analysis at Nagpur district of India. Weekly rainfall data for the 25 years were analyzed and rainfall curves were drawn at various probability levels. The values of weekly reference evapotranspiration were calculated by using weather data: relative humidity, temperature, wind speed and cloudiness. Crop coefficient values were determined to get water requirement of sugarcane. Annual reference evapotranspiration and crop water requirement were 2126 and 1982 mm, respectively.

The application of water balance studies in irrigated sugarcane from Karnataka – India has been reported by Rajegowd et al. [25]. It was observed that the crop coefficient value during initial stage was 0.5 and then gradually increased during vegetative phase and reached maximum during flowering and reproductive phase and decreased to 0.6 after the maturity. The available water holding capacity of the soil was approximately 116 mm for 100 cm soil depth. The total water requirement was 1346 mm. The results indicated that the irrigated sugarcane planted during the month of July with field capacity can grow without any further irrigation until December. Under the annual rainfall of 844 mm in the zone, the annual surplus was approximately 138 mm; the total water available for the crop growth from the rainfall was approximately 818 mm. The additional need of approximately 529 mm was met through supplement irrigation.

Kumar et al. [15] conducted experiment on probabilistic irrigation water requirement of major crops at Udham Singh Nagar district of Uttaranchal state – India. The water requirement was estimated using daily rainfall and observed Class A pan data. The irrigation water requirements of sugarcane were calculated at 50 and 80% probability of occurrence of the effective rainfall. The seasonal average water irrigation requirement was 1119 mm for sugarcane crops at 80% probability.

The water requirement for seasonal and annual sugarcane crop (per plant basis) using open pan evaporation method from Akola, Maharashtra was reported by Ingle [10]. Sugarcane crop water requirement (WR) increased from 0.52 L.day^{-1} to 11 L. day^{-1} on 21 MW (meteorological week), and then decreased gradually to 1.80 L day^{-1} at 26 MW. The maximum WR for Suru and Adsali sugarcane were 122 and 148 m^3·day^{-1}·ha^{-1}, respectively.

Larger yield in sugarcane depends on the availability of adequate quantity of water, Water is most important input in an assured sugarcane production system, especially in area, where sugarcane production suffers due to scarcity and or irregular distribution of rainfall specially from mid April to end of June (before the onset of monsoon). Micro irrigation is a method by which one can overcome scarcity of water for sugarcane crop.

9.1.2 ADVANTAGES OF MICRO IRRIGATION IN SUGARCANE

In micro irrigation system, water is applied in the form of drops directly to the plant through emitters from which it drops into the soil slowly and frequently to keep the soil moisture within the desired range for healthy plant growth so that the plant do not experience any moisture stress throughout the life cycle. It is particularly suited to soil with very low and very high infiltration rates under conditions of water scarcity and in area where drainage of excess water is difficult.

Advantages of micro irrigation in sugarcane are as follows: saving of irrigation water; low fertilizer/nutrient loss due to localized application and reduced leaching; high water application efficiency; grading of the field not necessary; ability to irrigate irregular shaped fields; allows better use of recycled water; moisture within the root zone can be maintained at field capacity; less weed growth; soil type plays less important role in frequency of irrigation; minimized soil erosion; highly uniform distribution of water, that is, controlled by output of each nozzle; lower labor cost; variation in supply can be regulated by valves and drippers; fertigation can easily be included with minimal waste of fertilizers; foliage remains dry thus reducing the risk of disease and decreased energy costs due to reduced pumping time to irrigate a given design area. It leads to optimum soil water air relations contributing to better germination, uniform field emergence and maintenance of optimum plant population with higher cane and sugar yields.

9.2 MICRO IRRIGATED SUGARCANE

9.2.1 INDIAN SCENARIO

Three levels of for surface irrigation (low, medium and high) were compared, based on IW/CPE ratio of 0.75, 1.00 and 1.25, respectively, and ratios of 0.40, 0.55 and 0.70, respectively, for micro irrigation by Ahluwalia et al. [1] for sugarcane. The medium level of irrigation for surface irrigation method (IW/CPE = 1.00) was the optimum level in terms of cane yield (81.4 t ha^{-1}) and WUE (0.484 tons per ha-cm). While the low level of irrigation was the optimum level of drip-irrigated sugarcane in terms of cane yield (80.6 t ha^{-1}) and WUE (0.779 tons per ha-cm). The micro irrigation resulted in saving of water by 38.0% with consequent increase of 60.9% in WUE over the surface irrigation. Juice extraction percentage in surface irrigated sugarcane was higher than in the drip-irrigated crop. The sugar yield with micro irrigation was higher than that with surface method.

Shinde and Jadhav [32] compared drip irrigated sugarcane with conventional irrigated in a study carried out at Pune, Maharashtra, India. Different surface and subsurface micro irrigation methods were compared with conventional irrigation. The results indicated that the automatically controlled micro irrigation used up to 56% less water, increased yield up to 52% and increased WUE by about 2.5–3 folds.

The effect of fertigation and planting techniques on yield and quality of sugarcane under micro irrigation on a Vertisol in Maharashtra was studies by Bhoi et al. [4]. Sugarcane was grown in paired rows (75 cm), four rows (90 cm) and drip irrigated. Fertilizer N (Nitrogen) was applied through micro irrigation in 4, 10 and 20 splits. Mean cane yield of 171.4 t ha^{-1} was found highest with four row planting and 20 splits. Paired row planting with 20 splits of N produced similar yield of 169.9 t ha^{-1}.

The influence of planting techniques and fertigation through micro irrigation on yield, quality and economics of sugarcane has been reported by Shinde et al. [33] from Rahuri, Maharashtra, on seasonal sugarcane and its ratoon. Planting technique of one skipped furrow (90 cm) after four planted furrows gave higher cane yield (mean 151.57 t ha^{-1}), net returns (Rs 44,451 ha^{-1}) and B:C ratio (1:75) than paired row planting. Fertigation of liquid fertilizers through micro irrigation gave 25% fertilizer saving and 20.74% higher yield. Kumar et al. [14] reported on evaluation of micro irrigation systems adopted for sugarcane crop in Vertisols. Among the two systems of micro irrigation evaluated for sugarcane crops in Karnataka, India (1994), both micro tube and bi-wall systems of micro irrigation were found to be suitable for sugarcane crops. The micro tube system was more efficient, recording higher growth and yield parameters, as compared to the bi-wall system of micro irrigation.

The field performance of pressure compensating, nonpressure compensating and inline micro irrigation systems in sugarcane crop has been compared with furrow irrigation by Shinde and Jadhav [31]. The results revealed that pressure compensating and inline micro irrigation systems saved irrigation water up to 50% along with 17 to 20% increase in cane yield. The commercial cane sugar production of more than 16 t ha^{-1} and 93% uniformity of water distribution were observed in pressure compensating and inline micro irrigation systems.

Raskar and Bhoi [21] studied effects of source and levels of fertigation with modified planting techniques on yield and quality of sugarcane under micro irrigation. The results of the first two trials indicated 19 to 25% higher cane yield in one row skip after four row planting with micro irrigation than conventional planting under surface irrigation. In the third trial sugarcane yield and commercial cane sugar (CCS) yield increased with increase in levels of fertilizer and was maximum in 125% fertilizer levels (157 and 17 t ha^{-1}). However, the yield obtained due to application of 75 and 100% recommended dose of water soluble fertilizer was on par, indicating 25% savings in fertilizer. Among the various sources, yield of water soluble fertilizer (146 t ha^{-1}) was at par with the yield obtained by fertigation of urea, di-ammonium phosphate and Muriate of potash. The overall increase in cane yield and total water savings by use of micro irrigation ranged from 20 to 30% and 42 to 52%, respectively. The WUE ranged from 10.17 to 14.03 100 kg per ha-cm as in micro irrigation compared to 4.82 to 6.00 100 kg per ha-cm in surface method.

The opinions of drip adopters about nature and extent of benefits due to micro irrigation system used for sugarcane crop of a total of 102 sugarcane farmers from 20 villages employing micro irrigation for sugarcane have been compiled by Chavai et al. [5] from Maharashtra. The major benefits reported by majority of the farmers were: savings in water ranging from 20 to 60%, in labor and fertilizer, reduction in weed intensity and increase in sugarcane yield.

The scheduling of drip irrigated sugarcane using the index tensiometer method has been reported by Muthy et al. [20]. An array of eight tensiometers was used to schedule irrigation and was compared to the water budget method over one crop cycle (plant cane and 8 ratoons). More water was applied with index tensiometer, on average an additional amount of 165 mm.yr^{-1}. This resulted in extra cane yields of 11.6 tons ha^{-1}.yr^{-1}. The efficiency of irrigation water use with tensiometer and water budget was 0.055 and 0.051 tons.(ha^{-1} mm^{-1}), respectively. In addition to this higher irrigation water use efficiency, the financial analysis showed net profit of approximately MUR 11,400 ha^{-1}.yr^{-1} in favor of the index tensiometer technique. Scheduling micro irrigation of sugarcane using tensiometers was found to be a viable practice.

Four specific irrigation schedules for sugarcane crop through drip system viz., I$_1$ (0.6, 0.8, 1.0, 0.8, etc.), I$_2$ (0.8, 1, 1.2, 0.6, etc.), I$_3$ (1.0, 1.2, 0.8, 0.6, etc.), through drip and I$_4$ (1.00 etc.), by gravity flow permuted with fertilizer treatments viz., application of RDF (F$_1$, 250:115:115 NPK) through soil, F$_2$ (100% RDF), F$_3$ (80% RDF) and F$_4$ (60% RDF) have been reported by Digrase et al. [7]. Results showed that drip irrigation produced significantly higher cane yield (162 tons.ha^{-1}) than gravity irrigation (96 tons.ha^{-1}). Irrigations applied through drip at I$_2$ schedule (0.8, 1.0, 1.2, 0.6, etc.) proved significantly superior to the rest of the schedules. It required 1981 ha-mm water against 2549 ha-mm in gravity flow to maximize the yield. Application of 80% recommended dose of fertilizers produced higher cane yields than its application at RDF through soil, resulting in 20% saving in fertilizer cost. The sugarcane crop responded quadratically to irrigation water and fertilizers. The resource use efficiency under drip was appreciably higher than gravity flow.

Narayanamoorthy [21] reported 23% higher yield, 44% water saving and 1059 kwh.ha^{1} electricity saving with micro irrigation in comparison with flood method of irrigation for sugarcane production in India.

The feasibility of micro irrigation for sugarcane in Haryana has been reported by Goel et al. [8]. A field experiment was conducted to effects of compare drip (1.0 and 0.8 IW/CPE ratio) and furrow (1.0 IW/CPE ratio) irrigation on water use, yield, juice quality and net returns of sugarcane. As compared to furrow irrigation, micro irrigation at 1.0 = IW/CPE ratio increased cane yield, number of millable cane, sugar yield, WUE and nutrient content in the index leaf of plant ratoon. The net profit under micro irrigation method was insignificant (Rs. 1610 ha^{-1}) for plant crop and Rs. 1820 ha^{-1} for first ratoon crop).

Sosa et al. [35] studied the response of sugarcane to distinct micro irrigation thresholds. Rainfall during the period was 1246 mm, and the irrigation treatments applied were none, and treatments that consumed 10, 30 or 50% of the water used (irrigation thresholds of 90, 70 and 50%). The results showed that all the irrigated plots gave better growth and yield than the nonirrigated plot, and better sugar contents and yields. The best treatment was the 10% irrigation (threshold 90%) but this was not significantly different from the 30% (threshold 70%) treatment, although it was significantly different from the 50% (threshold 50%) treatment, and the 30% (threshold 70%) and 50% (threshold 50%) treatments were not significantly different.

The effects of fertigation on emission uniformity of micro irrigation system for sugarcane have been reported by Kadam [12]. The recommended dose (RD) of NPK nutrients for sugarcane was applied in the form of commercially available water soluble fertilizers through micro irrigation system. The entire fertilizer dose of NPK was applied in ten equal splits at fortnightly interval. Each level of application was considered as separate treatment: T_1 = 10%RD, T_2 = 20%RD, T_3 = 30%RD, T_4 = 40%RD, T_5 = 50%RD, T_6 = 60%RD, T_7 = 70%RD, T8 = 80%RD, T9 = 90%RD, and T10 = 100%RD. The average discharges before and after application of fertigation was recorded and analyzed to assess the percent reduction in discharge and emission uniformity. The average reduction in initial discharge was 8.8%. The reduction in initial discharge is suggestive to have one acid treatment at the end of the season when the quality of irrigation water is C3S1 and water soluble fertilizer as acidic. The field emission uniformity values were in the range of 90–93% with an average value of 92% for the entire unit. The percent reduction in field emission uniformity was to the extent of 3.5% at the end of passing of 100% of RD. The reduction in discharge and the variation in EU were due to the variation in discharge of emitters due to clogging. It was found that the source of water was mainly responsible for clogging of emitters. The analysis of deposits in emitters and laterals revealed that the dissolved salts in water source dominated by carbonates, bicarbonates, chlorides and sulfates of calcium and magnesium are responsible for the emitter clogging. Thus, the reduction in discharge and emission uniformity was attributed to the water quality (C3S1) and not to the fertigation.

The socio economic analysis of micro irrigation in sugarcane at Tamil Nadu was reported by Shanthy and Kumar [30]. Results from the study concluded that micro irrigation for sugarcane cultivation is a valuable technology and that in the long run it was economical to lay a drip system for sugarcane.

Ravikumar et al. [28] from Coimbatore has evaluated a fertigation schedule for sugarcane using a vadose zone flow and transport model, HYDRUS-2D, and showed that the urea requirement can be reduced by 30% while at the same time providing enough N for its assimilation at all stages of crop growth.

9.2.2 WORLD SCENARIO

Brazil is the largest producer of sugarcane in the world. Nunes et al. [23] conducted a study on the effect of sub superficial micro irrigation system at three spacing of dripping tubes and four planting densities of sugarcane crop at North-east Brazil. The results indicated that smallest spacing between dripping tubes increased sug-arcane's yields. Highest yields were achieved when a 1.0×1.0 m^2 and 1.2×1.2 m^2 between plant rows and dripping tubes were used, attaining yields of 136.9 and 154.7 tons ha^{-1}, respectively, which corresponded to a 32.0 and 35.0% increase in relation to the same cropping system under dry farm conditions. The uses of double sugarcane rows (1.4×0.6 m^2) and (1.2×0.8 m^2), with 2.0 m spaced dripping tubes had attained the lowest yields, however, it still exceeded in 15.6 tons ha^{-1} the yield of non-irrigated plots.

Romero et al. [29] conducted a study on the effects of trickle irrigation (with drip lines buried in all the furrows, with drip lines buried in alternate furrows and with drip lines buried in alternate inter-furrows) for sugarcane crop at Argentina. The population, weight and height of stalks and crop yield increased under all the drip irrigated treatments. Trickle irrigation with drip lines buried in all the furrows resulted in the highest increase in crop production (55%).

The sensitivity analysis of sugar cane productions with drip irrigation and gravity irrigation has been compared by Torres et al. [36] in Mexico. Three variables were considered: cane production per hectare (made sensitive to low prices), selling price of sugarcane per ton (made sensitive to low prices), and variations in direct costs of production (made sensitive to high prices). The three variables showed significant differences, with favorable results for the micro irrigation system over the gravity irrigation system.

Kwong et al. [16] reported after conducting a study in Mauritius that drip fertigation may be used as a mean for reducing fertilizer nitrogen in sugarcane. Under the soil and climatic conditions that prevailed at the study site, fertilizer-N use was reduced by 30%. The growth pattern (as reflected by tiller density and leaf area development) and sugarcane yields from drip fertigation applied at the rate of 80 kg of N.ha^{-1} per year were not inferior to those obtained with the standard practice of burying 120 kg of N ha^{-1} per year along the cane rows.

Wei et al. [37] conducted a study on water requirement and effects of fertilizer on sugarcane in China. It was reported that distribution of rainfall in spring, autumn and winter did not satisfy the water requirement of sugarcane, as it provided only 74, 69 and 36% of the water requirement of sugarcane at the seedling, tillering and maturity stages, respectively. Fertigation enhanced sugarcane fertilizer and WUE.

The irrigated sugarcane production functions from South Africa were reported by Lecler [17]. The typical practice of increasing irrigation water application amounted to account for low irrigation uniformity. The results indicated that the maximum crop yields in Komatipoort required at least 1150 mm of irrigation water on shallow, 0.6 m deep sandy clay loam soils compared with only 900 mm on 1.2 m deep sandy clay loam soils.

The irrigation scheduling in sugarcane based on atmospheric evaporative demand (AED) from Australia has been reported by Attard et al. [3]. Two alternative scheduling techniques were developed that use simple tables and computerized systems based on AED, knowledge of crop response to water stress, and soil water holding capacity. The simple scheduling tables indicated that irrigation should be as frequent as 10 days prior to the wet season and as infrequent as 21 days after the wet season.

Brouwers et al. [5] compared the mineral nitrogen contents in cane cropped in vertisols of Sudan and Guadeloupe as influenced by urea application management. The effects of two methods of urea application management on soil mineral N levels in ratoon cane at two sites were measured and compared. Both experiment sites received different irrigation practices. In Sudan (site S), cane is grown under furrow irrigation, and urea is broadcast and then buried on the rows by hilling up. In Guadeloupe (site G), cane is grown rain-fed with complementary micro irrigation, and urea is broadcast on and near the cane rows. Site G results showed that one week after urea application, all applied N was recovered in the topsoil. At site S, however, only 70% N was recovered. The results also revealed that at site S, where yield was higher, the amount of mineral N in the topsoil was at a higher level than the preapplication amount for a far longer period than at site G. Of the application methods tested, the best commercial practice to extend the time that mineral N is at an adequate level in the rooting zone, and thus enhancing cane yield, appears to be hilling-up of the cane rows after broadcasting the urea.

9.3   DRIP FERTIGATION IN SUGARCANE

Micro irrigation systems provide a convenient method of applying fertilizers and chemicals with irrigation, which is called fertigation using special devices. The fertigation devices include pressure differential systems (fertilizer tank), suction produced by a venturi principle (venturi injectors) and pumps (diaphragm or piston or electrically operated). The fertilizer unit is an integral part of control head. Drip fertigation may be used as a mean for reducing fertilizers in sugarcane. For sugarcane crop, the Fertilizer N was reduced by 30% in a study conducted by Kwong et al. [16]. Shinde et al. [33] studied the influence of planting techniques and fertigation through micro irrigation on yield, quality and economics of sugarcane by conducting a field experiment in Maharashtra, on seasonal sugarcane and its ratoon. Planting technique of one skipped furrow (90 cm) after four planted furrows gave higher cane yield (mean 152 t ha^{-1}), net returns (Rs. 44,451 ha^{-1}) and B:C ratio (1:75) than the paired row planting. Fertigation of liquid fertilizers through micro irrigation gave 25% fertilizer saving and 20.74% higher yield.

Mahendran and Dhanalakhmi [18] reported effects of crop geometry and drip fertigation on growth and yield of sugarcane crop in a field experiment conducted at Madurai, Tamil Nadu. The experiment consisted of surface irrigation at 0.75 = IW/CPE ratio to a depth of 6 cm (T_1), micro irrigation in paired row (60:100) at 100% crop evapotranspiration (ETc) with and without fertigation (T_2 and T_3), micro irrigation in 80 cm spacing at 50, 75 and 100% ETc with fertigation (T_4, T_5, T_6), micro irrigation in 120 cm spacing at 50, 75 and 100% ETc with fertigation (T_7, T_8, T_9). Fertigation was given at biweekly intervals starting from 15 days after planting (DAP) onwards up to 150 DAP in 10 equal splits. Growth parameters such as germination percentage, plant height, tiller production, leaf area index and dry matter production were higher in 120 cm spaced sugarcane under micro irrigation at 100%ETc level with fertigation. The same treatment recorded higher yield attributes like number of millable canes, millable cane length and girth. Higher cane and sugar yield of 182 and 26 t ha^{-1} were recorded in drip and fertigation plots, respectively, at 100% ETc level fewer than 120 cm spacing.

Rajanna and Patil [24] studied the effects of fertigation on yield and quality of sugarcane in a medium black soil at Belgaum district of Karnataka – India. N and K were applied at recommended rates of 250 and 185 kg ha^{-1}, respectively. N was applied at 6-day intervals starting from 30 DAP to 240 DAP. Results showed that fertigation through micro irrigation produced a 24% higher yield, and saved 47% water compared to recommended fertilizer rate applied with surface irrigation (107 t ha^{-1}). Quality parameters such as brix, pol and percentage commercial cane sugar were not affected by fertigation.

Shukla et al. [34] reported that the potassium (K) fertigation in sugarcane increased the number of buds per stubble, number of stalks, dry matter accumulation, number of millable canes and individual cane weight in ratoon cane. Potassium content of stubble was increased by 16.7% with K fertigation. The content of reducing sugars in buds at the time of ratoon initiation was improved significantly with K fertigation. Ratoon cane yield increased by 15.2% (74.1 t ha^{-1}) while sugar yield increased by 13.9% (8.2 t ha^{-1}) as compared with control with K fertigation.

9.4   ECONOMIC VIABILITY OF DRIP IRRIGATED SUGARCANE

Economic viability of sugarcane has been established with micro irrigation as emphasized in this section. More and Bhoi [19] studied the economic analysis of Suru sugarcane (CO86032) and its ratoon under micro irrigation and wide row planting system. There were 12 treatment combinations involving three row spacing's (150, 180 and 270 cm), two planting techniques (single row and paired row), and two intercrops (cucumber and watermelon). Additional two control treatments of planting systems without intercrops (100×30 cm spaced normal planting with surface irrigation and 90×30 cm spaced four-row planting with micro irrigation) were also included for comparison. The irrigations were scheduled with drip system in all the treatments on alternate day on the basis of cumulative pan evaporation (CPE), while in surface irrigation, irrigations were scheduled at 75 mm CPE with 8 cm depth. In wide-spaced paired row planting at 75–150 and 90–180 cm, 19.92 and 12.97% higher cane yield with 54.50 and 54.24% saving in irrigation water over 100 cm spaced normal planting was recorded. The net profit/cm of water in paired row planting of 75–150 and 90–180 cm were Rs. 676.53 and Rs. 752.63 with benefit:cost (B:C) ratio of 2.09 and 2.18, respectively, which were higher than 100×30 cm normal planting (Rs 284.05 net profit and 2.31 B:C ratio) and 90×30 cm 4 row planting (Rs 526.32 net profit and 1.38 B:C ratio).

Raskar and Bhoi [28] conducted a study on response of sugarcane to planting materials, interrow spacing, and fertilizer levels under micro irrigation. The response of sugarcane to intrarow spacing (30, 60 and 90 cm), fertilizer treatments (75, 100 and 125% of the recommended fertilizer rate) and source of planting material (tissue culture plantlets and polybag settlings) were studied under micro irrigation. Sugarcane yield, commercial cane sugar, gross monetary net returns and WUE increased with increasing intrarow spacing and fertilizer rate, and were higher with the use of polybag settlings.

The economics of micro irrigation in sugarcane cultivation using data collected from a case study in Sivagangai district, Tamil Nadu, India has been reported by Narayanamoorthy [22]. Results showed productivity gains of 54% and water saving of 58% due to micro irrigation over flood irrigation. Discounted cash flow analysis suggested that investment in micro irrigation in sugarcane cultivation was economically viable even without subsidy. The benefit-cost ratio varies from 1.98 to 2.02 without subsidy and 2.07 to 2.10 with subsidy at different rates of discount. Similarly, Torres et al. [36] from Mexico have also reported favorable economic results for the micro irrigation system over the gravity irrigation system.

9.5   DISADVANTAGES OF MICRO IRRIGATION

There is a Chinese proverb: "You can't expect both ends of a sugarcane are as sweet." So, along with the advantages of drip-irrigated sugarcane, as mentioned in this chapter, there are certain disadvantages of micro irrigation like initial high installation cost. The other unfavorable factors are as follows; the sun can affect the micro irrigation components shortening the usable life; clogging may occur if the water is not properly filtered and the equipment not properly maintained; drip lateral causes extra clean up costs after harvest; waste of water, time and yield may occur if system is not installed properly; germination problems may occur in lighter soils as subsurface drip irrigation may be unable to wet the soil surface for germination. Most drip systems are designed for high efficiency, meaning little or no leaching fraction. Without sufficient leaching, salts applied with the irrigation water may build up in the root zone, usually at the edge of the wetting pattern. On the other hand, micro irrigation avoids the high capillary potential of traditional surface-applied irrigation, which can draw salt deposits up from deposits below.

9.6   CONCLUSIONS AND FUTURE PROSPECTS

Sugarcane is an important cash crop in India. Surface irrigation is the prevalent irrigation method. Micro irrigation in sugarcane is a relatively new innovative technology that can conserve water, energy and increase profits. Thus, micro irrigation may help solve three of the most important problems of irrigated sugarcane: water scarcity, rising pumping (energy) costs and depressed farm profits in India. The application of micro irrigation in sugarcane has convincingly shown that the technique results in high WUE, saves water, reduces fertilization requirement, provides better quality crop and higher yield. However, if not installed properly, it may result in waste of water, time and yield. Application of micro irrigation requires careful study of all the relevant factors like land topography, soil, water, crop and agro-climatic conditions, and suitability of micro irrigation system and its components. The subsidy and technical support to farmers may be an incentive to adopt this method in India on a large scale. Adoption of micro irrigation (surface or subsurface) system in sugarcane cultivation is technically feasible and economically viable and needs to be vigorously followed.

9.7   SUMMARY

Sugarcane (*Saccharum officinarum L.),* a major cash crop in India, has a unique role in sustaining agro industrial economic growth. Sugarcane being a long duration crop produces huge amount of biomass, and requires large quantity of water (1100–2200 mm) and is mostly grown as an irrigated crop using surface irrigation. The micro irrigation adoption in sugarcane increases WUE (60–200%), saves water (20–60%), reduces fertilization requirement (20–33%) through fertigation, produces better quality crop and increases yield (7–25%) as compared with conventional irrigation. However, if not installed properly, it may result in wastage of water, time, money and yield. The subsidy and technical support to farmers acts as an incentive to adopt this method on a large scale in India. Adoption of micro irrigation (surface or subsurface) system in sugarcane is technically feasible and economically viable and needs to be vigorously followed.

KEYWORDS

- available water holding capacity
- bagasse
- cane yield
- cash crop
- cattle feed
- climatological data
- drainage
- drip irrigation
- drippers
- fertigation
- fertilizer
- field capacity
- filter cake
- industrial alcohol
- infiltration rate
- IW/CPE ratio
- long duration crop
- micro irrigation
- nutrient loss
- open pan evaporation
- paper industry
- scarcity of water
- soil erosion
- sugar
- sugarcane
- surface irrigation
- valves
- water application efficiency
- water balance
- water requirement
- water use efficiency, WUE

REFERENCES

1. Ahluwalia, M. S., Singh, K. J., Singh, B., Sharma, K. P. (1998). Influence of drip irrigation on water use and yield of sugarcane. *Intern. Water and Irrig. Rev.*, *18*, 12–17.
2. Arulkar, K. P., Hiwase, S. S., Deogirikar, A. A. (2004). Water requirement estimation from climatological data by probability analysis for sugarcane crop in Nagpur district. *News Agric.*, *15*, 75–78.
3. Attard, S., Inman, J., Bamber, N. G., Engelke, J. (2003). Irrigation scheduling in sugarcane based on atmospheric evaporative demand. In: Proc. Australian Society of Sugar Cane Technologists, Queensland – Australia. pp. 35–38.
4. Bhoi, G., Bankar, M. C., Raskar, B. S., Shinde, S. H. (1999). Effect of fertigation and planting techniques on yield and quality of Suru sugarcane under drip irrigation. *Indian Sugar, 49(7)*, 487–492.
5. Brouwers, M., Vassal-Courtaillac, N., Osman, A. A. (2008). Mineral nitrogen contents in cane-cropped Vertisols of Sudan and Guadeloupe as influenced by urea application management. In: Proc. Annual Congress South African Sugar Technologists Association, South Africa. 343–347.
6. Chavai, A. M., Sadaphal, S. S., Bhange, S. B., Gurav, B. S. (2003). Opinions of the drip adopters about nature and extent of benefits due to drip irrigation system used for sugarcane crop. *Cooperative Sugar, 34*, 961–962.
7. Digrase, L. N., Sondge, V. D., Sawant, B. P. (2004). Optimization of irrigation water and fertilizer use for seasonal sugarcane (CO-7714) through drip irrigation system on Vertisols. *J. Soils. Crops, 14*, 273–277.
8. Goel, A. C., Kumar, V., Dhindsa, J. P. (2005). Feasibility of drip irrigation in sugarcane in Haryana. *Indian Sugar, 55*, 31–36.
9. Gulati, J. M., Nayak, B. C. (2002). Growth cane yield and WUE of sugarcane as influenced by irrigation and planting dates. *Indian J. Agron., 47*, 114–119.
10. Ingle, P. M. (2007). Estimation of water requirement for seasonal and annual crops using open pan evaporation method. *Karnataka J. Agric. Sci., 20*, 676–679.
11. Jain Irrigation System, Ltd, (2011). Modern irrigation and fertigation methodologies for higher yields in sugarcane. <http://jisl.co.in/PDF/ crop/sugarcane cultivation.pdf>.
12. Kadam, S. A. (2009). Effect of fertigation on emission uniformity of drip irrigation system. *International J. Agric. Eng., 2*, 72–74.
13. Kaushal, A., Patole, R., Kamal G. Singh, (2012). Drip irrigation in sugarcane: A review. <www.arccjournals.com>, *Agric. Reviews, 33(3)*, 211–219.
14. Kumar, B. M., Murthy, K. N., Chandrappa, H. (2000). Evaluation of drip irrigation systems adopted for sugarcane crop in Vertisols. *J. Environ. Ecology, 182*, 365–366.
15. Kumar, Y., Singh, J. K., Shashi, K. (2005). Probabilistic irrigation water requirement of major crops of Udham Singh Nagar district of Uttaranchal State. In: *Drainage and Irrigation Water Management* by Kumar, V., Singh, J., Bhakar, S. R. (eds), 80–88. Himanshu Publications, New Delhi.
16. Kwong, K., Paul, J. P., Deville, J., 1999). Drip fertigation a means for reducing fertilizer nitrogen to sugarcane. *Expl. Agric., 35*, 31–37.
17. Lecler, N. L. (2009). Irrigated sugarcane production functions. In: Proc. of the Annual Congr. South African Sugar Tech Association, Darban, South Africa. pages 604–607.
18. Mahendran, S., Dhanalakhmi, M. (2003). Effect of crop geometry and drip fertigation on growth and yield of sugarcane crop. In: Souvenir 65 Annual Convention of the Sugar Technologists Association of India. Bhubaneshwar, Orissa, India. pp. 80–87.
19. More, S. M., Bhoi, P. G. (2004). Economics analysis of Suru sugarcane (CO-86032) and its ratoon under drip irrigation and wide row planting system. *Indian Sugar, 54*, 447–452.

20. Muthy, N. K., Piat, B., Cheong, L. R. (2003). Scheduling drip irrigation of sugar cane using the index tensiometer method a viable practice. Revue Agricole et Sucriere de l' Ile Maurice, *82*, 31–36.
21. Narayanamoorthy, A. (2004). Impact assessment of drip irrigation in India the case of sugarcane. *Develop. Policy Rev., 22*, 443–462.
22. Narayanamoorthy, A. (2005). Economics of drip irrigation in sugarcane cultivation case study of a farmer from Tamil Nadu. *Indian J. Agric. Econ., 60*, 235–248.
23. Nunes, F. J., Simoes, A. L., Sousa, A. R., Maiciel, G. A., Franca, J. G. (2003). Sugarcane yields at several crop spacing and drip irrigation rows under the dry rain forest climate conditions of Pernambuco State North-east Brazil. *Pesqu. Agropec. Pema., 14*, 71–79.
24. Rajanna, M. P., Patil, V. C. (2003). Effect of fertigation on yield and quality of sugarcane. *Indian Sugar, 52*, 1007–1011.
25. Rajegowd, M. B., Muralidhara, K. S., Ravindrababu, B. T. (2004). Application of water balance studies in irrigated sugarcane. J. Agrometeorology, *6*, 138–141.
26. Raskar, B. S., Bhoi, P. G. (2003). Response of sugarcane to planting materials, inter row spacing's and fertilizer levels under drip irrigation. *Indian Sugar, 53*, 685–690.
27. Raskar, B. S., Bhoi, P. G. (2001). Productivity of sugarcane as influenced by planting techniques and sources of fertigation under drip irrigation. *Indian Sugar, 50*, 801- 810.
28. Ravikumar, V., Vijayakumar, G., Simunek, J., Chellamutthu, S., Santhi, R., Appavu, K. (2011). Evaluation of fertigation scheduling for sugarcane using a casdose zone flow and transport model. *Agric. Water Manage., 98*, 1431–1440.
29. Romero, E. R., Scandaliaris, J., Sotomayor, L., Alonso, L. (2003). Results of the first experience of drip irrigation on sugarcane in Tucuman Argentina. Revista Industrial Agricola de Tucuman, *80*, 5–9.
30. Shanthy, T. R., Kumar, S. R. (2010). Socio economic analysis of drip irrigation in sugarcane. Cooperative Sugar, *41*, 41–44.
31. Shinde, P. P., Jadhav, S. B. (2000). Water management with drip irrigation system for sugarcane. In: Proc. 62 Annual Convention Sugar Technologists Association of India, Agra, India. pp. 36–41.
32. Shinde, P. P., Jadhav, S. B. (1998). Drip in sugarcane an experience in India. In: Proc. International Agricultural Engineering Conference, Bangkok, Thailand. pp. 734–747.
33. Shinde, S. H., Dahiwalkar, S. D., Berad, S. M. (1999). Influence of planting technique and fertigation through drip on yield quality and economics of sugarcane. J. Maharashtra Agric. Univ., *24*, 276–278.
34. Shukla, S. K., Yadav, R. L., Singh, P. N., Singh, I. (2009). Potassium nutrition for improving stubble bud sprouting dry matter partitioning nutrient uptake and winter initiated sugarcane (*Saccharum spp. hybrid complex*) ratoon yield. *Europ. J. Agron., 30*, 27–33.
35. Sosa, F. A., Roberto, F. L., Hernandez, C. F., Morandini, M., Agustin, S. G., Hasan, A. J., Romero, E. R. (2008). Response of sugarcane to distinct drip irrigation thresholds. *Advanced Agroindustrial, 29*, 36–39.
36. Torres, S. G., Prado-Vazquez, V. H., Rivera-Espinoza, M. P. (2010). Sensibility analysis of sugar cane productions with two irrigation technologies (drip and gravity) in Zapotiltic Jalisco Mexico. *Revista Mexicana de Agronegocios, 14(26)*, 193–201.
37. Wei, T. H., Qiang, Z. L., Rulin, Z., Meifu, H. (2008). Water requirement of sugarcane and effect of application of fertigation on sugarcane. South-west China *J. Agric. Sci. 21*, 1381–1384.
38. Wikipedia, (2011). *Sugarcane*. <http://en. wikipedia.org/wiki/Sugarcane>.
39. Zade, M. R., Hiwase, S. S., Deshmukh, M. M. (2001). Probability estimation of water requirement of major crops of Akola. *PKV. Res. J., 25*, 30–34.

CHAPTER 10

DRIP IRRIGATION DESIGN FOR SUGARCANE

POOJA BEHAL and KAMAL G. SINGH

10.1 INTRODUCTION

Drip irrigation is one of the most recent developments in irrigation methods. It was originally developed in Israel by Simca Bloss in 1959. This method proved very economical in water use and also produced higher yield with good quality water. The water demand has increased many folds due to improvement in agricultural production technology in the last few decades. The pumping cost has increased substantially with the inflation of machinery prices and enhancement in diesel and electric energy price. This has made the water not only scarce but also a costly resource. Because of the anticipated shortage of this vital agriculture input, we have always been endeavoring to increase our water resources.

The choice of any irrigation system depends on number of factors such as topography (land surface configuration), soil type (i.e., texture, structure, depth), stream size (i.e., flow rate of irrigation water available), economic viability (benefit – cost rates), farmer's capacity to meet initial investment and social acceptance of the system.

Drip irrigation is one of the latest methods of irrigation, which is becoming increasingly popular in areas with scarcity of irrigation water and salt problem. This is also called trickle irrigation system or localized irrigation system. Drip irrigation is very well suited for widely spaced high value crops. The required quantity of water is applied daily at the root zone of plants through a network of piping system. The key to the success of drip irrigation in the arid region is the fact that both water and nitrogen fertilizers are applied frequently in small quantities to meet the needs of the plant. This is a continuous balance of water, air and fertilizer in the entire life cycle to the plant; resulting in the optimum growth, better fruiting and early maturity of crop.

The main characteristics of drip system consist of uniform, small continuous flow and reduced surface wetting around plants. Greater application efficiency is

---

*In this chapter, the currency is expressed in Indian Rupees (1.00 US$ = Rs. 60.93; 1.00 Rs. = 0.02 US$).

obtained due to better placement of water near the roots in precise quality. The reduced surface wetting reduces surface evaporation. For widely spaced crops like fruit trees, the system may be even more economical than other method of irrigation.

The comparative merits and demerits of surface, sprinkler and drip irrigation methods reveal that drip irrigation method is most efficient method of irrigation having irrigation efficiency of about 90% compared to sprinkler method with about 70–80% and conventional method of surface irrigation having about 40–50%. Although drip method is the most efficient method yet it can be recommended and utilized only under particular situation in certain pockets of Punjab because of high initial cost and other limitations. Even under most suitable and favorable situations for drip irrigation method, the farmers have not adopted drip irrigation method. The reasons for this are that farmers do not realize the importance and value of irrigation water available from canal and tubewell source, as they get electric powers for motors at a cheap flat rate.

Keeping in view the water scarcity, topography and water quality, Kandi areas and south western districts of Punjab are most suitable areas especially, where fruit trees can be grown for drip irrigation system [6, 7, 10, 14]. The existing paddy – wheat cropping system in subtropical parts of northern India leads to hastening of water table declines, deterioration of soil health, increased incidence of insect/pests, and weeds and reduced availability of protein rich legumes and oil seed crops. Sugarcane is emerging as an alternative to the paddy-wheat cropping system and has a unique role of play in sustaining agro-industrial economic growth of our country [17]. The potential sugarcane production depends upon optimum availability of water throughout the crop growth period. Sugarcane is the most important cash crop sugarcane is adapted to a range of tropical and subtropical climates and can be successfully grown on all types of soil ranging from sandy to clay loam.

The total area under sugarcane in the world was 19.90 million ha during 1998 with the production of 1,254.94 million tones of sugarcane. The area under sugarcane increased almost 1.5 times from 156,000 ha in 1966–1967 to 235,000 ha in 1996–1997. Sugarcane is the most important sugar crop contributing more than 62% of the world sugar production. Sugarcane occupied about 96 thousand hectares in Punjab during 2005–2006. The average cane yield is 601 quintals per hectare. The average sugar recovery was 9.8%. Before recommending the adoption of drip irrigation system for sugarcane it is necessary that drip system be properly designed and its economic viability is worked out.

Keeping the above facts in consideration, the present study was taken up with following objectives. To design the drip method of irrigation for sugarcane; and to determine the cost estimation of drip irrigation.

10.2 REVIEW OF LITERATURE

Hapase et al. [4] revealed that the surface and subsurface drip (B1-wall) with daily irrigation and paired row planting is effective to save water upto 50–55%, better crop growth response and about 2.7 times increase in water-use efficiency.

Shih [16] in a study conducted at University of Florida compared the difference in sugarcane yield and water-use efficiency between drip irrigation and subirrigation using a lysimeter system. The average sugarcane yield under drip irrigation was significantly high than that under subirrigation.

Jones et al. [5] reported an increase in cane yield by more than 20% while reducing the water use of almost 40% with drip as compared with surface irrigation method. Drip irrigated fields were reported to have produced 22% more sugar yield per hectare.

Parikh et al. [12] in field experiments on sugarcane in Gujarat compared drip irrigation method with the furrow irrigation. Drip irrigation, at IW/CPE ratio of 0.3 and 0.6, gave cane yield of 106.4 t/ha and 120.5 t/ha, respectively, compared with 79.4 t/ha under furrow irrigation at IW/CPE ratio of 0.75. Studies also revealed that drip irrigation consumed 7.6–50% less water than furrow irrigation.

Batta and Singh [2] Studied the efficiency of water use investigation in the Jordan Valley project were determined by the evaluation of project efficient (Ep). This included the irrigation system efficiency (Es) and field application efficiency (Ea). Evaluation included the comparison of open causal with surface irrigation versus pressurized pipes with sprinkles on drip irrigation system. Ep for open surface causals with surface irrigation under citrus was 53% for vegetables, it was 42% for pressurized pipe system, Ep was 68% and 70% for sprinkler and drip irrigation methods, respectively. Es varies from 65% for an open causal to 77% for pressurized pipe projects. The Ea was 82% for surface irrigation on citrus and 64% for vegetables. Under drip irrigation it was found to be 91% for vegetables. The low Es value for the causal was attributed to water theft, measurement loss and leakage.

Ahire et al. [1] using two irrigation methods, that is, trickle and surface observed that trickle irrigation system produced higher number of tubers per plant (5.58), larger size of tubers (16.08 cm), more weight of tubers per plant (145.68) and higher tuber yield (20.43 t/ha) as compared to surface irrigation. Also trickle irrigation saved 46% irrigation water over surface irrigation.

Tumbare et al. [19] studied the liquid fertilizer through trickle system to Okra crop. The treatments consisted of five fertigation levels, that is, 125, 100, 75, 50 and 25% of recommended dose of solid fertilizer with surface irrigation method. They found that 75% of recommended dose of liquid fertilizer through trickle system was equally to produce yield equal to that of conventional method of irrigation resulting in 25% saving in fertilizer than band placement of fertilizer.

Singh and Sharma [18] on water economy and yield aspects of the crop and little attention has been paid to the study of plant growth parameters under K fertigation and they failed to relate them with the yield variation. These parameters help in assessing the performance of a crop under various agronomic and management practices. Accordingly this study was taken up on economizing K through a suitable irrigation method like trickle irrigation using fertigation.

Sharma and Singh [15] studied the effects of NPK fertigation and planting patterns on yield and economics of potato through drip irrigation. Fertigation using trickle irrigation where fertilizer is delivered along with water to the growing crops is an innovative technique to use fertilizers in the most effective manner. Phosphorus has not been generally recommended for application by trickle irrigation system because of its supposedly high clogging potential and limited movement in the soil.

Singh et al. [8] Studied the effect of P fertigation on potato. Crop using di-ammonium phosphate as P fertilizer. Trickle fertigated treatments were compared with conventional method of fertilizer application in terms of biometric parameters as well as available P. It was found that plant performance parameters like leaf area, dry matter accumulation and yield attributes had much higher values in P-fertigated crop than furrow irrigated and conventionally fertilized crop. Available P was also more in the entire soil profile under trickle-fertigated treatment.

Kaur et al. [9] Studied K-fertigation with 100, 75 and 50% of the recommended dose of muriate of Potash (200 kg/ha) in potato using trickle and furrow method of irrigation. The results showed leaf area index, dry matter accumulation and yield attributes to much higher in trickle method. Also, the trickle-fertigated gave higher proportion of (A&B) grade tubers than furrow irrigated and conventionally fertilized crop. For the same fresh tuber yield, about 50% of potassium fertilizer could be saved adopting trickle fertigation instead of conventional fertilization.

10.3   COMPONENTS OF THE MICRO IRRIGATION SYSTEM

A drip irrigation system consists essentially of a main line, submains, laterals, and emitters. The main line delivers water to the submain, and the submains in to the laterals. The emitters, which are attached to the laterals, distribute water for irrigation. The mains, submains and laterals are usually made of black, PVC (Poly Vinyl Chloride) tubing's. The emitters are also usually made of PVC material. PVC materials are preferred for drip system as it can withstand saline irrigation water and is also not affected by chemical fertilizers. Auxiliary components are valve, pressure regulator, filter, pressure gauge, and fertilizer application components. Brief description of the components of typical drip irrigation systems is given in the following sections.

10.3.1   PUMPING SET

It should be able to cater the required amount of water at desired operating pressure. The pressure necessary to force water through the components of the system, including the fertilizer tank, filter unit, mainline, lateral and the nozzle, is obtained by a pump of suitable capacity. Centrifugal pumps are generally used in drip system.

10.3.2 PIPELINE

Mainline delivers water to submain and submain to lateral. The main and submain are made up of black; PVC (Polyvinyl chloride) and lateral lines are made up of PE (Polyethylene) Plastics with carbon black, which makes it water resistant. They are also not affected by the reaction of any chemical fertilizer. The variation is lateral pressure is 0.15 to 1.75 kg/cm^2 pressure goes on decreasing as it move from submain to lateral and their emitters.

10.3.3 EMITTERS

Emitters are fixed on lateral line for discharge of water to plants. It should be in expensive, durable reliable and provide uniform discharge. Some of the more distinctive emitters designs are the short path long path, small orifice, vertex, pressure compensating self flushing, perforated single and double chamber tubing or minimum sprinklers used in spray irrigation depending upon emission point, mode of operation. Depending upon the emission point above or below the soil surface, the drip irrigation system is classified as: surface trickle irrigation system; and subsurface trickle irrigation system.

10.3.3.1 SURFACE TRICKLE IRRIGATION SYSTEM (DRIP)

These are the emitters and lateral lines that are laid on the soil surface. Surface trickle has been primarily used on widely spaced plants but can also be used for row crops. Generally in this case discharge rates are less than 12 L/hr for single outlet point source emitter and less than 12 L/hr/m for line source emitter. The advantages of trickle irrigation system includes the case of installation, inspection, changing and cleaning emitters, plus possibility of checking soil surface wetting patterns and measuring individual discharge rate.

10.3.3.2 SUBSURFACE TRICKLE IRRIGATION SYSTEM

In this system water is applied slowly below the surface through emitter's with discharge rate in the same range as a surface trickle system. The advantages of subsurface trickle irrigation include freedom from achieving of the lateral lines at the beginning and removing them at the end of growing season, little interference with intercultural and other cultural practices.

Now-a-days, the combination of subsurface and surface irrigation system has been tried, where lateral lines are buried and emitters are located above the soil surface through the use of riser fittings.

10.3.4 CLASSIFICATION OF DRIP IRRIGATION SYSTEM

10.3.4.1 POINT SOURCE

Point source emitter discharge water flows through individual or multiple outlet, which are spread in 1 m open point source and generally used for trees.

10.3.4.2 LINE SOURCE

The line source, emitter having perforated holes or pores which discharge water at close spacing are used usually for small fruit, trees, vegetables or closed spaced row crops.

10.3.4.3 SPRAY

Spray emitters apply water to soil in the form of small spray. The air is instrumental in distributing the water. They have high discharge rate (175 L/hr) and is used for irrigating widely spaced crops. The main disadvantages of spray system are high evaporation loss particularly when plants are young.

10.3.4.4 FILTERS

Filter is used to remove impurities for preventing blockage. A two-stage filter is usually provided. It consists of one coarse filter and one fine filter.

10.3.4.5 SAND FILTER

Pressure type, high flow sand or mix bed filters are the more popular ones and gravity operating sand filters has low flow rates and thus require large surface area to produce equivalent volume of the filtered water and pressurized filter and are not widely used. Filter capacity is designed in terms of volume of flow per unit time per unit bed area. In pressurized filters full depth of sand is used.

10.3.4.6   SCREEN FILTERS

These are most commonly used filters. In this water comes from inside of the screen and exist from the periphery. Others are designed with water flowing in opposite direction when properly sized and maintained, screen filters do an adequate job of separating suspended particular from water, but there are limited in their load capacity. For removing large organic debris such as leaves and weeds nonpressurized gravity screen filters are installed in an irrigation canal before the pump in take or delivery system. In case of extremely high-suspended load smaller screen openings have been used to remove gravel, sand and silt.

10.3.4.7   CYCLONIC OR CENTRIFUGAL FILTERS

They are used in the line system, which are used to remove suspended particles (having specific gravity greater than water). Their operational principle is entirely different from the screen or sand filters.

10.3.4.8   VALVES

Valves are the integral part of trickle irrigation system. The nature of the valves for a given installation will depend upon the automation degree of pressure regulator and number of sets required. Types of valves used in drip irrigation system: Manual valves, automatic valves, check valves, and air release valve.

10.3.4.8.1   MANUAL VALVES
They include gate, globe, butterfly and ball valves. Butterfly valves are commonly used in the larger size. Manual valve are generally troubles free and require little maintenance or service throughout the life of the system.

10.3.4.8.2   AUTOMATIC VALVES
These are found at pump and filter stations to regulate mainline pressure, control backwash cycle in the filters or control flow through branching mainlines. Automatic valve require periodic maintenance to assure satisfactory operation. Maintenance schedules depend on the use of the valves and the cleanliness of the water.

10.3.4.8.3   CHECK VALVES
Check valves are normally used only at pump station. Air release and vacuum relief valves are located at high points on main, submain and lateral.

10.3.4.8.4   AIR RELEASE VALVE

Air release valves are generally placed at high points in mainline, submain and pump station. They release entrapped air as system. Startup and allow air to enter the pipeline under conditions of negative pressure. They are of two types: (i) large orifice type, which seals completely when the pipeline is pressurized; (ii) combination of large and small orifice type, which release entrapped air while the system, is operation.

10.3.4.9   FERTILIZER TANK

Fertilizer tank is provided at the head of drip irrigation system for supplying fertilizers in solution directly to the field along with irrigation water. By connecting fertilizers tank with the irrigation water supply we can save the cost of labor required for applying fertilizer to the field.

10.4   METHODS AND MATERIALS

10.4.1   DESIGN PROCEDURE OF INLINE DRIP IRRIGATION SYSTEM

10.4.1.1   SELECTION OF SYSTEM CAPACITY

Drip Irrigation System shall have the capacity adequate to fulfill 90% of daily peak water requirements within the designed area and in a stipulated time period usually not more than 16 h of operation per day. The system capacity should include an allowance for losses of time in related operations. The capacity of the system should match with the available quantity of water in the source. In case of a bore well (tube well), system should be designed to match the recharging rate of water. Similarly, in case of Pond, Canal, Reservoir, system should be designed to match the available flows.

10.4.1.2   ESTIMATION OF PEAK WATER REQUIREMENT OF CROP (PWR)

$$PWR = ET_0 \times K_c \times AW \tag{1}$$

where: ET_0 = reference evapotranspiration during peak water demand period, mm/day; Kc = crop coefficient; and AW = wetted area as a percentage of total area at maturity. Wetted area is the area which is shaded due to its canopy cover when the sun is over head, which depends on the stage of crop growth.

10.4.2   SELECTION OF EMITTING DEVICES OR DRIPPERS

There are numerous varieties of drippers/inline tubes available, with different discharge rates, features, functions, characteristics and suitability to different crops. The selection of emitting devices are based on peak water requirements of crop, age and root zone, soil type, topography, soil water holding capacity and infiltration rate, hydraulic conductivity, life expectancy and cost economy. However, an ideal or perfect emitting devices or drippers should meet the following requirements:

a.   Should be inexpensive, durable and serviceable.
b.   Should have relatively low discharge rate to keep the system cost low.
c.   Should have discharge variation not more than ±10%.
d.   Should have relatively large cross sectional area and flow path to avoid clogging
e.   Should preferably have turbulent flow path and pressure compensation action in undulated and sloping areas.
f.   Should not create runoff within the immediate application area.

10.4.2.1   SELECTION AND DESIGN OF LATERALS

Laterals are the conduits, which carry water from submain, and feed the emitters and are available in different sizes 12 mm, 16 mm, 20 mm, etc. Emitting devices/Drippers can be fitted at determined spacing as per the requirements of crops. In case of Inline Drip Tubing, dripper intervals and discharge has to be determined before designing of system. As the laterals are significant with regard to pressure variation, proper evaluation of fractional head losses in the laterals is essential to achieve higher uniformity. Pressure variation 5 to 20% and discharge variation 5 to 10% throughout the system should be maintained in a range. The size, allowable length and frictional losses of laterals have to be determined by nomograms or charts and design guidelines provided by the manufacturer for their specific emitting devices.

10.4.2.2   SELECTION AND DESIGN OF SUBMAINS

Submain (PVC/HDPE) carries water from main line and distributes among the laterals. The size, length and frictional head losses of submain have to be determined by nomograms or charts & design guidelines provided by the manufacturer or by using Hazen William equation and limiting the frictional head losses within the limit of design tolerance of the particular emitting devices as specified by the manufacturer.

10.4.2.3   SELECTION AND DESIGN OF MAIN LINE

Mainline carries water from source to submain. The size of mainline is determined by considering the quantity of water flowing through it, length and path or mainline, elevation of ground, velocity, safety parameters, cost economy and nomograms provided by the manufacturer. While designing the mainline following points should be kept in mind:

a.   Permissible velocity: Should not exceed 1.5 meter per second.
b.   Friction head losses: Should be limited to 5 to 20 meters per 1000 m length of pipe.
c.   Economic size: Should include low initial investment; Low annual and low power cost.
d.   Elevation and class of pipe: Use of optimum pressure rating (class) of Pipes. Run the mainline as straight and shortest as possible with due consideration for the field slopes.
e.   Control measures: Provide air release cum vacuum breaker; non-return valves or check valves; and pressure sustaining valves, etc., at appropriate locations based on the engineering requirements.

10.4.2.4   SELECTION AND DESIGN OF FILTRATION UNIT

Drip irrigation systems are characterized by large numbers of emitting devices, having fairly smaller flow paths. Because of the smaller flow path and orifice, they are more susceptible to clogging due to presence of physical impurities, organic/inorganic precipitates, silts, clay, sand, suspended particles, etc. A properly designed filtration system is the key to long-term successful operation of drip irrigation systems. The selection and design of filtration system is based on:

(a) Source and quality of water.
(b) Type, size and concentration of physical impurities.
(c) Design system flow (filtration capacity).
(d) Type of irrigation system.
(e) Workability of filtration system.
(f) Ease for handling, cleaning, maintenance and repairing.
(g) Filtration media and low frictional losses.
(h) Initial and maintenance costs including power costs.

Normally three types of filtration systems are used for drip irrigation system:

10.4.2.4.1   HYDRO CYCLONE FILTER

It is also known as Centrifugal Filter or Sand Separator, and is used to separate the sand, silt or particles heavier than water but cannot remove algae, fibers, clay, etc. It is used immediately after the pump deliver as a primary filtration system for waters with heavy loads of suspended impurities and particulate matter in water sources such as Canals, Dam or Reservoir, Rivers, Ponds, Open Wells, etc.

10.4.2.4.2   SAND MEDIA FILTERS

Sand Media Filters are the most effective in removal of all types of physical impurities – organic/inorganic, algae, silt, clay and suspended particles, etc., as they are three-dimensional filters. Sand filters are recommended for all types of drip irrigation system. Especially it is recommended in case of open well, river, canal, dam, water sources and where the quality of water in source is expected to vary with seasonal rainfall and runoff.

10.4.2.4.3   SCREEN/MESI I FILTER

Screen/mesh filter is the simplest of the entire filter and also costs less compared to all other filters. But these filters cannot filter algae, fine silt, trash, etc. It is normally used in case of clean water source with no presence of algae; clay and suspended particles usually for bore wells.

10.4.2.5   SELECTION AND DESIGN OF PUMP

Pump Unit is an electromechanical device, which lifts water from one level to another level and provides the required pressure. Pump unit must be capable of supplying required pressure and discharge for efficient functioning of system. Pump unit can be selected and designed by calculating total had and discharge required for efficient operation of the system. Total head required for the system is calculated as:

Total head = (Suction + Delivery) + Filtration Losses + Frictional Losses in Main Line + 10 m Operating Pressure + Fitting Losses + Venturi Head + Elevation (if any)   (2)

where: Suction = vertical distance between water level to center to of pump; Delivery = vertical distance between center of pump to ground level; Filtration Losses = friction head losses in different types of filters (It is assumed 2 m for type of filter unit, that is, Hydro cyclone, Sand and Screen Filters. If these are connected in series, then sum of frictional head losses of these three filters is 6 m); Friction Head Losses accruing in mainline; Operating pressure is the recommended operating pressure given by the manufacture at which system has to be operated and designed (in general, 10 m or 1 kg/cm² of pressure is required for nonpressure compensating drippers and 15 m or 1.5 kg/cm² for pressure compensating drippers); Fittings Losses (friction head losses occur in fittings like bends, elbows tees, reducers, valves, other fittings, etc. and is assumed to 2 m for overall operation); Venturi head is required to operate venturi type fertilizer applicator, as recommended by the manufacturer (it is assumed to be 5 m for manually operated venturi system); and Elevation = vertical distance between ground level near the source (as considered above) to the highest level of ground. With total head and discharge required, we can calculate horse power of the pump for efficient operation of drip irrigation system.

$$HP = [Q \times H]/[75 \times a \times b] \tag{3}$$

where: Q = discharge required in liters per second; H = total head required in meters; a = efficiency of motor (assumed 85%); and b = efficiency of pump (assumed 80%).

10.5   RESULTS AND DISCUSSION

10.5.1   DESIGN OF INLINE DRIP SYSTEM

The system has been designed for two hectares under sugarcane.
 Area of farm: 200 m × 100 m = 20,000 m² = 2 ha
 Location: Ludhiana, Punjab, India
 Soil: Clay
 Row-to-row spacing: 90 cm (0.9 m)
 Sufficient quantity of water is available for irrigation.
 Step 1: Calculate peak water requirement of sugarcane, use Eq. (1).
 ET_0 = 10.03 mm/day (From Table 10.1)
 Crop Coefficient = 1.10 (From Table 10.2)
 PWR = 10.03 × 1.10 × 0.6 = 6.61 L per day per m².
 Step 2: Water availability check and number of operations.
 Total water requirement for entire field of 200 m × 100 m = 200 × 100 m² × 6.61 L per day per m² = 132,200 L per day.
 Since sufficient quantity of water is available, entire field can be irrigated at a time.
 Step 3: Selection and Design of Drip line: We will use inline drip tubing of 16 mm class 2.
 Dripper Spacing: 50 cm
 Discharge: 2 L per hour per dripper
 Operating pressure: 10 m or 1 kg/cm²
 Outer Diameter: 16 mm
 Inner Diameter: 14 mm
 Maximum length of drip line is 96 m at 10% discharge variation and 121 m at 20% discharge variation. Actual length in field is 100 m, the discharge variation will be 10.50%.
 Total length of inline tubing = (Area in m²)/(Spacing between adjacent tubes or laterals)

$$= (200 \times 100 \text{ m}^2)/(0.90 \text{ m}) = 22{,}222 \text{ m}$$

Add 2% for temperature stress and snacking = $0.02 \times 22{,}222 = 444$ m
Total length of inline tube = $22{,}222 + 444 = 22{,}666$ m
Total flow of water = Total length (m) × discharge per meter
  = $22{,}666 \times 4$ L per hour per meter = 90,664 L per hour
Total flow Q, liters per second = Flow in liter per hour/3600
  = $90{,}664/3600 = 25.2$ L per second

TABLE 10.1 Normal Monthly Pan Evaporation (mm) Data For Ludhiana, India [11]

Month	2001	2002	2003	2004	2005	Average
Jan.	37.1	47.5	24.2	57.8	39.2	41.14
Feb.	76.0	60.6	66.3	65.9	50.5	63.86
Mar.	132.6	127.3	108.2	130.6	86.5	117.04
Apr.	214.0	234.0	211.1	236.3	215.5	222.38
May	303.9	319.2	324.8	323.9	283.1	310.98
Month	2001	2002	2003	2004	2005	Average
Jun.	188.8	241.0	279.0	236.0	302.8	249.52
Jul.	114.3	230.3	138.5	213.4	139.6	167.22
Aug.	140.9	168.1	125.0	125.1	141.6	140.14
Sep.	151.4	115.0	115.5	142.6	121.4	129.18
Oct.	106.1	102.6	106.2	876.6	102.2	100.94
Nov.	80.7	65.4	65.2	66.0	78.6	71.18
Dec.	49.7	42.6	39.0	38.3	43.6	42.64

TABLE 10.2 Peak Period Crop Coefficients of Selected Crops, India

Crop	Crop coefficient
Beans, green	0.90
Cabbage	0.80
Citrus, kinnow	0.60
Corn, Sugarcane	1.10

TABLE 10.1 *(Continued)*

Corn, sweet	0.95
Cotton, beans, potato, tomato peas	1.00
Grapes	0.70
Melons, peanuts, lettuce	0.95
Onion, Green	0.80
Onion, dry	0.90
Pea	0.95
Pecans, almonds, apricots, peaches	0.75
Peeper	0.80
Walnuts apples	0.85
Watermelon	0.85

Source: Rajput and Patel [13].

Step 4: Design of Submain: Submain is a conduit, which carries water from main line and distributes among the laterals or dripper lines. Assume the submain is placed along 200 m side. Calculate standard discharge rate (SDR) in the submain in liters per hour per meter.

$SDR_{submain}$ = (Total flow in submain × 3600)/(length of submain in m)

= [25.2 × 3600]/200 = 543 L per hour per meter = Total flow in submain (lps).

The required size of the submain line is 75 mm diameter. Here the submain length is three segments each of 66.66 m in length. Friction head losses in submain = 1.4 m for 66.66 m length and 75 mm size.

Submain Design Check:

Friction Head loss in submain should not exceed 20% of operating pressure (i.e., 10 m × 0.20 = 2 m). It is also known as design tolerance. Since the friction head in submain is less than design tolerance, Design is safe. As three submains are of same length and flow, therefore size and friction head loss for each submain will also be same.

Step 5: Design of Mainline: Main line is a conduit, which feeds water from water source to submains. In this design, mainline is connected to three submains from water source as far as possible in a straight line. For ease of operation of system, we prefer to operate the each submain separately.

Flow, lps = (453 × 66.6)/3600 = 8.38 lps

The required size of the mainline is 90 mm. The results are summarized as follows:

Operation section	From	To	Distance (m)	Flow (lps)	Main Size (mm)	Friction head loss (m)
I	S/M3	S/M2	67 m	8.38	90	0.804
II	S/M2	S/M1	67 m	8.38	90	0.804
III	S/M1	Well	100 m	8.38	90	1.200
Total frictional head loss, m						2.808 m

Step 6: Pump Design (Eqs. (2) and (3)):

(Suction + Delivery) head = 5 m (assumed).

Friction head required loss in filter unit = 2 m for screen filter only.

Friction head loss in main line = 2.8 m

Operating Pressure = 10 m given by manufacturer

Friction head loss in fitting = 2 m (assumed

Pressure head required for venturi = 5 m (assumed).

Elevation (up ward if any) = Field is flat = 0 m

Total head required for the system, H, Eq. (2) = 5 + 2 + 2.8 + 10 + 2 + 5 + 0 = 26.8 = 27 m

From Eq. (3):

HP = $(Q \times H)/(75 \times a \times b) = (8.38 \times 27)/(75 \times 0.80 \times 0.85) = 4.19 = 4$ HP

KW = 0. 746 × HP = 0.746 × 4 = 2.98 KW

(*Note*: Please check the performance chart of pump to know available or manufacturing HP of pump set).

Application rate of system = (Total flow, lps × 3600)/(total area, m²)

= $(25.18 \times 3600)/(200 \times 100) = 4.53$ lph per m²

Irrigation duration, hours = (Crop water requirement)/(System application rate)

= $(8.38$ L per day per m²)/(4.53 lph per m²) = 1.84 h

Total irrigation hours at peak water requirement stage = 1.84 hrs. × 3 times operation

= 5.52 h = 6 h.

10.6   COST ESTIMATION OF DRIP IRRIGATION SYSTEM

Table 10.3 indicates installation cost of different components of a drip irrigation system in this study for 200 × 100 m² of sugarcane. All package practices were according to the bulletin by Punjab Agricultural University, Ludhiana [11].

TABLE 3   Cost estimation of drip irrigation system.

Item	Specification	Unit price Rs.	Qty.	Total Cost Rs.
Main line	90×10 kg/cm^2	Rs.81.05/m	234 m	18965.7
Submain line	75×6 kg/cm^2	Rs.40.40	200 m	8080
Lateral	16 mm dia.	Rs.9.00 per m	22666	203,994
Control valve	63 mm dia.	636.75 each	1	636.75
Sand filter	25 m^3/hour	17,500 each	1	17,500
Fertilizer tank	FT (mini)	7300 each	1	7300
Pumping unit	Motor (Mono block pump) + starter and accessories	10,000 each	1	10,000
Miscellaneous charges (Fitting charges)	Fitting and accessories @ 10% of total cost of system			26,647.6
Govt. VAT @ 4%				11,724.9
Installation cost of drip irrigation system for 200 m $\times$ 100 m (2 ha) of sugarcane				304,848.9
Installation cost of drip irrigation system for of sugarcane, Rs./ha				152,424.45

10.7   SUMMARY

Drip irrigation system is one of the most advance methods of irrigation. It is very economical in area having undulating topography where cost of land leveling is high. As the initial cost of drip system is very high, therefore this system is most economical for widely spaced orchards and crops having high value. A drip irrigation system for sugarcane (200 m × 100 m) has been designed. The cost of installation of drip irrigation system for 1 ha of sugarcane was found to be Rs. 152,424.45. The cost of installation of drip irrigation system is on higher side because the water source is assumed at one end of field, which increases the length of main and submain and the total cost includes the cost of pumping unit also.

KEYWORDS

- application efficiency
- drip irrigation
- economic viability
- electric energy
- emitter
- Israel
- okra
- pumping unit
- Punjab
- soil depth
- soil structure
- soil texture
- stream size
- submain
- sugarcane
- topography
- widely spaced orchards

REFERENCES

1. Ahire, N. R., Bhoi, P. G., Solanke, A. V. (2000). Effect of row spacing and planting system on growth and yield of potato under surface and drip irrigation *J. Indian Potato Assoc.*, 27(1–2):59 60.
2. Batta, R. K., Singh, N. (1998). Drip irrigation to sugarcane: Indian experience. Proc. National Seminar on Micro-Irrigation Research in Indian Status and perspective for the twenty-first century, Bhubaneswar – India. pp. 133–147.
3. Lenka, D. (1991). Irrigation and Drainage, Kalyani Publishers, New Delhi.
4. Hapase, D. G., Deshmukh, A. S., Gynjal, B. B., Shinde, P. P. (1988). Evaluation of Micro Irrigation System in Sugarcane Agriculture. Proceedings of XI International Congress on the use of Plastics in Agriculture.
5. Jones, C. A., Santo, L. P., Kigton, G., Gascho, G. (1990). Irrigation of agricultural crops. *Am. Soc. of Agronomy*, 835–854.
6. Michael, A. M. (1978). *Irrigation – Theory and Practices.* Vikas Publication House Pvt. Ltd. Masjid Road, Jangpura, New Delhi.
7. Michael, A. M., Ojha, T. P. (1966). *Principles of Agricultural Engineering,* Volume II. Jain Brothers, Jaipur.
8. Singh, Harjinder, Narda, N. K., Chawla, J. K. (2004). Efficacy on phosphorus through trickle fertilation of potato. *Indian Journal of Agricultural Sciences, 74(8),* 476–478.

9. Kaur, Mandeep, Narda, N. K., Chawla, J. K. (2005). Irrigation and potassium management in trickle fertigated potato. *Indian Journal of Agricultural Sciences, 75(5),* 290–291.

10. Nakayama, F. S., Bucks, D. S. (1986). *Trickle Irrigation For Crop Production.* American Society of Agricultural Engineers, St. Joseph – MI – USA. 383 pages.

11. *Package of Practices for Kharif Crops of Punjab* (2006). Punjab Agricultural University, Ludhiana, India. 25–40.

12. Parikh, M. M., Srivastava, P. K., Sawani, N. G. (1992). Response of Sugarcane crop of drip method of irrigation. Co-operative Sugar, *23(10),* 673–677.

13. Rajput, T. B. S., Patel, Neelam, (2001). *Drip Irrigation Manual.* Water Technology Centre, Indian Agricultural Research Institute, New Delhi.

14. Reddy, K. S., Reddy, G. P. (1995). Micro irrigation for water scarce areas. Yojaina Magazine, June, 39.

15. Sharma, R. C., Singh, B. P. (2003). Effect of NPK fertigation and planting patterns on yield and economics of potato under drip irrigation. *Journal of Indian Potato Association, 30(1–2),* 69–70.

16. Shih, S. F. (1988). Sugarcane yield, Biomass and water use efficiency. *Transaction of ASAE, 31(1),* 142–148.

17. Sidhu, G. S., Waraitch, K. S. (2002). *Sugarcane Production and Protection Technologies.* Punjab Agricultural University, Ludhiana.

18. Singh, N., Sharma, R. C. (2002). Effect of drip irrigation on yield, quality and economy of potato. In: Proceedings of International Agronomy Congress on Balancing Food and Environmental Security -A Continuing Challenge, held during 26–30 November at New Delhi – India. 347–348.

19. Tumbare, A. D. (1999). Effect of liquid fertilizer through drip Irrigation on growth and yield of Okra. *Indian Journal of Agronomy, 44,* 176–178.

CHAPTER 11

DESIGN AND COST ESTIMATION OF MICRO-SPRINKLER IRRIGATION SYSTEM FOR CHILI

SATWINDER SINGH and KAMAL G. SINGH

11.1 INTRODUCTION

Micro irrigation is one of the most recent developments in irrigation methods [1, 2]. It was originally developed in Israel by Simca Bloss in 1959. This method proved very economical in water use and also produced higher yield. The water demand has increased many folds due to improvement in agricultural production technology in the last few decades. The pumping cost has increased substantially with the inflation of machinery prices and enhancement in energy prices. This has made the water not only scare but also a costly resource. Because of the anticipated shortage of this vital agricultural input, we have always been endeavoring to increase the water resources.

The choice of any irrigation system depends on number of factors such as topography (land surface configuration), soil type (i.e., texture, structure, depth), stream size (i.e., flow rate of irrigation water available), economic viability (benefit–cost ratio), farmer's capacity to meet initial investment, and social acceptance of the system.

Micro irrigation is one of the latest methods of irrigation and is becoming increasingly popular in areas with scarcity of water and salt problem. This is also called tickle irrigation system or localized irrigation system.

Micro sprinkler irrigation is very well suited for widely spaced high value crops and trees. The required quantity of water of water is applied daily near the root zone through a network of piping system. The soluble fertilizers are fertigated frequently in small quantities to meet the needs of the crop. There is a continuous balance of water, air and fertilizer in the entire life cycle of plant; resulting in the optimum growth, better fruit quality and early maturity of crop.

---

*In this chapter, the currency is expressed in Indian Rupees (1.00 US$ = Rs. 60.93; 1.00 Rs. = 0.02 US$).

The main characteristics of micro sprinkler irrigation system consist of uniform, small continuous flow and reduced surface wetting around plants. Greater application efficiency is obtained due to better placement of water near the roots in precise quantity. The reduced surface wetting reduces surface evaporation. For widely spaced crops like fruit trees, the system is even more economical than other methods of irrigation.

The comparative merits and demerits of surface, drip and micro sprinkler irrigation methods reveal that micro sprinkler irrigation method is most efficient method of irrigation having irrigation efficiency of about 90% as compared to ordinary sprinkler method with about 70–80% and conventional method of surface irrigation having about 40–50%. Although micro sprinkler method is the most efficient method, yet it can be recommended and used only under particular situations in certain pockets of Punjab because of high initial cost and other limitations. Even under most suitable and favorable situations for micro sprinkler irrigation method, the farmers have not adopted micro sprinkler irrigation. The reasons for this are that farmers don't realize the importance and value of irrigation water available from canal and tube well source, as they get electric power at a cheap flat rate.

Keeping in view the water scarcity, topography and water quality, south western districts of Punjab are most suitable areas, especially where fruit trees can be grown under micro sprinkler irrigation system. The existing paddy-wheat cropping system in subtropical parts of northern India leads to hastening of water table declines, deterioration of soil health, increased incidence of insect/pests, weeds and reduced availability of protein rich legumes and oil seed crops. Sugarcane is emerging as an alternative to the paddy-wheat cropping system and has a unique role to play in sustaining agro-industrial economic growth of India. The potential sugarcane production depends on optimum availability of water throughout the growth period. Sugarcane is the most important cash crop. Sugarcane is adapted to a range of tropical and subtropical climates and can be successfully grown on all types of soils ranging from sandy to clay loam.

The total area under sugarcane in the world was 19.90 million ha during 1998 with production of 1254.94 ha in 1966–1967 to 235,000 ha in 1996–1997. Sugarcane contributes more than 62% of the world sugar production. Sugarcane occupied about 96 thousand hectares in Punjab during 2005–2006. The average sugarcane yield is 60,100 kg per ha. The average sugar recovery was 9.8%. Before recommending the adoption of micro sprinkler irrigation system for sugarcane, it is necessary that drip system is properly designed and its economic viability is evaluated.

At Mahatma Phule Krishi Vidyapeeth, Rahuri (Maharashtra – India), studies indicated that the efficiency of the sprinkler system is higher than the border method of irrigation. The water application, distribution, storage and water use efficiency (WUE) were higher for sprinkler irrigation with 13.9% increase in yield. Indian National committee on the use of plastics in agriculture has also reported supremacy of sprinkler over conventional irrigation methods for certain crops.

Keeping above facts in consideration, the present study was taken up: to design the micro sprinkler irrigation system for chili crop; and to estimate the cost of micro sprinkler irrigation system.

TABLE 11.1 Water Use in Sprinkler Versus Surface Method

Crop	Water depth, cm		Water saved, %
	Sprinkler irrigation	Surface irrigation	
Bajra (Pearl millet) Barley Gram	7.82	17.78	56
Cotton	29.05	40.64	29
Jowar (Sorghum)	11.27	25.40	56
Potato	30.00	60.00	50
Wheat	14.52	33.02	56

11.2 ADVANTAGES OF MICRO SPRINKLER IRRIGATION SYSTEM

1. Micro-sprinkler applies less water per unit area than drop emitters because they spread water over a large area.
2. Micro-sprinkler prevents runoff and erosion in heavy soils.
3. Even distribution of water since droplet size is small.
4. Less evaporation losses as compared to conventional sprinkler due to low height.
5. Simple assembly and operation, needs little or no maintenance.
6. Useful for fruits trees and orchards below the canopy.
7. Can be used in green house and nurseries.
8. Soils having high infiltration rate can be efficiently irrigated.

11.3 DISADVANTAGES

1. High initial cost.
2. Micro-sprinkler requires larger diameter of lateral than drip system.
3. Cannot be used for delicate flowering plants.
4. There are chances to soft fruits in dwarf plants.

11.4   COMPONENTS OF MICRO-SPRINKLER IRRIGATION SYSTEM

The basic components of all micro/mini sprinkler irrigation system are: main line, submain, microsprinklers, laterals, micro sprinkler head (sprinkler nozzle), stake for sprinkler, ball valve, flush valve, air release valve, pressure gauge, gromet and take off for LDPF lateral, fertilizer applicator, check valve, pumping unit, and filtration system.

11.5   DESIGN OF MICRO SPRINKLER SYSTEM

Steps in design of microsprinkler are:
1.   Determination of water requirement.
2.   Selection of sprinkler.
3.   Layout of sprinkler system.
4.   Selection and design of lateral.
5.   Selection and design of sub main.
6.   Selection and design of mainline.
7.   Selection of pump.
8.   Calculation of irrigation time.

11.5.1   DETERMINATION OF WATER REQUIREMENT

Determination of water requirement is one of the basic needs for crop planning and for designing of sprinkler irrigation system. Water requirement includes the losses due to evapotranspiration (ET) or consumptive use (CU) plus losses during the application of irrigation water. By using sprinkler system, we are applying water directly to the soil, so that we can reduce conveyance losses. Water requirement for orchards (tree crops) is estimated as:

$$WR = \{(\text{crop area } x \text{ P.E } x \text{ P}_c x \text{ K}_c) \times (\% \text{ wetted area})\}/U_c \qquad (1)$$

where: crop area = plant-to-plant spacing multiplied by row-to-row spacing, m^2; P.E.= pan evaporation, mm/day; P_c = crop coefficient, depends on foliage characteristics, stage of growth; and U_c = uniformity coefficient of microsprinkler. Percent wetted area is the area, which is shaded due to canopy cover when the sun is overhead. Table 11.2 shows average values of percent wetted area for different orchards.

TABLE 11.2 Percent Wetted Area of Various Crops Under Drip Irrigation

Crop	Wetted area (%)
Banana	50 to 60
Chilies	75
Grapes	50 to 60
Lime	20
Orange	30
Pomegranate	20
Sugarcane	60 to 70
Vegetables	60 to 70

For row crops, the water requirement is calculated per unit area per day as follows:

Method I

Volume of water required per unit area per day = Net water depth x percent wetted area (2)

where: net depth of water or crop evapotranspiration of the crop $(ET_c) = PE \times P_c \times K_c$; P.E. = pan evaporation, mm/day; P_c = pan coefficient, average value is 0.72; and K_c = crop factor.

Method II

Water requirement (lit/day/plant)= [crop area $\times$ P.E $\times K_c \times$ % wetted area]/U_c (3)

Method III

Volume of water (lit/day/dripper) = $S_1 \times S_s \times PE \times P_c \times K_c$ % wetted area (4)

where: S_1 = lateral spacing along the submain, m; and S_s = sprinkler spacing along the lateral, m. If data on evaporation rate is not available, then the average evaporation can be assumed between 6 to 8 mm/day depending on the season and location. In case of row crops, generally paired row or skip row plantation method is most suitable for sprinkler irrigation. In this method, the entire strip of paired row needs to be irrigated and hence the wetted area should be calculated for this paired row strip excluding the area in between the paired rows. In case of narrow row crops, the following crop factors are recommended:

Crop growth stage	Crop factor
Emergence to early growth	0.3 to 0.5
Initial period of growth	0.6 to 0.7
Major crop growth period	0.8 to 0.9
Flowering and fruiting stage	1.0 to 1.15
Harvesting stage	0.8

11.5.2   SELECTION OF MINI SPRINKLER NOZZLE

The selection of sprinkler nozzle depends upon the crop water requirement and soil infiltration rate. The sprinkler nozzle should supply enough water to the plant. Generally the spray pattern of sprinkler is affected by operating pressure. Therefore, before selecting a particular type of microsprinkler using existing pump, it must be ensured that: the sprinkler will get enough pressure, the sprinkler should give uniform and constant discharge and it should not vary significantly because of minor differences in pressure head. The nozzle cross sectional area should be large in order to reduce the clogging problem.

11.5.3   LAYOUT OF MINI SPRINKLER SYSTEM

The design of sprinkler system mainly depends on the layout of laterals, submains and mainline. The layout is considered in relation with dimensions, area topography of the field. The length of submain and mainline depends on dimensions and shape of the field. The layout of the system should be such so that it should require minimum length of submain, it should be laid at the center of the plot. As a thumb rule, if the land is undulating, the submain and mainline should be laid along the slope and laterals should be laid along the contour line, that is, across the major slope to have minimum variation in pressure and discharge. The layout is also dependent on the location of the water source and quantity of water available for irrigation.

Wind speed and direction has direct effect on uniformity coefficient. Wind direction decides the positioning of lateral as well as the spacing between the laterals and sprinkler. Usually laterals should be positioned across the direction of wind, when slope is not constraint.

11.5.4 SELECTION AND DESIGN OF LATERALS

For microsprinkler system, the lateral may be of small diameter, made up of low density polyethylene (LDPE) or linear low density polyethylene (LLDPE) of 16 or 20 mm in diameter. The lateral may also be made up of PVC pipes with small riser heads. The LDPE laterals are connected to submain by using rubber grommet and take off. The size and length of lateral is decided by the discharge of the sprinkler and number of sprinkler on one lateral.

11.5.4.1 CALCULATION OF FLOW RATE

$$Q_1 = ns \, x \, q_s \tag{5}$$

where: Q_1 = flow rate of one lateral, lit/hr; ns = number of sprinklers on one lateral; and q_s = nominal discharge of sprinkler, lit/hr.

11.5.4.2 LATERAL HEAD LOSS CALCULATIONS

The diameter of lateral is usually selected such that the difference in the discharge between two extreme sprinklers operating simultaneously should not exceed 10% and pressure head difference should not exceed 20% of average operating head. The allowable head loss for level area should be divided between the lateral and submain lines as follows:

$$\Delta H_1 = 55\% \text{ of } \Delta H_s \text{ and } \Delta H_{sm} = 45\% \text{ of } \Delta H_s \tag{6}$$

where: ΔH_s = total allowable difference in pressure head in submain and lateral, m; ΔH_1 = head loss in lateral, m; and ΔH_{sm} = the head loss in submain, m. For example, if microsprinkler has operating pressure of 2 kg/cm² (20 m), then total allowable head loss is 20% of 20 m = 4 m. Then head loss in lateral should is 55% of 4 m = 2.2 m and head loss in submain should be 45% of 4 m = 1.3 m. The Hazen Williams equation is most commonly used for calculation of head losses in pipes:

$$\Delta H_1 = K \times (Q_1/c)^{1.852} \times D_1^{-4.871} \times L_1 \times F \tag{7}$$

where: ΔH_1 = head loss in lateral, m; K = constant = 1.21×10^{10}; Q_1 = flow rate in the lateral, lit/sec; C = the friction loss coefficient, depends on the roughness of pipe material; D_1 = inside diameter of lateral, mm; F = the outlet factor depending on number of outlets; and L_1 = length of lateral, m.

11.5.5 SELECTION AND DESIGN OF SUBMAIN

Submain is generally made up of PVC (polyvinyl chloride) pipes of 32 mm, 40 mm, 50 mm, 63 mm or 75 mm in diameter. The design of submain is based on both capacity and uniformity. Capacity implies that the submain size should be large enough to deliver required amount of water to irrigate the subsequent part of the field. Uniformity means that the submain should be designed to maintain the allowable pressure variation, so that flow into all lateral lines will have an acceptable variation. Submain supplies water to individual lateral. Design of submain is similar to that of lateral, however it differs in that the spacing between outlets is greater and large flow rates are involved. The size and length of submain is determined by number of laterals and distance between laterals. Usually in a flat field, the position of the submain should be located at the center of the plot. On slopping field, the lateral lines should be laid along the contour lines and the submain along the slope, as far as possible. Flow rate and length of submain is calculated as follows:

$$Q_{sm} = N_1 \times Q_a \qquad (8)$$

$$L_{sm} = N_1 \times S_1 \qquad (9)$$

where: Q_{sm} = flow rate in the submain, lit/sec; N_1 = number of laterals on the submain; Q_a = average flow rate of laterals, lit/sec; L_{sm} = length of sub main, m; and S_1 = average spacing between laterals on the submain. The friction loss for the submain is calculated as:

$$\Delta H_{sm} = K \times (Q_{sm}/c)^{1.852} \times D_{sm}^{-4.871} \times L_{sm} \times F \qquad (10)$$

where: F = outlet factor; and L_{sm} = length of submain, m.

11.5.6 SELECTION AND DESIGN OF MAINLINE

Generally the size of mainline is one size higher than submain. The size of main line is decided by flow rate of all the submains. The sizes of mainline are 40 mm, 50 mm, 63 mm, 75 mm, 90 mm or 110 mm, etc. The head loss is calculated by using Hazen and Williams equation:

$$\Delta H_m = 15.27 \{Q_m^{1.852}/D_m^{4.871}\}L_m \qquad (11)$$

$$\Delta H_m = K (Q_m/C)^{1.852} D_m^{-4.871} \times L_m \qquad (12)$$

where: Q_m = total discharge of mainline, lit/sec; D_m = inside diameter of mainline, cm; L_m = length of mainline, m; ΔH_m = head loss in mainline, m; and for C = 150; K

= constant = 1.21×10^{10}; C = friction coefficient for continuous section of pipe and depends on pipe material; and D_m = inside diameter of mainline, mm.

11.5.7 PUMP SIZE

To determine the pump size, consider the following two equations:

Total head of pump (H) = suction head + delivery head + filter losses + mainline loss + operating pressure + fitting loss + venture head loss + elevation difference

(13)

$$H.P. = (Q.H./75.\eta_{motor} \cdot \eta_{pump}) \qquad (14)$$

where: filter losses are assumed to be 2 m for screen filter (disc filter) and 2 m for sand filter; operating pressure is about 1.5 kg/cm² (15 m); fitting loss = 2 m; Venturi head loss = 5 m; Q = maximum flow rate of system, lit/sec; H = total head of the system, m with Eq. (13); η_{motor} = motor efficiency = 80%; and η_{pump} = pump efficiency = 75%; and H.P. = pump horse power.

11.5.8 CALCULATION OF IRRIGATION TIME

Irrigation rime for tree crop (hours) = water requirement (lit/day/tree)/application rate (mm/hr)

Irrigation time for row crops (hours) = water requirement (li/day/area)/application rate (mm/hr) (15)

Application rate is calculated by dividing the discharge of sprinkler to the sprinkler spacing along the lateral and submain.

11.6 DESIGN OF MICRO SPRINKLER SYSTEM: CASE STUDY

11.6.1 DESIGN DATA

1. Size of the field = 1 ha
2. Assume square plot of 100 m × 100 m
3. Crop is chili
4. Crop spacing = 30 cm × 45 cm × 75 cm
5. Pan evaporation = 8 mm/day
6. Area to be wetted as a percentage of total area = 75%

7. Availability of electricity per day = 12 h
8. Soil type is sandy loam
9. Topography of field is slightly undulating with 1% slope
10. Uniformity coefficient = 85%
11. Water source is well at the corner of the field
12. Well depth = 10 m
13. Harzen Williams constant, C = 150 for PVC pipe and 140 for LLDPE pipes
14. Assume crop coefficient as 0.8 and pan coefficient as 0.7, for entire growing season

11.6.2 DESIGN PROCEDURE

Step 1. Volume of water = [crop area (m²) × P.E. × P_c × K_c % wetted Area]/U_c = (0.3 × 0.45 × 8 × 0.7 × 0.75)/0.85 = 0.53 lit/day/plant, volume of water needed.

   Step 2. *Selection of Microsprinkler:* For economic design of microsprinkler system, the paired row planting is adopted, in which one lateral for two rows of plants is provided. From the wide range of micro sprinkler available with the manufacturer, the microsprinkler having maximum radius of throw with minimum discharge can be selected. From Table 11.3, select micro sprinkler of 83 lit/hr discharge having radius of throw of 3 m, considering 100% coverage and 50% overlapping between the microsprinkler, the spacing between microsprinkler of 3 m.

   Then, application rate = 83 /(3×3) = 9.2 mm/hr.

   Step 3. *Selection and Design of Lateral:* To divide the flow into two sections, install the submain at the center of the field. For economic design, divide the area into four sections.

1. Length of lateral on one side of submain = 25 m
2. No. of microsprinklers on one lateral at a spacing of 3 m = 25/3 = 8.3 = 8 m (Approx.)
3. Actual spacing between the microsprinkler = 3.12 m
4. Discharge rate of lateral = 83 x 8 = 664 lit/hr = 0.184 lit/sec
5. Select 16 mm diameter lateral

Using Hazen Williams equation, head loss due to friction is:

$\Delta H_1 = K \times (Q_1/c)^{1.852} \times D_1^{-4.871} \times L_1 \times F = 1.21 \times 10^{10} (0.184/140)^{1.852} \times 13.8^{-4.871} \times 25 \times 0.41 = 1.62$ m

   As the calculated head loss is nearly equal 10% of operating pressure, the lateral of 16 mm size is sufficient to carry the flow.

   Step 4. *Selection and Design of Submain:* Use four submains of 50 m length each.

1. Length of submain = 50 m
2. Distance between two lateral on one side of submain = (2 × 0.45) + (2 × 0.225) + 1

= 2.35 m

TABLE 11.3   Size of Sprinkler Nozzles and Mini Sprinklers, India

Operating pressure Kg/cm²	Nozzle color and size													
	Black 0.85 mm		Blue 1.0 mm		Green 1.2 mm		Red 1.4 mm		White 1.6 mm		Violet 1.8 mm		Yellow 2.0 mm	
	Flow lps	Rad m	Flow lps	Rad m	Flow lps	Rad m	Flow lps	Radm	Flow lps	Radm	Flow lps	Rad m	Flow lps	Radm
1.5	33	3.0	44	3.25	64	3.5	90	4.0	116	4.4	150	4.5	185	4.7
2.0	39	3.3	50	3.75	75	4.2	100	4.4	135	4.6	175	4.8	215	5.2
2.5	43	3.3	56	3.85	83	4.5	115	4.7	150	5.0	195	5.2	238	5.4
Available sizes of micro sprinklers in the market, India													
1.0	28	1.9	98	3.3	67	2.84			43	2.4				
1.5*	36	2.2	122	3.7	83	3.35			54	2.9				
2.0	41	2.5	148	3.9	97	3.65			63	3.2				

*Recommended operating pressure = 1 5 kg/cm²; Rad = Radius of throw in meters.

3. No. of laterals on one side of submain = 50/2.35 = 21.27 = say 21
4. No. of laterals on both side of submain = 21 × 2 = 42
5. Discharge rate of submain = 42 × 0.184 = 7.73 lit/sec
6. Outlet factor (for 42 outlets) for submain = 0.361.

Select 63 mm diameter of submain, that is, inside diameter of submain = 58.8 mm. Using Hazen and Williams equation, head loss due to friction is:

$$\Delta H_{sm} = K \times (Q_{sm}/c)^{1.852} \times D_{sm}^{-4.871} \times L_{sm} \times F = 1.21 \times 10^{10} \, (7.73/150)^{1.852} \times 58.8^{-4.871} \times 50 \times 0.361 = 2.15 \text{ m}$$

As the calculated head loss is within desired limit, the design is accepted. Submain is a conduit, which carries water from main line and distributes among laterals of dripper lines. Assume the submain has to be placed along 100 m side. Calculate SDR (standard discharge rate is submain in liter per hour per meter)

SDR of submain = 3600 × (Total flow in submain, lps)/(Submain length in meter) = (7.73 × 3600)/100 = 278.28 L per hour per meter

The required size of the submain line is 75 mm in diameter. Here the submain length consists of 4 segments each of 25 m length. Friction head loss in submain = 1.4 m for submain for 25 m length and 75 mm size.

Submain Design Check:

Friction head loss in submain should not exceed 20% of operating pressure (10 m × 0.20 = 2 m). It is also known as design tolerance. Since the friction head in submain is less than design tolerance, design is safe.

Step 5. *Selection and Design of Mainline:* Operating two submains at a time.
1. Length of mainline = 75 m
2. Discharge rate of mainline = 7.73 × 2 = 15.46 lit/sec
3. Select main line of 90 mm in diameter.

Main line is a conduit, which feeds water from water source to submains. In this design, mainline is connected to submains from water source as far as possible in a straight line. To operate the system, we prefer to operate the each submain separately.

$$\text{Flow (lps)} = (278.28 \times 66.6)/3600 = 5.14 \text{ lps}$$

Therefore, the required size of the mainline is 90 mm with Inside diameter of 85.8 mm. Using Hazen Williams equation, head loss due to friction is:

$$\Delta H_m = K \, (Q_m/C)^{1.852} \, D_m^{-4.871}. \, L_m = 1.21 \times 10^{10} \, (15.46/150)^{1.852} \, 85.8^{-4.871} \times 75 = 5.11 \text{ m}$$

This calculated head loss of mainline should be added, while selecting the total head of the pump.

Step 6. *Pump Size:*

Total head of pump (H) = suction head + delivery head + filter losses + mainline loss + operating pressure + fitting loss + venture head loss + elevation difference = $10 + 15 + 5.11 + 2 + 2 + 5 = 39.11$ m

For Q = 15.46 lit/sec, h = 39.11 m, motor efficiency = 80%, pump efficiency = 75%, we have:

H.P. = $(Q.H./75.\eta_{motor} \cdot \eta_{pump}) = (15.46 \times 39.11)/(75 \times 0.80 \times 0.75) = 13.43$

Adding 10% for unforeseen losses,

$$H.P. = 13.43 \times 1.1 = 14.77 = \text{say 15 H.P.}$$

Step 7: *Irrigation time:* It is the ratio of volume of water applied for four plants to the discharge rate dripper.

IR = $(0.53 \times 16)/9.2 = 0.92$ hrs. 55 min for one shift of two submain.

Total irrigation time = 2×55 min = 110 min = 1 hr, 50 min.

11.7 COST ESTIMATION OF MICRO-SPRINKLER FOR ONE HA AREA

The cost of micro sprinkler system depends on the spacing between micro-sprinklers, discharge of microsprinklers and other factors. However, maximum care should be taken to reduce the cost of the system without affecting plant population and uniformity coefficient. The calculations are shown in Table 11.4.

TABLE 11.4   Cost of Installation of Micro Sprinkler Irrigation System For One Hectare Under Chili Cultivation

Items	Quantity	Rate (Rs./-)	Amount (Rs./-)
PVC pipe 90 mm	180 m	80 per m	14,400/-
PVC pipe 63 mm	204 m	40 per m	8,160/-
LLDPE lateral 16 mm	4500 m	7 per m	31,500/-
Micro-sprinkler,	1350 each 1000	9 per No.	12150/-
83 lit/hr extension tube, 6 mm	m	3.5 per No.	3,500/-
Screen/ disc filter, 40 m³/hr	1 each	5000 per No.	5,000/-
Ball valve, 63 mm	4 each	350 per No.	1,400/-
Flush valve, 63 mm	4 each	200 per No.	800/-
Filter and accessories, @ of 10% of above.		@ 10%	7,691/-
		Total	84,601/-

KEYWORDS

- design
- flow rate
- irrigation time
- micro sprinkler irrigation system
- mini sprinkler nozzle selection
- mini sprinkler system layout
- pump size
- water requirement

REFERENCES

Goyal, Megh R. (2014). *Management of Drip/Trickle or Micro Irrigation.* Oakville, ON, Canada: Apple Academic Press Inc.

Goyal, Megh R., (2015). *Research Advances in Sustainable Micro Irrigation, Volumes 1 to 10.* Oakville, ON, Canada: Apple Academic Press Inc.

APPENDIX I: PHOTOS OF DRIP IRRIGATION COMPONENTS

CHAPTER 12

DEVELOPMENT OF LOW PRESSURE FERTIGATION INJECTOR

SANTOSH KUMAR, KAMAL G. SINGH, and CHETAN SINGLA

12.1 INTRODUCTION

There has been a tremendous growth in fertilizer use throughout the world in the twentieth century. By the end of twentieth century, developing countries had increased the utilization of fertilizers to 60% of the world's fertilizer use and produced 55% of total nitrogenous fertilizer [9]. Drip irrigation offers several advantages like high irrigation efficiency, labor and energy savings, improvement in quality and yield of produce and opportunity to manipulate inputs, as per the crop demands, over the conventional irrigation method. Black [1], Miller et al. [8], Kaneworthy [6] and Smith et al. [12] indicated that there was a saving of up to 50% of the fertilizer, when applied through a drip irrigation system.

Fertigation is the technique of supplying water-soluble fertilizers to crops through an irrigation system. It has become a common practice in modem agriculture. Fertigation through drip irrigation can save fertilizers from 25 to 50% [5]. Fertilizers can be injected into irrigation system by three principal methods: (i) bypass tank, (ii) venturi, and (iii) pumps. The venturi is considered as the best method of fertigation. The main limitations of venturi are large pressure drop and high head requirement for its operation. Lewitt [7] gave the mathematical equation for measurement of flow by pipe orifice. Ree [10] indicated that shop- built orifices, when used without correcting to true diameter, may yield errors in a discharge estimate as large as 15% for the 2.54 cm diameter orifice and 6% for the three larger (4.44, 6.35, 8.89 cm) orifices [13].

Eisenhauer and Bockstadter [4] studied injection pump flow considerations for center pivots with corner watering systems and concluded that injecting chemicals at constant rate into center pivot equipped with either guns or swing-booms resulted in systematic chemical application rate errors due to the variable wetted radius of the irrigation system. Replogle and Wahlin [11] constructed a special venturi meter

*In this chapter, the currency is expressed in Indian Rupees (1.00 US$ = Rs. 60.93; 1.00 Rs. = 0.02 US$).

from a plastic pipefitting and attempted to address the economic limitations, and fouling water management requirements. Thirty venturi meters were constructed. It was established that individual calibrations were not required for effective application in irrigation practice. Camp et al. [2] conducted an experiment to develop a variable rate, digitally controlled metering device to permit variable flow of a fluid. The device consisted of a reservoir that was alternatively filled and emptied at a rate dependent upon a digital pulse from an external source.

These studies show that there are number of methods such as by-pass tank, venturi and injection pump for fertigation. The most commonly used device for fertilizer application is the venturi that requires1 kg/cm² to 4 kg/cm² for operation. Therefore, the authors developed low head injector for fertigation through a drip irrigation system. It will be useful for small and marginal farmers and green houses.

12.2   MATERIAL AND METHODS

The orifice meter was developed based on Bernoulli's Theorem, which is defined as an increase in the water speed occurs simultaneously with a decrease in pressure or a decrease in the potential energy [3, 13, 14], due to total constant energy at any point along the stream line:

$$P1/w + V_1^2/2\,g + Z1 = P2/w + V_2^2/2\,g + Z2 = \text{Constant} \tag{1}$$

Substitution for $Z1 = Z2$ for a horizontal orifice and $V_1 = (a2\,V_2)/a1$ in Eq. (1), we get:

$$h = V_2^2/2\,g\,[1 - (d/D)^4] \tag{2}$$

$$V_2 = \{(2\,gh)/(1 - (d/D)^4)\}^{0.5} \tag{3}$$

The theoretical discharge $(Q_{th}) = a2\,V_2 = a1\,V_1$ is:

$$Q_{th} = a2\,\{(2\,gh)/(1 - (d/D)^4)\}^{0.5} \tag{4}$$

For an orifice of area a2:

$$Q_{act} = (C_d)\,a2\{(2\,gh)/(1 - (d/D)^4)\}^{0.5} \tag{5}$$

$$Q_{act} = Q_e \times N \text{ or} \tag{6}$$

$$C_d = Q_{act}/Q_{th} \tag{7}$$

In Eqs. (1) to (7), a_1 is the cross-sectional area of inlet pipe (mm²); a_2 is the cross-sectional area of orifice meter (mm ²); D is the diameter of inlet pipe (mm); d is the diameter of orifice meter (mm); V_1 is the velocity of water approaching orifice meter (m/s); V_2 is the velocity of water through orifice meter (m/s); p_1 is the pressure of fluid before orifice meter (kg/cm²); p_2 is the pressure of fluid after orifice meter (kg/cm²); h is the measured difference of pressure (cm); w is the unit weight of water (kg/cm³); Q_{th} is theoretical discharge rate without the equipment (lph); Q_{act} is the actual rate of discharge rate with equipment (lph); C_d is the orifice coefficient of discharge; N is the total number of emitters in a given area of fertigation; and Q_e is the emitter discharge rate (lph).

For losses due to passage of the liquid through the orifice, the actual rate of discharge is obtained by Eq. (5). Due to drop in pressure in the orifice meter, water-soluble fertilizer was injected the orifice meter. Then, it was mixed with the water flowing through the orifice meter. The discharge rate at the outlet is given by Eq. (6). The coefficient of discharge for an orifice meter is defined in Eq. (7).

TABLE 12.1   Various Sizes of the Developed Orifice Meters

Orifice size	Thickness	Diameter of orifice	Diameter of fertilizer injection hole
	mm	mm	mm
O1	25.40	3.18	1.59
O2	19.05	3.18	1.59
O3	12.70	3.18	1.59
O4	9.53	3.18	1.59
O5	12.70	2.38	1.59
O6	12.70	1.59	1.59

FIGURE 12.1   Different thickness of fabricated orifice meters.

12.3   DESIGN OF EQUIPMENT FOR FERTIGATION

Various sizes of orifice meters in this study are given in Table 12.1. The figures of different thicknesses of orifice meter are given in Fig. 12.1. With decrease in diameter, the injection rate was increased (continuity equation in one dimension). By keeping the diameter of orifice meter constant (3.18 mm), the thickness of the orifice meter was varied to get the optimal value. Hole of 1.59 mm of fertilizer injector was drilled just at the edge of the orifice meter. A plastic tube was connected from the orifice meter to the fertilizer tank, which was placed at a height of one meter above the ground. The pressure drop and fertilizer injection rates values were recorded. The difference between inlet pressure and outlet pressure was measured by the difference in height of a mercury manometer with two limbs. This pressure gradient is related by the following equation:

$$h = R\{(Sm/S) - 1\} \tag{8}$$

where: h = pressure head in cm; R = reading of the U-tube manometer, cm; S_m = specific gravity of mercury; and S = specific gravity of water.

The tests on this equipment were conducted to determine the injection rate through the different sizes of the orifice meter. After preparing the fertilizer solution, the water flowing through orifice meter was allowed through the orifice area. All developed orifice meters were tested under different inlet discharge rates and rates of fertigation. All orifice meters were tested for areas of 450 m², 300 m² and 150 m².

12.4   RESULTS AND DISCUSSIONS

The developed orifice meters were evaluated for inlet pressure of 0.42 to 0.944 kg/cm^2 and three areas of orifice of 150 m^2, 300 m^2, and 450 m^2. The orifice meter having maximum injection rate and the corresponding operating pressure for different orifice meters are given Table 12.2. The maximum injection rate depends on the outlet pressure for a specific inlet pressure. The best fit equation of average injection rate and outlet pressure for six orifice meters (0_1–0_6) are given below:

For orifice meter 0_1:
$S = 2292.1\ P^2 - 924.74\ P + 99.326,\ R^2 = 0.9448$
For orifice meter 0_2:
$S = 45937\ P^2 - 12{,}322\ P + 833.65,\ R^2 = 1$
For orifice meter 0_3:
$S = 1498.1\ P^2 - 573.94\ P + 64.788,\ R^2 = 0.9778$
For orifice meter 0_4:
$S = 1541.7\ P^2 - 519.09\ P + 58.239,\ R^2 = 0.9956$
For orifice meter 0_5:
$S = 1061.9\ P^2 - 409.21\ P + 47.007,\ R^2 = 0.9517$
For orifice meter 0_6:

$$S = 4610.1\ P^2 - 1105.1\ P + 84.857,\ R^2 = 0.9952 \qquad (9)$$

where: S = average injection rate, lph; and P = outlet pressure, kg/cm^2.

TABLE 12.2   Maximum Injection Rates and the Corresponding Operating Pressures For Different Orifice Meters (1 to 6)

| Orifice meter | Size | | Maximum injection rate | Operating pressure | |
| | Thickness | Diameter | | Inlet pressure | Outlet pressure |
	mm		lph	kg/cm^2	
1	25.4	3.18	24.82	0.772	0.117
2	19.05	3.18	12.00	0.784	0.124
3	12.7	3.18	45.00	0.702	0.037
4	9.53	3.18	27.70	0.795	0.102
5	12.7	2.38	15.00	0.830	0.108
6	12.7	1.59	27.67	0.944	0.078

12.4.1   COST ESTIMATION OF THE BEST DEVELOPED ORIFICE METER

The materials used were two pieces of flanges with thickness 12.7 mm (Rs.150.00), 900 mm length of GI pipe with diameter of 12.7 mm (Rs. 30.00), and two 12.7 mm diameter elbows (Rs. 100.00). The total cost of the system including the best-developed orifice meter was Rs. 300.00.

12.5   CONCLUSIONS

The performance of all developed orifice meters were tested at a pressure < 1 kg/cm^2, while the venturi operates at 1 to 4 kg/cm^2. The orifice meter with 12.7 mm thickness and 3.18 mm diameter was found the best with maximum fertilizer application rate of 45 lph at an inlet pressure of 0.72 kg/cm^2. The orifice meter was found economical than the most commonly used devices (the venturi). The cost of the orifice meter was around Rs. 300.00, whereas the cost of the venturi is about Rs. 1048.00 for same purpose of fertigation.

12.6   SUMMARY

The methods of fertilizer application through drip irrigation are by-pass tank, venturi and pump system. The venturi is most commonly used for fertigation. The main limitations of the venturi are large pressure drop and high-required pressure head. Therefore, an orifice meter was developed for fertigation, which operates at inlet pressure <1 kg/cm^2.

Six different sizes of orifice meter were developed and evaluated at the Research Farm of Department of Soil and Water Engineering, Punjab Agricultural University, Ludhiana – India. Four orifices had a diameter of 3.18 mm and thickness of 25.4, 19.05, 12.7, and 9.53 mm. Two orifices with diameters of 2.38 mm and 1.59 mm and with a thickness of 12.7 mm were evaluated at pressure of 0.42 to 0.944 kg/cm^2.

The maximum injection rate was 45 lph with the orifice meter having a diameter of 3.18 mm and thickness of 12.7 mm at an inlet pressure of 0.702 kg/cm^2. This was found to be best. The developed equipment was also suitable for farmers having small land holdings or in a greenhouse.

KEYWORDS

- by-pass tank
- center pivot
- drip irrigation
- fertigation
- fertilizer application
- fertilizer injector
- fruit tree
- greenhouse
- injection pump
- injection rate
- metering device
- nitrogen
- orifice meter
- pipe orifice
- pump system
- small land holding
- trans critical flow
- trickle irrigation
- venturi meter

REFERENCES

1. Black, J. D. F. (1976). Trickle Irrigation – A Review. *Hort. Abstract, 46(1),* 69–74.
2. Camp, C. R., Sadler, E. J., Evans, D. E., Usrey, L. J. (2000). Variable rate, digitally controlled metering device. *Applied Engineering in Agriculture, 16(1),* 39–44.
3. Chanson, H. (2009). Trans-critical flow due to channel contraction. *Journal of Hydraulic Engineering ASCE, 135(12),* 1113–1114.
4. Eisenhauer, D. E., Bockstadter, T. L. (1990). Injection pump flow considerations for center pivots with corner watering systems. *Transactions in Agriculture, 33,* 162–166.
5. Haynes, R. J. (1985). Principles of fertilizer use for trickle irrigated crops. *Fert. Res., 6,* 235–255.
6. Kenworthy, A. L. (1979). Applying nitrogen to fruit trees through trickle irrigation system. *Acta. Hort., 89,* 107–110.
7. Lewit, E. H. (1952). The mathematical equation for measurement of flow by pipe orifice. Hydraulics and Fluid Mechanics Handbook. 752 pp.

8. Miller, R. J., Rolston, D. E., Rauschkolb, R. S., Wolfe, D. W. (1976). Drip application of nitrogen. *California Agri., 30(11),* 16–18.
9. Rajput, T. B. S., Patel, N. (2002). *Fertigation Theory and Practices.* Publication No. IARI/ WTC/2001/2 by Water Technology Centre, IARI, New Delhi, 56 pages.
10. Ree, W. O. (1977). How accurate are shop-made orifice plates. *Transactions of the ASAE, 20(2),* 298–300.
11. Replogle, J. A., Wahlin, B. (1994). Venturi meter constructions for plastic irrigation pipe lines. *Applied Engineering in Agriculture, 10(1),* 21–26.
12. Smith, M. W., Kenworthy, A. L., Bedford, C. L. (1979). The response of fruit trees to injection of nitrogen through trickle irrigation. *Jr. Amr. Soc. Hort. Sci., 104,* 311–313.
13. Thomas, J. T. (1985). Orifice plates for furrow flow measurement: Part 1: Calibration. *Transactions ASAE, 24(1),* 103–111.
14. Yuan, Z., Choi, C. Y., Waller, P. M., Colaizzi, P. (2000). Effects of liquid temperature and viscosity on venturi injectors. *Transactions of ASAE, 43(6),* 1441–1447.

APPENDIX I: INJECTORS FOR FERTIGATION

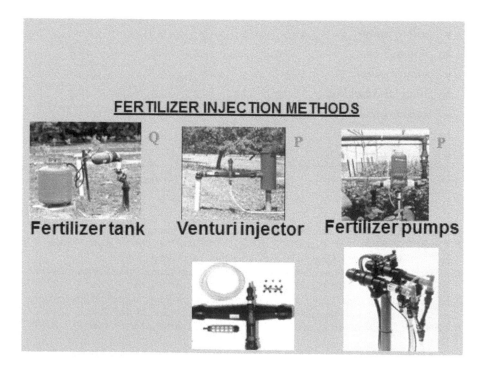

PART II
MICRO IRRIGATION SCHEDULING

CHAPTER 13

SIMULATION OF SALT DISTRIBUTION AND MOISTURE WETTING PATTERNS IN DRIP IRRIGATED TOMATO

D. D. NANGARE and KAMAL G. SINGH

13.1 INTRODUCTION

Water scarcity is becoming one of the major limiting factors to economic development and welfare in most parts of the semiarid regions of the world. In India, about one third of geographical area has been classified as arid and semiarid regions. The arid regions are mostly confined to western and northwestern Rajasthan and extend to some parts of the states of Haryana, Punjab and Gujarat. The entire arid and semiarid region is characterized by low rainfall and has the problems either of water scarcity or poor quality ground water. The regions identified for poor quality water are major parts of Rajasthan, Gujarat, Haryana, North Western UP and South Western parts of Punjab.

The use of alkali ground waters constitutes a major threat to irrigated agriculture in semiarid parts especially south Asia [66]. High incidence (30–50%) of these waters is found in semiarid parts (annual rainfall 500–700 mm), which are the most intensively cultivated areas in the Indo-Gangetic plains. Soil properties and permeability are adversely affected by irrigation with sodic water [71] in the arid and semiarid region. Poor quality water constitutes 32–84% of ground water surveyed in different parts of India is related either saline or alkali [65]. Farmers of these regions are compelled to use poor quality water to irrigate crops due to inadequate availability of good quality water. In southwestern Punjab, the quality of underground water is marginal and unfit for irrigation [18]. The 22% ground water is fit, 31% marginal and 47% water is unfit for irrigation due to poor quality. Brackish groundwater with high EC (0.2–12.6 dSm^{-1}) and RSC ranging from 0.3 to 35.1 mel^{-1} has been observed in this zone [45]. Saline water irrigation is practiced in several regions of the world [87], where water scarcity prevents the use of freshwater for irriga-

---

*In this chapter, the currency is expressed in Indian Rupees (1.00 US$ = Rs. 60.93; 1.00 Rs. = 0.02 US$).

tion. Utilization of saline water for irrigation is associated with salt accumulation in the soil, which might be harmful to plants, and reduces yields [7, 17, 56, 82, 94, 115]. The salt effects on physiological process result from lowering of the soil water potential and the toxicity of specific ions. On the other hand, it has been reported that nontoxic highly saline water has an agricultural potential. By using low quality water for irrigation, it would be possible to extend the normal growing season, thus targeting markets when the demand is especially high. The use of saline drainage water for irrigation has the environmental advantages of reducing the nonsaline water requirement for salt-tolerant crops and decreasing the volume of drainage water requiring disposal or treatment. When water resources are limited and the cost of nonsaline water becomes prohibitive, crops of moderate to high salt tolerance can be irrigated with saline water especially at later growth stages, provided appropriate irrigation methods and management practices are used. There are also a number of social and economic benefits that encourage the use of saline water for agricultural purposes. Saline water improved the welfare for local communities, as the use of a nonconventional water resource for irrigation will reduce the stress that exists on conventional freshwater supplies. Moreover, using saline water is a way to mitigate shortages of irrigation water within local communities and reduce conflict over water resources. Additionally, it might also allow irrigated agriculture to take place in areas where it is currently not possible due to the lack of good quality water.

In irrigated agriculture, any shift from freshwater to saline water irrigation incurs a yield reduction, but it enables more freshwater to be allocated for other purposes. The greatest risk for the farmer using saline water is that of salinization, which may lead to crop failure. The yield of crops irrigated with saline water could be enhanced substantially, if an additional source of good quality water were available for use at critical times during the season. The successful use of low-quality water requires the selection of salt tolerant crops, the application of a suitable water management strategy, the choice of the most appropriate irrigation systems, water management practices, irrigation system, soil type and salinity distribution affect crop productivity. There is a necessity of development of proper irrigation management practices so that poor quality water can be used in conjunction with good quality water with minimum adverse effect on crop yield. The irrigation water can be used as a mixture of saline water with fresh water (blending or mixing) or saline water can be applied in cycles with fresh water. The practical strategies include the mixing of water of varying quantity or cyclic use of fresh and saline water [86] and proper management and use of drip irrigation system [65].

The use of drip irrigation system has great potential in arid and semiarid regions particularly for light textured soils. Drip irrigation has been shown to be the most useful irrigation technique when irrigating with saline water [29] as it avoids the leaf injury to plants and improves the yield, WUE and quality of vegetables [108]. If irrigation can be managed in a way such that it provides high soil moisture content and consequently high soil water potential within the whole root zone then the osmotic effects will be damped. Moreover, when saline water is skillfully used for irrigation, it can be beneficial for agricultural production, particularly in fruits and vegetables. A regular and frequent water supply is possible with the drip system for crop production. Also the use of drip system raises the threshold limits of salt tolerance in crops by modifying the pattern of salt distribution and maintaining the higher matric potential in the root zone [64].

In India, crop yield is about doubled and water productivity increase ranges from 40 to 250%, when conventional irrigation is replaced with drip irrigation [76]. In addition, as competition for fresh water increases, water of better quality is primarily used for domestic purposes, whereas water of lower quality is often used for irrigation [51]. Therefore, one of challenge for the future was to maintain or even increase crop production with less water that may often be of poor quality, that is, saline water.

The level of salinity that can be tolerated in the soil water (hence in the irrigation water) will depend not only on the salt tolerance of the crop to be grown, but also on the initial salt content and distribution of salinity in the soil profile, amount and frequency of irrigation, the extent to which the soil water is depleted between irrigations and on the water content and hydraulic properties of the soil. In drip irrigation increased irrigation frequency typically results in a decreased depth of rooting and an increase in the mean salt concentration in the upper, main part of the root zone. Thus, the net result of increasing irrigation frequency may increase soil salinity and causes its deleterious effects on crop growth. The net overall effect of saline water irrigation on osmotic and matric-potentials is not easy to predict. This is an area of understanding that needs improvement. Additional research should be carried out to predict as to how much irrigation frequency can be increased to reduce salinity and water stresses on crop production. Studies on moisture and salt distribution under drip irrigation with different salinity levels and discharge rates of emitters and their influence on crop growth, yield and quality are also important. These studies are essential towards the development of improved water management technology in areas having problem of water scarcity and poor water quality. Therefore, there is a need to develop criteria for the use of saline water under drip irrigation with different mixing ratios of saline ground water and fresh canal water to improve quality of vegetables with minimum reduction in yield [17, 56, 82, 94, 115].

Keeping the above in view, this chapter discusses the research results to: (i) study the salt and moisture distribution patterns with different salinity levels of irrigation water under drip irrigation and its effect on yield and quality of tomato crop; (ii) simulate salt distribution, moisture wetting pattern and crop yield for tomato crop; and (iii) validate the model with experimental data.

13.2   LITERATURE REVIEW

13.2.1   DRIP IRRIGATION WITH SALINE WATER

Rhoades et al. [86, 87] recorded an increase in the quality of wheat, melon and al-falfa with the use of saline irrigation water. Shelf-life of fruits was not affected by saline treatments, whereas the yield and fruit size were found lower than fruits those from the control treatments. Further, increase in salinity increased the soluble solids concentration and slightly improved the appearance of muskmelons. Shannon and Grieve [97] examined the salinity problems which occur in irrigated agriculture and studied salt tolerance of fruits and vegetables, and cyclic reuse of saline water to determine the potential for growing leafy vegetables using relatively saline irrigation water. They found that crop yields were less affected by saline water if it was applied later in the season. In most cases vegetables grown at high salinity were found darker green, which would enhance their marketability. Irrigation with saline water with high in sulfates seemed to enhance the flavor of leafy cruciferous vegetables.

Amor et al. [8] suggested that that the response of melons to salinity depends on the duration of exposure to saline water. Salinity treatments increased fruit reducing sugars, acidity, and total soluble solids. Fruit yield reduction at each salinization time was correlated with salinity levels and concluded that brackish waters can be used for growing melon with minimum yield losses if concentration and duration of exposure are carefully monitored. Cuarter and Rafael [26] reviewed the effect of salinity on tomato plant growth, fruit production, cultural techniques which can be applied to alleviate the negative effects of salinity on tomato crop. They found that an increase of EC of irrigation water 1 dS/m resulted in reduction of about 9–10% yield. Salinity enhanced tomato fruit taste by increasing both acids and sugars. Firmness was found unchanged or slightly lowered. They reviewed in details effect of salinity on seed germination, root development, shoot development, and yield related characteristics and fruit quality under different salinity levels of irrigation water.

Cucci et al. [27] found that tomato yield generally decreases if a threshold salinity level is exceeded and further subsequent fruit deterioration in quality (increase in fiber content, size reduction, shorter storing period) causes economic damage. Salinity can induce some qualitative improvements such as better color and taste, increased soluble solids, reduced sugars with positive effects on dry matter and fruit cracks but got reduced yield and size of fruits. In Spain, Franco et al. [34] indicated that tomato yield per plant was found higher and fewer fruits were affected by blossom-end rot (BER) at the higher irrigation rate. No significant effects were found on macronutrient content in leaves and fruits. N concentration in the fruit was significantly increased in the water stressed plants. The Ca concentration in the stylar portion of mature fruits, which is related to the incidence of BER, was not significantly affected by irrigation rate. As regards micronutrients, only the Fe (in leaves and fruits of the first truss), Cu (in leaves of the first truss), Zn (in leaves and fruits of the first truss, and leaves of the fifth truss) and Mn (in leaves of the first truss) concentrations differed significantly. The total free amino acid leaf content was similar in both irrigation treatments. Pascale et al. [73] found that electrical conductivity of the saturated-soil extract (ECe) between 1.9 dS m^{-1} (treatment 0%) and 4.2 dS m^{-1} (treatment 1%), the marketable yield decreased by about 65% and the marketable root number m^{-2} by 57%. The N concentration was increased from 2.53 to 3.32 g 100 g^{-1} dry matter (leaf) and from 1.28 to 1.47 (root) with increased soil salinity. Na and Cl concentrations of leaf and root were doubled in plants grown on soil with the highest salinity. The threshold value was found 2.0 dS m^{-1} and yield was reduced at the rate of 28% per unit increase in soil salinity. The rate of yield reduction was higher than those reported in literature for sensitive crops.

Cuartero et al. (1999) found that salinity reduced tomato seed germination and lengthens the time needed for germination. A yield was reduced when plants were grown with a nutrient solution of 2.5 dS/m or higher. An increase of 1 dS/m above the 3.0 dS/m nutrient solution resulted reduction in yield of about 9–10%. At low EC, yield reduction was caused mainly by reduction in the average fruit weight. At high EC, yield reduction was caused mainly by declining number of fruits. Root growth, which slowed down when salinity reaches 4–6 dS/m. Salinity raised Na+ concentration in roots and leaves of tomato plants. Salinity enhanced tomato fruit taste by increasing both sugars and acids, fruit shelf life and firmness were unchanged or slightly lowered, but the incidence of blossom end rot was found much higher in salinity treatment. Pascale et al. [74] found that salinity and water stress limited plant gas exchanges and reduced growth, and increased the concentrations of Na$^+$ and Cl$^-$ ions in the plant's vegetative and reproductive tissues. However, the effects of salinity stress were found mostly apparent at higher salinity levels (0.5 and 1% NaCl), regardless of irrigation levels.

In South Portugal, Beltrao et al. [15] showed that for lower water salinity levels (from 1 up to 3 dS/m), cabbage yield decreased about 40% at the higher plant population density. However, at the lower plant density, cabbage yield decreased only about 20%. Yield decreased was found almost linearly with the increase of salinity; however, between 1 and 6 dS/m, yield reduction was found higher, as well as from 9.6 to 11 dS/m. Wang and Shanon [113] determined the characteristics of salt distributions under drip and sprinkler irrigation regimes and correlated these distributions and irrigation methods with salt concentrations in soybean plants. They observed that differences in irrigation methods affect plant salt uptake during irrigation with brackish saline water. The results showed that more salt accumulated in the soil profile in the drip than in the sprinkler irrigation plot. Under salinity treatment, higher concentrations of Na$^+$ were found in roots than in leaves or stems, and significantly higher concentrations of Ca^{2+} were found in plant leaves than in the stems or roots. Higher concentrations of Cl$^-$ were found in leaves and stems in sprinkler than drip irrigation plots, indicating strong foliar uptake of Cl$^-$. For irrigation with saline water, drip irrigation was found more suitable for salt sensitive plant species susceptible to foliar salt damage. Sprinkler irrigation can be used for more salt toler-ant plants and can create more leaching, leaving less residual salt in the soil profile at the end of the growing season.

Sharma [100] compared the surface methods with the high-energy pressurized irrigation methods and found sprinkler and drip are more efficient methods for ap-plication of marginal quality of water, as it can be adequately controlled. Sprinkler irrigation causes leaf burn but helps in uniform distribution of water on undulating soils with the increase in efficiency of salt leaching. The regular and frequent sup-ply is possible with drip irrigation, as it will not cause the leaf injury to plants with application of saline water. Drip irrigation enhances the threshold limits of salt tol-erance by modifying the pattern of salt distribution and maintenance of constantly higher matric potential. In Maharashtra – India, Kadam and Patel [48] found that the yield of tomato decreased with increased in saline water level from 0.21 dS/m to 9.5 dS/m, the reduction in yield ranged from 3.13 to 24%. The acidity and total soluble solid and lycopene contents were found increased with increased in saline water level. However, the pH of fruit juice decreased with increased water salinity. For young lemon trees at Jorden valley, Abu Awwad [1] found that the increase in irrigation water salinity by 3.7 times increased the crop root zone salinity by about 3.8–4.1 times. The higher salt concentration at soil surface was due to higher evapo-ration rates for wetted areas. The salt concentration followed bulb shape of wetted volume under trickle irrigation.

In Brazil, Amorin et al. [9] indicated that garlic plants were found relatively tol-erant to salinity at the bulb formation stage and initial growth up to 30 days of plant-ing (DAP). During the final stage (90–120 DAP), wetting of the leaves affected the growth of aerial parts and the number of garlic cloves. The salinity levels affected aerial parts during the period 30–60 DAP, while the bulb was affected only between 60–90 DAP. The most sensitive phase of bulb growth to salinity was observed the last 30 days of the crop cycle. Malash et al. [62] showed that salinity at 4 and 6 dS/m decreased total yield, marketable yield, number of fruits and average fruit weight in all genotypes of tomato. However, irrigation with saline water up to 6 dS/m in-creased fruit total soluble solids (TSS), vitamin C, and dry matter contents in all genotypes. Fruit acidity was also enhanced by salinity although 4 dS/m level gave the most acid fruits. Salinity had no significant effect on fruit firmness and storage ability. It was also observed that the reduction in yield in the salt tolerant genotypes, was mainly due to the reduction in fruit weight rather than fruit number.

Oron et al. [70] found reasonable yield of pear by using the saline water through subsurface drip irrigation (SDI). Moisture distribution under SDI was better ad-justed to root pattern in order to counteract osmotic effects of the soil salinity in comparison to commercial drip irrigation. Saline water particularly tends to increase sugar content and acidity of the Pear fruits using SDI. For a tomato crop (Marmande Raf) growing in an unheated plastic greenhouse, Restuccia et al. [85] showed that the total yield accumulated over the harvest periods was significantly higher for 1.6 dS/m than 6.0 dS/m and for 100% ETM+Lr than 100% ETM. The plants irrigated with both high saline concentration of water and high water supply (100% ETM+Lr), crop yield was found similar to that observed on plants grown under low salinity (1.6 dS/m) and 100% ETM but without considering Lr requirements. Fruit quality (dry matter content, reducing sugar concentration, and firmness) was improved with greater salinity, but was unaffected by water irrigation regime with the exception of dry matter. It was found that using an irrigation regime equivalent to 100% ETM+Lr reduces the negative influence induced by high salinity of the irrigation water on yield, while maintaining the beneficial effect on fruit quality.

For two sun-cured tobacco genotypes, Angelino et al. [10] indicated that the soil EC was increased with increase of the salinity of the irrigation water. The soil water content increased with increasing salinity and during the growth season. Increasing salinity progressively reduced the leaf turgor pressure and enhanced the cellular osmotic adjustment. In Italy for Sicilian winter melon, Incalcaterra et al. [44] found that the vegetative parameters at 60 and 70 days after transplanting (DAT) were slightly influenced by the treatments. At 80 DAT, plant length was positively affected by increasing amount of irrigation water, but was significantly reduced as salinity increased. Plants irrigated with 40 L of saline water recorded higher early and total yields, and fruits of better quality than those irrigated with lower volumes of good quality water.

Lei et al. [53] conducted the experiment on drip irrigation for watermelons with saline water in the Hetao Region, China and showed that the yield was increased and the quality improved under drip irrigation, as compared with control, with the highest increases in both yield and quality in the 60% treatment. The water production efficiency was 39.2 kg m^{-3}, respectively. The results suggested that drip irrigation of watermelon with saline water was feasible in the region. Lei et al. [54] studied the suitability of honeydew melons drip-irrigated with saline groundwater in Changsheng experimental Station in Nei Menggu (China). Results showed that the yield and the quality increased those under treatments of drip irrigation, as compared with control. The highest increase in yield was found in 60% Epan treatment. The water production efficiency was 18.0 kg/m^3 at 60% Epan. It was observed that a higher amount of salts were leached beyond the root zone under the treatments with 60 and 90% of evaporation, which suggested the feasibility of drip-irrigating honeydew melons can be grown in the saline soils with saline water.

Olympios et al. [69] found that salinity negatively affected the plant size and total fruit weight. There were a 20.3, 30.2 and 49.0% reduction in the yield, and 2.9, 12.2 and 20.1% reduction in the plant height as compared with the control for 3.7, 5.7 and 8.7 dS/m, respectively. In contrast, fruit number was significantly reduced only at 8.7 dS/m. The average fruit weight was reduced at highest salinity level especially when applied at an early growth stage. They observed that when good quality water was applied at the beginning of growth, followed later by salinity, the negative effect on plant height, shoot fresh and dry weights, leaf area, yield, average fruit weight and the percentage of fruits with BER was less severe. In contrast, when saline water was applied at the early stages of growth, followed by good quality water, the effect on plant parameters (total yield, average weight of fruits and blossom-end rot) was more severe. The longer was the duration of saline application, the more severe was the reduction in leaf area than the control. Increased salinity (various concentrations) increased soluble solids, Na, K, Cl contents of the fruits, irrespective of the time of application.

Sharma et al. [102] conducted an experiment in Bikaner, Rajasthan, India, during 1999–2002 on saline and BAW mixed irrigation indicated that the salinity of mixed water had significant effect on groundnut pod yield. The maximum average pod yield (42 per 100 kg/ha) was observed with canal water, while 34 per 100 kg/ha pod yield was obtained with BAW (ECiw 2.5 dS/m). Approximately, 18.3, 51.1, 59.7 and 79.3% reduction in groundnut pod yield was observed at ECiw 3.75, 5.0, 6.25 and 7.5 dS/m, respectively, compared to BAW. Similarly, kernel weight per plant and number of kernels per plant were reduced as the EC of the mixed water increased from 2.5 to 7.5 dS/m. For wheat, the maximum grain yield (36 per 100 kg/ha) was recorded with canal water. There was a reduction of 12.5, 22.9, 35.5 and 46.7% in wheat grain yield at ECiw 3.75, 5.0, 6.25 and 7.5 dS/m, respectively, compared to BAW. Plant height, ear length and number of tillers per plant reduced with increasing the salinity of mixed water. The soil EC also increased from 0.16 to 1.26 dS/m with an increase in the salinity of irrigation water after three years of rotation of experimentation, whereas there was little increase in pH.

Singh and Sanwal [107] tested the suitability of high frequency drip irrigation with the depths of irrigation as 25%, 50%, 75% and 100% of the maximum crop water requirement using saline/sodic drainage effluent as the irrigation water supply, and compared the results with traditional methods of surface irrigation for tomato using drip irrigation and traditional surface irrigation (rectangular check basin). High frequency drip irrigation gave better results with saline/sodic water, compared to other methods of irrigation. This can be attributed to its frequent water application and specialized salt and water distribution in the crop root zone. The highest crop yield of 22.05 t/ha was obtained for good quality canal water with daily drip irrigation of water application. The highest crop yield for surface irrigation was found 19.82 t/ha for good quality canal water with 82.7 cm depth of water application.

Yazar et al. [114] evaluated the maize yield and WUE in relation to salt concentration level of irrigation water applied with trickle irrigation in Turkey. Saline irrigation water with ECw of 3.0, 6.0, 9.0, and 12.0 dS/m along with canal water of 0.5 dS/m was used. In addition, three treatments were included by applying 10% leaching fraction to 0.5, 6.0, and 12.0 dS/m treatments after flowering. There was no significant difference in maize grain yield among the treatments. The highest yield of 8875 kg/ha was obtained in plots irrigated with canal water. In soil profile, salt concentration increased with increasing salinity of irrigation water. Higher salt concentration on the top layer was due to higher evaporation rate from the wetted surface. Applying a leaching fraction of 10% after flowering did not affect the salt distribution profile significantly in the treatments. There were no significant difference in dry matter production levels, WUE, 1000-grain weight, and harvest index among the salinity treatments. Gawad et al. [35] assessed the effects of management of crop production to obtain a better understanding of irrigation with saline water (8 ds/m) and role of crop tolerance to salinity using mixing and cyclic irrigation management with furrow and drip irrigation methods. It was found higher WUE with drip irrigation over traditional methods and higher sugar content in tomato fruit grown using saline irrigation water compared with that irrigated with nonsaline (1ds/m) water.

Malash et al. [61] studied the performance of two water management strategies, that is, alternate and mixed supply of fresh (canal water 0.5 ds/m) and saline / drainage water (4.2–4.8 ds/m) applied through drip and furrow method on tomato yield and growth in Nile Delta, Egypt. The mixed water management practice gave higher growth and yield than alternate irrigation. Sharma and Minhas [101] gave the strategies for managing saline/alkali water for sustainable agricultural production in South Asia. Rui et al. [93] evaluated tomato rooting patterns, yield and fruit quality in a field trial under three irrigation regimes 0.6 (D1), 0.9 (D2) and 1.2 ETc (D3)] and three drip irrigation depths surface (R0), subsurface at 20 cm depth (R1) and subsurface at 40 cm depth (R2). The behavior of the root system in response to the irrigation treatments was evaluated using mini-rhizotrons installed between two plants, near the plant row. Root length intensity (La), length of the root per unit of mini-rhizotron surface area (cm/cm^2) was measured at four crop stages. It was found that most of the root system was concentrated in the top 40 cm of the soil profile for all treatments, where the root-length density was found ranged from 0.5 cm/cm^3 to 1.4 cm/cm^3. The response of tomato fruits to an increase in the water applied was similar in quantitative and qualitative terms for the different drip irrigation depths. Water applied by drip irrigation had the opposite effect on commercial yield and soluble solids (°Brix), however, yield in terms of total soluble solids was the same for the 0.9 and 1.2 ETc.

Badr and Taalab [14] found that the yield reduction due to saline water use was minimal under 2 l/h, which gave better water and salt distribution in the root zone. The growth and yield performance of tomatoes irrigated through subsurface drip was found lower when compared with surface drip irrigation. The maximum tomato yield was obtained under surface drip irrigation applied at 2 l/h while the lowest yield was observed when water applied at 8 l/h with subsurface irrigation. Tomato production was found almost 16.4% lower with the highest discharge rate compared with the lowest rate while the yield reduction was found more pronounced with subsurface drip irrigation and found that 24.7% lower than with surface drip irrigation. Chauhan et al. [21] evaluated the response of potato (*Solanum tuberosum*), sunflower (*Helianthus annus*) and Sesbania (*Sesbania sesban*) to green manure and the combined use of good quality canal water (CW, ECcw 1.1 dS/m, RSC nil, SAR 1.8) and an alkali water (AW, ECaw 3.6 dS/m, RSC 15.8 me/L, SAR 12.4) on a well-drained sandy loam soil (ECe 2.5 dS/m, pH 7.9, exchangeable sodium percentage, ESP 5.3). They found that increase in soil pH (8.9–9.1), salinity (4.7–5.1 dS/m) and sodicity (ESP 25–41) as a consequence of irrigation with alkali water affected the growth and yields of all crops. The sustainability yield index (SYI) when irrigated with AW was 0.063 and 0.133 for potato and sunflower, respectively, indicating that these crops should not be irrigated with such high alkalinity waters. The SYI of potato improved to 0.703, 0.642, 0.442 and 0.579, respectively, with the cyclic 1CW:1AW, 2CW:2AW, 2AW:2CW and CWp:AWs treatments. The values of SYI were found 0.633 and 0.415 for potato when irrigated with blends of CW and AW in the ratio 2:1(2CW:1AW) and 1:2 (1CW:2AW), respectively. The SYI for sunflower was found ranged between 0.481–0.736 and 0.512–0.592 for cyclic use and blending but was reduced to 0.394 with 4AW:2CW in cyclic use mode. The relative yields was found (compared to CW) ranged between 65–85 and 61–94% for cyclic use and 66–83 and 71–81% for blended waters in potato and sunflower, respectively. In terms of potato grade, considerable deterioration in produce quality was observed. Weight loss on storage as well as the smaller seeds and lower oil content was observed in sunflower [22].

Dehghanisanij et al. [30] conducted field experiments using saline water for maize at the Arid Land Research Center, Tottori University, Japan, Variables affecting soil water content (theta w) and soil solute salinity (ECw), including time after irrigation and crop growth stages; early (ES), mid (MS) and late (LS) stages, were investigated at different radius from the emitter (lateral pipe) and found that theta w and ECw increased in the order: ES<MS<LS. The interaction between theta w and ECw for a given radius and different crop growing stages showed that the influence of theta w on ECw was restricted to a small radius of approximately 20 cm from the emitter, which decreased further to only 10 cm during LS. Beyond this range, increasing theta w was not enough to significantly affect ECw. The information obtained from this research was essential for the design, operation and management of saline water use with drip irrigation system in sand dune fields. Kadam et al. [47] conducted a study in Maharashtra, India to evaluate the effects of saline water (S_0, best available canal water, EC 0.21 dS/m; and S1, saline water, EC 2.0 dS/m); and fertigation (100, 75 and 50% recommended urea applied) through drip irrigation on the yield and yield components of eggplant. N content in the plant and fruit at harvest was found highest under S_0 + F1 (1.60 and 2.81%). P content was found highest in plant and fruit under S_0 and F1 (0.46 and 0.77%, respectively). K content was higher under S_0 (2.46 and 3.56%) than under S_0 (2.87 and 3.96%) in the plant and fruit, respectively. K content observed decreased with decreasing fertilizer level.

At the Hatfield Experimental Farm in Pretoria – South Africa, Karlberg et al. [49] used two low-cost drip irrigation systems with different emitter discharge rates (0.2 and 2.5 l h⁻¹) to irrigate tomatoes with three different salinity levels (0, 3 and 6 dS m⁻¹) of water along with plastic mulch to minimize soil evaporation was compared to a "bare soil" or uncovered treatment. An average yield of 75 Mg.ha⁻¹ was recorded for all treatments and seasons, which can be compared with the average marketable yield for South Africa of approximately 31.4 Mg ha^{-1}. Even at the highest irrigation water salinity (6 dS/m), a yield above the average marketable yield was achieved, indicating that low-cost drip irrigation works well in combination with saline water. The study showed that the choice of drip irrigation system with regard to discharge rate was of minor importance when irrigating with saline water. However, combining low-cost drip irrigation with plastic mulch increased the yield on an average 10 Mg ha^{-1} for all treatments. Nagaz et al. [68] conducted field study to determine the effect of irrigation regimes with saline water (3.25 ds/m) on soil salinity, yield and WUE of potato in southern Tunisia. Highest EC values were found at about 20 and 10 cm from emitters, respectively for 100 and 40 L. The yields were highest under 100 L although no significant difference was observed with daily irrigation regime at 100%. Full irrigation with daily application resulted in lowest WUE values because of highest evaporative losses.

Badr et al. [13] compared the effect of drip irrigation over other irrigation methods with fresh and saline water on the performance of field-grown cantaloupe. The fresh water was applied at crop water requirement, 1.0 ETc (as a control) and saline water (3.8 dS/m) applied at amounts equivalent to 1.0 ETc, 1.2 ETc and 1.4 ETc, respectively. Saline water significantly depressed cantaloupe total yield but the reduction was minimal under 1.4ETc irrigation regime. Total yield with saline water was almost 18–32% lower than with fresh water but offered several benefits as the absolute exportable yield was equaled to that of the control but the export rate was observed 90% versus 72%, respectively. However, saline water provided an attractive compromise between fruit size and quality. Plants irrigated with saline water were extremely high export rates. Saline water contributed markedly to the improvement of fruit quality by increasing total soluble solids and sugar contents. The use of mild saline water for irrigation is an attractive approach to optimize cantaloupe production with taking advantage of saline water effects on crop quality. Malash et al. [60] studied the use of nonconventional water resources including saline drainage water, brackish groundwater and treated wastewater to irrigate tomato using drip and furrow irrigation systems. They found that the growth parameters, yield, and WUE were greater for drip irrigated tomato plants than furrow-irrigated plants. However, furrow irrigation produced higher individual fruit weight. The electrical conductivity of the soil solution (extracted 48 h after irrigation) showed greater fluctuations when cyclic water management was used compared to those plots irrigated with blended water. Soil moisture was at a minimum in the root zone (20–40 cm layer), but showed a gradual increase at 40–60 and 60–90 cm and was stable at 90–120 cm depth. Soil water content was decreased gradually as the distance from the irrigation water source increased.

Harbi et al. [41] investigated the effects of water quality and subsurface drip irrigation management on the growth and yield of tomato under greenhouse conditions. Two water qualities (fresh and saline water with EC of 1.2 and 4.7 dS m^{-1}, respectively) along with two different irrigation frequencies (three and six days/ week) and three irrigation rates (2, 4 and 6 lph) were used. It was found that the saline irrigation water having an EC of 4.7 dS/m significantly decreased average fruit weight and total yield which was decreased by 62.3 and 24.3%, respectively, compared to irrigation with fresh water having an EC of 0.5 dS/m. Irrigating tomato using subsurface drip irrigation for six days/week significantly stimulated number of leaves, enhanced fruit size and increased total yield 26.4% (from 98 to 133 t/ha) compared to irrigation for three days/week. No significant differences were found in total yield when irrigated with rate of 4 and 6 lph. The irrigation with fresh water using subsurface drip irrigation for six days week with a rate of 6 lph gave the best results in respect to total yield.

13.2.2   SALT AND MOISTURE DISTRIBUTION IN DRIP IRRIGATION WITH SALINE WATER

Shalhevet [96] studied major issues regarding the use of marginal quality water for crop production. In irrigation requirement, crop water production function-relating yields to evapotranspiration are not influenced by water salinity and reduction of water uptake with increasing salinity is the result of reduction of growth. Singh et al. [104] developed a simulation model for assessing water quality to judge its suitabil- ity for irrigation. The model, which classifies the water, determines the potential of water for direct application and also evaluates management strategies based on con- junctive use of canal and saline water. This developed model is useful for predicting irrigation schedules as well as crop root zone salinity on land irrigated with saline water. Coelha and Ore [24] concluded that analytical solution for study and transient flow conditions predicted reasonably well measured distribution of water content and matrix potential in the absence of plants and that the study flow conditions were seldom realized under practical drip irrigation. The transient flow solutions provide more realistic distribution of soil water dynamics and introducing uptake terms to transient analytical solutions provided a simple and effective framework for captur- ing soil water dynamics in presence of pant roots [106].

Philip [75] studied the effect of root water extraction on wetted regions from continuous trickle sources. A model was introduced with moisture dependent root water extraction rate, so that the spatial distribution of uptake is no longer a datum but emerges as part of the solution. For a diversity of conditions, he obtained simple steady solutions describing interactions between continuous sources and regions of wetting and water extraction. It was shown that as root uptake increases, the region decreases markedly in size and becomes more symmetrical in the vertical; for sources at depths that are small relative to the sorptive length, interaction with the surface becomes important; surface evaporation losses significantly reduce the region and increase asymmetry; with nonevaporating surfaces, the region is larger and more symmetrical. He also observed that in wet regions close to the source we may expect that root density is large and that roots take up water readily, while the converse is true for the drier regions far from the source. Jinguan et al. [46] developed root water extraction model to incorporate the effect of soil water deficit and plant root distribution on plant transpiration of annual crops. They used ratio of actual to potential cumulative transpiration to determine leaf area index at optimum soil water conditions. The developed numerical model was applied to simulate soil water movement with root water uptake and found that simulated soil matrix potential, soil water content and cumulative evapotranspiration had reasonable agreement with measured data. Khumoetsile and Ore [52] reviewed dynamics of water uptake and solute in-situ soil with introducing nutrient in the root zone with different irrigation methods and quality of irrigation water.

Lubana et al. [57, 106] presented an extensive review of research on modeling of various processes associated with moisture distribution patterns under point source trickle emitters. The review promoted better understanding to facilitate more rational analysis of soil water dynamics process. He presented review pertaining to the modeling of infiltration, wetted soil volume in trickle irrigation, emitter spacing, moisture distribution pattern and plant water uptake, which is very useful for further research study.

Fares et al. [32] used a water flow and solute transport numerical model, HYDRUS-2D, to demonstrate the performance of drip irrigation with three different soil types. Results showed that with sandy soil, the waterfront moves vertically; with loam and clay soils, water front movement was a multidirectional process. Compared to sands, the same drip system can cover twice and 1.5 times as much horizontal area in clay and loamy soils, respectively. It was found that drip irrigation systems work better with loamy and clay soils than sandy soils.

Assouline [12] compared three-emitter discharges 0.25, 2 and 8 lph on different aspects of water regime for corn drip irrigated daily. He observed that drip irrigation tends to increase the yield and higher relative water content values at 0–30 cm soil depth. The relative yield was found statistically insignificant under field conditions. Numerical solution with HYDRUS-2D showed that micro drip irrigation leads to smallest wetted volume with the less extreme water gradients both in horizontal and vertical directions. The driest profile was obtained at lowest application rates compared to higher application rates.

LuDian et al. [58] studied infiltration simulation to study water movement and solute transport in a film drip irrigation system as affected by dripping rate, accumulated infiltration, initial soil moisture content, initial soil salt content, etc. Results showed that water content around the dripper increased with increasing dripping rate, which deferred the formation of a desalinization zone where crops can grow. It was also found that an increase in accumulated infiltration increased the volume of wetted soil and favored the formation of a desalinization zone. The increase in initial soil salt content decreased the desalinization coefficient. This study provided useful data for the design of a film drip irrigation system. Ragab [77] developed SALTMED model, which forms an integrated approach that accounts for water, crop/soil and field managements. The developed model can be used for a variety of irrigation systems, soil types, soil stratifications, crops and trees, water management strategies (blending or cyclic), leaching requirements and water quality. Water and solute transport, evapotranspiration and crop water uptake phenomenon were considered in developing the model. The model successfully illustrated the effects of the irrigation system, the soil type, the salinity level of irrigation water on soil moisture and salinity distribution, leaching requirements, and crop yield. Claire et al. [23] highlighted the impacts of soil properties on water and solute transport from buried trickle emitters. They analyzed effects of soil hydraulic properties, soil layering, trickle discharge rate, irrigation frequency, and timing of nutrient application on wetting patterns and solute distribution; and showed that trickle irrigation can improve plant water availability in medium and low permeability fine-textured soils, providing that design and management are adapted to account for their soil hydraulic properties. The results demonstrated the need to account for differences in soil hydraulic properties and solute transport, which will help in designing irrigation and fertigation management strategies.

Cote et al. [25] incorporated soil water and solute transport properties and soil profile characteristics in the design and management of trickle systems. The analysis addresses the influence of soil hydraulic properties, soil layering, trickle discharge rate, irrigation frequency, and timing of nutrient application on wetting patterns and solute distribution. It was observed that: (1) trickle irrigation can improve plant water availability in medium and low permeability fine-textured soils, providing that design and management are adapted to account for their soil hydraulic properties; (2) in highly permeable coarse-textured soils, water and nutrients move quickly downwards from the emitter, making it difficult to wet the near surface zone if emitters are buried too deep; and (3) changing the fertigation strategy for highly permeable coarse-textured soils to apply nutrients at the beginning of an irrigation cycle can maintain larger amounts of nutrient near to and above the emitter, thereby making them less susceptible to leaching losses. Feng et al. [33] used computer simulation model to evaluate the consequences of different management strategies on crop yield and salt distribution in the soil profile. The simulated salt distribution from ENVIROGRO model was compared with measured distribution at the end of growing season from an experiment having different EC of irrigation water and irrigation interval. The agreement between the measured and simulated salt distribution were found better for larger irrigation intervals than shorter ones. Huang et al. [43] studied water and salt movement in soil at Takelamakan desert of China, after irrigation with saline groundwater with an EC value of 6.3 mS cm^{-1}. They found that EC value of the soil solution reduced obviously in the upper part of the layer and was almost the same as that of before irrigation in the lower part of the layer. Twenty-four hours after irrigation, the amount of salt contained in the soil solution in 0–150 cm layer was less than that of before irrigation, while 144 h after irrigation, it was found only 53.46% of that of before irrigation.

Li et al. [55, 56] used HYDRUS-2D software for water and nitrate transport in soil from a point source of ammonium nitrate. In the model, simulated wetting dimensions of soil volume and distributions of water content and nitrate concentrations in soil were compared with data obtained from laboratory experiments conducted on a loam and a sandy soil. An excellent agreement was obtained between the simulated results and the measured data. Then, the verified model was used to simulate water and nitrate distributions for various initial conditions and fertigation strategies. They suggested strategies of first applying water for one-fourth of the total irrigation time, then applying fertilizer solution for one-half of the total irrigation time, followed by applying water for the remaining one-fourth of the total irrigation time will useful for keeping the most nitrate close to the source. Ragab et al. [78, 79] calibrated SALTMED model in Egypt and Syria using 100% fresh water treatment with the drip irrigation. The calibration was focused on yield prediction instead of calibration against water uptake/transpiration and soil moisture and salinity contents as yield finally depends on water uptake and moisture distribution. The model proved its ability to handle several hydrodynamic processes acting at same time. The model successfully predicted the impact of salinity on yield, water uptake, soil moisture and salinity distribution in the Floradade variety of tomato, which is salt tolerant and grown in the Mediterranean region. It was found that the relation between both yield and water uptake as a function of salinity of irrigation water was nonlinear and was a polynomial function of the fourth order. The model gave better research tool to field mangers to manage their water, crop and soil in an effective way to save water and protect the environment.

Razuri et al. [83] modeled soil salinity in a loamy soil in Valley of Quibor, Venezuela, using the prediction model LEACHM. The experiment was conducted in a heavy fine, isohyperthermic Typic Haplocambids soil, having an electric conductivity (EC) of 6.5 dS/m, with dominancy of calcium sulfates. The saline water classified as C3S1 was applied with drip irrigation with high frequency of application. The experimental results showed that maintaining soil water near to field capacity minimizes the EC in the crop root zone by means of the prediction model LEACHM.

Ben Asher et al. [16] evaluated soil water atmosphere and plant (SWAP) and SWAPd model for various salinity of irrigation water levels for grapevines under arid conditions in drip irrigation. They tested salinity treatments of 1.8, 3.3 and 4.8 dS/m on production parameters of grapevines. SWAP simulated higher water contents than SWAPd. Singh et al. [105] simulated soil wetting depth and width under subsurface drip irrigation with line source of water application. They found no significant difference between predicted and observed values of wetting width and depth. Karlberg et al. [50] developed a drip-irrigation module and included in an ecosystem model and tested on two independent datasets, spring and autumn, on field grown tomato. Simulated soil evaporation correlated well with measurements for spring (2.62 mm.d^{-1} compared to 2.60 mm.d^{-1}). Changes in soil water content were less well portrayed by the model (spring $r^2 = 0.27$; autumn $r^2 = 0.45$). Simulations showed that saline water irrigation (6 dS m^{-1}) caused reduced transpiration, which led to higher drainage and soil evaporation, compared with fresh water. They also found that covering the soil with plastic mulch resulted in an increase in yield and transpiration. The simulations indicated that the emitter discharge rate did not have any impact on the partitioning of the incoming water to the system. The model proved to be a useful tool for evaluating the importance of specific management options.

Ouda et al. [72, 94] used a "wheat-Stress" model to predict wheat yields under the following hypothetical situations: (1) reducing the amount of irrigation water by 20%; (2) using saline water for irrigation. Results showed that the model prediction gave good agreement with actual yield, for both varieties, that is, Sakha 93 and Giza 168 over the two growing seasons. Results indicated that the yield of both varieties was reduced under deducting 20% of irrigation water by 8.62 and 8.76% for Sakha 93 and Giza 168, respectively. Furthermore, using saline water for irrigation reduced wheat yield by 4.14 and 4.38% for Sakha 93 and Giza 168, respectively. Water consumptive use under total irrigation was reduced by 18.54 and 11.64% for both growing seasons, respectively, under saving 20% of total water irrigation. Whereas under saline water irrigation, water consumptive use was reduced by 15.02 and 11.93% for both growing seasons, respectively.

Raine et al. [80] studied water and associated solute movement within the root zone under precision irrigation applications, in Australia. They estimated that 10% of the irrigated land area (producing as much as 40% of the total annual revenue from irrigated land) can be adversely affected by root zone salinity resulting from the adoption of precision irrigation. A review of soil–water and solute movement under precision irrigation systems highlights the gaps in current knowledge including the mismatch between the data required by complex, process-based soil–water or solute simulation models and the data that is easily available from soil survey and routine soil analyzes. Other major knowledge gaps identified were: (a) effect of root distribution, surface evaporation and plant transpiration on soil wetted patterns; (b) accuracy and adequacy of using simple mean values of root zone soil salinity levels to estimate the effect of salt on the plant; (c) fate of solutes during a single irrigation and during multiple irrigation cycles; and (d) effect of soil heterogeneity on the distribution of water and solutes in relation to placement of water. Opportunities for research investment wee identified across a broad range of areas including: (a) requirements for soil characterization; (b) irrigation management effects; (c) agronomic responses to variable water and salt distributions in the root zone; (d) potential to scale or evaluate impacts at various scales; (e) requirements for simplified soil–water and solute modeling tools; and (f) the need to build skills and capacity in soil–water and solute modeling.

Wallender et al. [89, 110, 111] developed a flow and reactive-salt conceptual model for drip irrigated table grapes for understanding the complexity in the processes involved in the system as well as to provide management guidelines for farmers, district water management personnel and consultants. The developed model predicted the distribution of salts along the vine row and between the rows during the growing season, as affected by reactivity of salts of the applied irrigation water as well as rate and duration of drip application. The model predicted remarkable agreement between measured and simulated salinity. The developed model can predict soil water quantity and quality outcomes for possible land and water management scenarios.

Wang Dan et al. [112] investigated distribution characteristics of different salt ions in the soil under drip irrigation with saline water in North China. The experiment included five treatments, in which controlled mineralization degrees of irrigation water were 1.1 dS/m(K1), 2.2 dS/m(K2), 2.9 dS/m(K3), 3.5 dS/m(K4) and 4.2 dS/m(K5), respectively. It was found that the transferring velocity and distribution characteristic of salt ions were affected by water distribution in the soil under drip irrigation condition. Ca^{2+} Mg^{2+} and SO_4^{2-} were easily leached by irrigation water and distributed mainly outside the moist volume, while HCO_3-, Na^+ and Cl^- mainly distributed in the moist volume. During the growing period, HCO_3 , Na^+ and Cl^- increased slightly though the total salt content within the soil profile did not increase.

Zhou et al. [116] studied a two-dimensional (2D) root water uptake model based on soil water dynamic and root distribution of grape vine. The soil water dynamic model of APDI (APRI-model) was developed based on the 2D root water uptake model and soil evaporation function combined with average measured soil moisture content at 0–10 cm soil layer. Soil water dynamic in APDI was simulated by Hydrus-2D model and APRI-model. The simulated soil water contents by two models were compared with the measured value. The results showed that the values of root-mean-square-error (RMSE) range from 0.01 to 0.022 cm3/cm3 for APRI-model, and from 0.012 to 0.031 cm^3/cm^3 for Hydrus-2D model. The average relative error between the simulated and measured soil water content is about 10% for APRI-model, and from 11% to 29% for Hydrus-2D model, indicating that two models performed well in simulating soil moisture dynamic under the APDI. The APRI model was more suitable for modeling the soil water dynamics in the arid region with greater soil evaporation and uneven root distribution.

Ainechee et al. [5] investigated water movement in three soil types from a point source. They found that the surface wetted radius increased with an increase in application rate. A good relationship was found between the surface wetted radius and the volume of water applied. Predictability of model was 96.8 and 95.3%, respectively, for prediction of wetted width and depth. The results showed good agreement for all application rates. Golabi et al. [37] verified SALTMED model in irrigation management in semiarid area of south west of Iran to predict soil salinity changes due to irrigation water quality in sugarcane. They compared simulation results with actual results using maximum error, root mean square error, coefficient of determination, modeling efficiency, coefficient of residual mass and ANOVA table. The results indicated that there was no significant difference between simulation and measurement data. The overall results of this research confirmed that SALTMED model could be used successfully as tool for irrigation and drainage of sugarcane farms in Khuzestan province of Iran.

Roberts et al. [88] conducted study at the University of Arizona Maricopa Agricultural Center for predicting salt movement and accumulation with SDI using HYDRUS-2D model for crops of cantaloupe, melon and broccoli with two tape depths (18 and 25 cm), two different germination practices (germination with SDI and with sprinklers) along with two water salinity (1.5 and 2.6 dS/m). Predicted saturated paste electrical conductivity (ECe) values from HYDRUS-2D were significantly correlated with actual ECe data obtained from field experiments (r^2 = 0.08 to 0.93). After season one, the correlation coefficients were found highly variable, with the majority of model ECe values being higher than field data. Second season results indicated a much stronger relationship, with R^2 values gave as high as 0.93. Model predictions for season two showed under prediction of ECe when compared with actual ECe. Relationships between predicted ECe and actual ECe resulted in a slope of nearly 1.0 for all treatments and y-intercept was −1 dS/m.

The review of the past work done on drip irrigation with saline water found major gap in generating information on influence of discharge rate of emitters, irrigation levels and blending of fresh and saline water on growth and quality of fruits and vegetables for the semiarid regions of Punjab. Also, soil moisture and salinity distribution resulting from these influences have not been established under drip irrigation with saline water for tomato crop in the region.

Therefore, the research study in this chapter was conducted with a view to simulate salt and moisture distribution; and to study its effect on yield and quality of drip irrigated tomato with saline water.

13.3 MATERIAL AND METHODS

13.3.1 LOCATION AND CLIMATE

A field experiment was conducted at research farm of Central Institute of Post-Harvest Engineering and Technology (CIPHET), Abohar – Punjab – India during December 2008 to June 2009. Abohar is located at south-western part of Punjab with latitude of 30°4′ N and longitude of 74°21′ E and with mean sea level of 185 m. The climate of Abohar is semiarid with hot summer and cold winter. The average annual rainfall is 300–400 mm.

13.3.2 SOIL CHARACTERISTICS

The soil samples were taken 0–15, 15–30, 30–45, 45–60 and 60–90 cm depth for determining the soil characteristics. The soil at the experimental site is sandy loam. The physicochemical properties of soil at the research site are given in Table 13.1.

TABLE 13.1 Physio-Chemical Properties of Soil At Experimental Site

Depth cm	Sand %	Silt %	Clay %	EC dS/m	pH	Bulk density g/cm³
0–15	76.57	8.02	15.41	0.15	8.52	1.69
15–30	77.92	7.69	15.39	0.13	8.63	1.71
30–45	78.21	7.35	14.44	0.13	8.70	1.82
45–60	76.96	8.36	14.68	0.14	8.70	1.79
60–90	78.28	8.18	13.54	0.15	8.63	1.76

13.3.2.1   SOIL MOISTURE RETENTION CURVE

The moisture retention curves were developed for the soil at 0–30, 30–60 and 60–90 cm depth with pressure plate apparatus. The pressure plate apparatus run for the pressure ranging from 1/3 to 15 atmospheres. The developed moisture retention curves are shown in Fig. 13.1. The field capacity (1/3 atm) and wilting point (15 atm) of soil at 0–30 cm depth were 13.30% and 3.52% on dry weight basis, respectively.

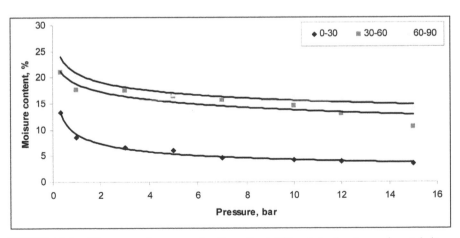

FIGURE 13.1   Water retention curves for the soils at different depths at experimental site.

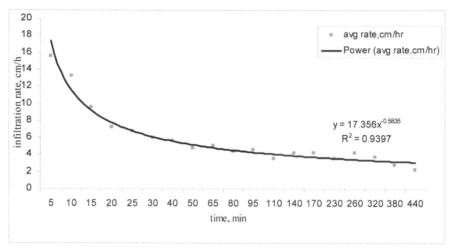

FIGURE 13.2 Infiltration rate of soil at experimental site.

13.3.2.2 BULK DENSITY

The soil bulk density was determined by core cutter method. The bulk densities of soil at different depths are presented in Table 13.1.

13.3.2.3 INFILTRATION RATE OF SOIL

The soil infiltration rate was determined by using double ring cylinder infiltrometer. The infiltration rate of soil against time is depicted in Fig. 13.2. The basic infiltration rate of soil at experimental site was 2.43 cm/hr. The saturated hydraulic conductivity for sandy loam soil was determined by auger-hole method and was found 620 mm/day. The saturated and residual moisture content for sandy loam soil were found 0.41% and 0.065%, respectively. The K-q and D-q relations developed for sandy loam soil at experimental site and are given in Table 13.2.

TABLE 13.2 Soil Hydraulic Characteristics at Experimental Site

Depth	Diffusivity	Hydraulic conductivity
0–30 cm	$D(q) = 13.53 \exp(12.84*q)$	$K(q) = 1.5E-5 \exp(34*q)$
30–60 cm	$D(q) = 12.65 \exp(11.64*q)$	$K(q) = 1.3E-5 \exp(33.6*q)$
60–90 cm	$D(q) = 14.12 \exp(13.06*q)$	$K(q) = 1.6E-5 \exp(34.1*q)$

13.3.3 EXPERIMENTAL LAYOUT

The drip irrigation system was installed at research farm of Central Institute of Post Harvest Engineering and Technology (CIPHET), Abohar (Punjab). The total area of the experimental site was 60×30 m^2. The area was divided into five subplots for irrigating crop with fresh and saline water with five different salinity levels, separately. Each plot was subdivided into nine lines with three replications (three discharges × three irrigation levels). Thus, each plot was 5 m^2. There were 135 combinations with three replications for the whole experiment. The treatment details of the experiment presented as follows:

13.3.3.1 DIFFERENT SALINITIES OF WATER

T1 100% fresh water.   T2 75% fresh + 25% saline water.
T3 50% fresh + 50% saline water.   T4 25% fresh + 75% saline water.
T5 100% saline water.

13.3.3.2 THREE DISCHARGE RATES OF INLINE EMITTERS

Q1 = 1.2 lph.   Q2 = 2.4 lph.   Q3 = 4.2 lph.
Three IW/CPE ratios: I1 = 0.6, I2 = 0.8, and I3 = 1.0

13.3.3.3 INSTALLATION OF DRIP IRRIGATION SYSTEM

The drip irrigation system was installed as per layout. Three inline dripper discharges were used for the study. After the installation of drip irrigation system (Fig. 13.3), it was tested for design discharge, uniformity of emitters and for clogging problem. At a pressure of 1 kg/cm^2 the average discharge per emitter was measured and the Christiansen uniformity coefficient was worked for analyzing the uniformity of emitter discharge. The Christiansen uniformity coefficient was calculated as given below:

$$Euc = \left(1 - \frac{\Delta \bar{q}}{\bar{q}}\right) \tag{1}$$

where: Euc = Christiansen uniformity coefficient; $\Delta \bar{q}$ = Mean absolute deviation of the emitter flow from the mean value (l/min); and $\bar{q}$ = Average discharge (l/min). The average uniformity coefficient was observed to be 98.65% at pressure of 1 kg/cm^2.

FIGURE 13.3   View of drip irrigation system at the site.

13.3.3.4   PREPARATION OF WATER/EXPERIMENTAL SETUP

The canal (fresh) and tube well (saline) water were both available at the experimental site. The fresh water was used from the water tank constructed at CIPHET, Abohar farm. The groundwater in the CIPHET farm was saline in nature. The tube well saline water was pumped for mixing as per given ratio with fresh water and made five different salinities of water. The view of experimental setup for mixing and storing of mixed water in storage tank are shown in Figs. 13.4 and 13.5. One tank of 2000-liters capacity was used for preparing mixture of fresh and saline water as per given ratio and then this mixed water was transferred to storage tank. Four storage tanks of 2000-liters capacity were put on the platform near the experimental site and these are used storing the water of different ratio. The separate valves were used for these tanks and these are connected to the inlet of pump. During the irrigation, only one valve of the storage tank was opened at a time for irrigating crop with given mixture of water. The characteristics of mixed irrigation water at different salinity ratios are given in Table 13.3.

TABLE 13.3 Characteristics of Water Used For Irrigation

Treatment	EC dS/m	pH	Co₃⁻	HCo₃⁻	Cl⁻	Ca²⁺	Mg²⁺	Na⁺	K⁺	SAR
					me/L					
100% F	0.38	7.51	Nil	2.00	1.00	1.60	0.80	1.60	0.17	1.46
75% F + 25% S	6.30	7.66	Nil	3.00	30.50	7.00	12.00	12.24	0.49	3.97
50% F + 50% S	9.10	7.77	Nil	3.00	46.50	9.20	19.80	14.40	0.64	3.78
25% F + 75% S	14.70	7.84	Traces	4.00	76.00	16.00	32.80	21.84	0.88	4.42
100% S	19.50	7.79	Traces	4.00	107.00	20.60	46.60	29.12	1.09	5.02

F = Fresh canal water, and S = Tube well saline water.

13.3.4 RAISING OF CROP

13.3.4.1 NURSERY PRODUCTION

The nursery of Golden seed hybrid tomato (GC1500) was raised in the poly house constructed at CIPHET, Abohar according to the package of practices of PAU, Ludhiana [11]. Tomato seeds were sown in the month of November 2008 after treating seeds with thiram @ 3 gm/kg. After one week, the plants were drenched with thiram @ 4 gm/L. Adequate moisture was maintained in the seed bed throughout growing period.

13.3.4.2 FIELD PREPARATION

The field operation like plowing and harrowing was done. The recommended dose of farm yard manure (FYM) was added in the soil. As per recommendation of PAU packaging practices for vegetable crops, NPK applied in band in form of urea and DAP (Di-ammonium Phosphate) before transplanting.

13.3.4.3 TRANSPLANTING

The tomato seedlings were transplanted in the field on 31st December of 2008. The row-to-row and plant-to-plant spacing was 1 m and 0.30 m, respectively. The tomato seedlings were covered with polyethylene sheets forming low tunnels to protect the crop from cold winter during January to mid-February 2009.

13.3.4.4   IRRIGATION APPLICATION

Fresh water was applied through drip irrigation to tomato crop. At early stage, fresh water with equal quantity applied to all plots. After establishing seedlings at early stage, the fresh water is applied to all plots as per three irrigation levels. To avoid the water stress to plants due to high salinity of water and heavy cold in early stage after transplanting, the mixed water treatment was started after 45 days. Treatment wise volume of irrigation water applied per plant through drip irrigation was estimated on the basis of spacing, pan evaporation, pan factor and crop coefficient. Based on discharge capacity of emitters, the drip system was operated for determined time to apply given volume of water per plant. In drip-irrigated treatments, water was applied for three different irrigation levels IW/CPE ratio of 0.60, 0.8 and 1.00. Drip irrigation water was applied after 10 mm cumulative pan evaporation. Total volume of water applied per plant was calculated [63] as below:

$$Vd = \frac{Ac \times CPE \times kp \times kc \times (IW/CPE) \times Aw}{Euc} \qquad (2)$$

where: Vd = volume of water applied per plant in drip irrigation system [l]; Ac = cropped area (m²) which is calculated by row-to-row spacing (m)x plant-to-plant spacing (m); CPE = cumulative pan evaporation (mm) after which the drip irrigation was applied; Kp = pan factor of 0.7; Kc = crop coefficient (the crop coefficient curves for tomato indicated in Fig. 13.6); IW/CPE = ratio of irrigation water to cumulative pan evaporation; Aw = fractional wetted area which was taken as 75% (0.75); and Euc = christiansen uniformity coefficient with Eq. (1).

FIGURE 13.4   View of experimental setup for mixing of saline and fresh water.

FIGURE 13.5   View of experimental setup for storing mixed water.

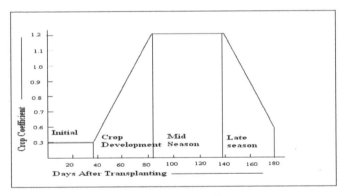

FIGURE 13.6   Crop coefficient curve for tomato [104].

The time of irrigation for operating drip system per application was calculated as given below:

$$T(drip) = \frac{Vd}{qe} \tag{3}$$

where: T (drip) = drip irrigation time, hours; Vd = volume of water applied per plant in drip irrigation system; and qe = average emitter discharge, lph. The irrigation duration in each treatment was determined. The starting and end time of operation of system in each treatment was recorded and it was used as input parameter in the model.

13.3.4.5   FERTIGATION

The 50% N was applied at the time of transplanting and remaining 50% N in the form of urea was fertigated in 10 splits.

13.3.4.6   CULTURAL PRACTICES

The weeding operation was carried out at every 30 days interval after transplanting. To protect the plants from harsh winter, the crop was covered with polyethylene sheet in the form low tunnel during evening/night hours (Fig. 13.7). When there was clear sunshine, these covers were removed during day time. To protect the plants from early blight disease due to high humidity inside plastic cover, alternate spray of M-45 and Bavistin was done at 10 days interval. When plant showed the late blight symptoms in the last week of March, the Redomil spray followed by 3 sprays of M-45 at weekly interval was done to protect the crop. When fruiting started, the fruits were protected from stem borer larva attack by spraying Endosulphon as per recommendation.

13.3.4.7   HARVESTING

The fruits were harvested from plants, when fruit color changed from orange to red. The pickings were done at 3–4 days interval.

13.3.5   EXPERIMENTAL OBSERVATIONS

13.3.5.1   SOIL MOISTURE DISTRIBUTION

The soil moisture distributions were studied in plots, where water was applied at IW/CPE ratio of 0.8 through three emitters. The soil moisture (m^3/m^3) at depth of 10, 20, 30, 40, 60 and 100 cm was determined by using Delta T device moisture meter (Fig. 13.8). The PR-1 access tubes were installed at 0, 10, 20 and 30 cm from the plant or emitter source in plots T1 to T5 (Fig. 13.8). The soil moisture distribution profile was studied at 30 days interval during growth period.

13.3.5.2   SALT DISTRIBUTION

The EC distribution was studied in plots where water was applied at IW/CPE ratio of 0.8. The soil sampling was done from 0, 10, 20 and 30 cm horizontally and at depth of 15, 30, 45, 60 and 75 cm. The soil sampling was done at 30 days interval to study the effects of different salinity of water application in soil profile. The EC was determined by using portable soil water analysis kit using the standard procedure.

13.3.5.3 GROWTH PARAMETERS

For comparing the plant performance, various biometric observations like plant height, ground cover were monitored at regular interval.

Plant Height: The five plants in each treatment were selected at random and height of plant was measured from base of plant to axil of the last unfolded leaf after 30 days interval.

Ground Cover: The area covered by the crop measured by counting squares of rectangular frame. By viewing the canopy of the plants through wooden frame and then counting the sections more than half filled with leaf, percent ground cover was calculated by counting number of sections.

FIGURE 13.7   View of tomato field with low tunnels during winter.

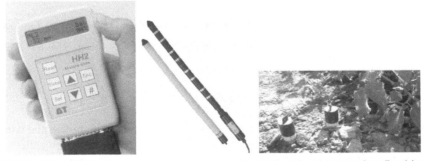

FIGURE 13.8   Delta-T device moisture meter with PR-1 probe; Access tubes fixed in the field near plant for measurement of soil moisture (Extreme right).

Plant Dry Matter Accumulation: The dry matter accumulation of plants was determined for each treatment. The three plants were cut from the surface and these were first air dried and then oven dried at 60°C to get constant weight. The dry matter accumulation of plants was determined at 30 days interval.

Tomato Yield: The three plants were selected for yield from each replication. The average yield after last picking was taken for each plot. The data on yield of all the plants of each treatment plot were recorded and total yield in t/ha worked out.

13.3.5.4   FRUIT QUALITY PARAMETERS

Total soluble solids for each treatment were determined by digital handheld refractometer (ERMA, made in Japan) with the scale of 0–32 °Brix and expressed as °Brix at room temperature. The three fully matured fruits were selected from each replicated treatment. Juice was extracted from these fruits and TSS was determined.

The acidity of fruits was determined for each treatment. The three fully matured fruits were selected from each replicated treatment and juice was extracted from these fruits. About 5 mL of this juice was taken and titrated against N/10 NaOH solution with phenolphthalein as indicator and pink color as end point. The volume of NaOH recorded and acidity was computed as follows [81]:

$$\text{Acidity (gm/100 ml of juice)} = \left(\frac{0.0064.X.100}{Y} \right) \tag{4}$$

where: X = mL of N/10 NaOH used; and Y = mL of sample taken for titration.

The ascorbic acid in tomato was determined by using 2–6 dichlorophenol-indophenol (dye) visual titration method, described by Ranganna [81]. The three fully matured fruits were selected from each treatment and juice was extracted from fruits. The ascorbic acid was determined by using the following formula.

Ascorbic acid = [(DF × volume make up × Titration reading × 100) ÷ (Sample weight × aliquote taken for estimation)]   (5)

where: DF = dye factor.

Firmness of Fruit: The textural properties like fruit firmness of matured tomato were determined using textural analyzer. The representative samples of fruits were selected from each treatment.

13.3.6   METEOROLOGICAL DATA

The meteorological data included: rainfall, temperature, relative humidity, and wind speed and pan evaporation during crop period (Table 13.4). The data were collected from mini observatory at CIPHET, Abohar and PAU regional fruit research station, Abohar.

TABLE 13.4   Monthly meteorological data at experimental site (2009).

Month	Average temperature		Pan evaporation	Average relative humidity	Rainfall
	Maximum	Minimum			
	°C		mm	%	mm
January	19.9	6.8	34.5	75.3	17.4
February	22.8	8.8	50.2	71.6	34.1
March	30.9	10.8	70.9	64.4	17.9
April	42.2	15.7	134.8	41.5	13.8
May	44.3	22.2	291.6	35.6	4.9
June	42.4	27	250.2	36.7	0

13.3.7 SIMULATION OF SALT AND MOISTURE DISTRIBUTION

The SALTMED model was used for simulating the data of the experiment [77]. Model predictions provide a glimpse into the future. SALTMED model offers an integrated water management scheme for crop, water and field using saline water. It is used to test the model performance and its ability to predict the final yield, plant water uptake, soil moisture and salinity profiles under different irrigation water salinity levels, water management strategies, irrigation systems.

13.3.7.1 THE BASIC EQUATIONS FOR SALTMED MODEL

The SALTMED model includes different key processes like: evapotranspiration, plant water uptake, water and solute transport under different irrigation systems, and drainage. A brief description of each process is presented below:

13.3.7.1.1 EVAPOTRANSPIRATION
The evapotranspiration has been calculated using the Penman–Monteith equation according to the modified version of FAO-56 [7], as below:

$$ET_o = \frac{0.408\Delta(R_n - G) + \gamma\dfrac{900}{T + 273}u_2(e_s - e_a)}{\Delta + \gamma(1 + 0.34u_2)} \tag{6}$$

where: ETo is the reference evapotranspiration, (mm); R_n is the net radiation, (MJ m^{-2} day^{-1}); G is the soil heat flux density, (MJ m^{-2} day^{-1}); T is the mean daily air temperature at 2 m height, (°C); Δ is the slope of the saturated vapor pressure curve (kPa C^{-1}); g is the psychometric constant, 66 Pa° C^{-1}; e_s is the saturated vapor pressure at air temperature (kPa); e_a is the prevailing vapor pressure (kPa); and u_2 is the wind speed at 2 m height (m s^{-1}). The crop evapotranspiration ETc is calculated as,

$$ETc = ETo \ (Kcb + Ke) \tag{7}$$

where: K_{cb} is the crop transpiration coefficient also known as basal crop coefficient; and K_e is the soil evaporation coefficient. The values of K_{cb} and K_c, for each growth stage and the duration of each growth stage for different crops, were taken from FAO-56. K_e is calculated according to FAO-56 [7], and K_{cb} and K_c are adjusted according to FAO-56 [7] for wind speed and relative humidity different from 2 m s^{-1} and 45%, respectively.

13.3.7.1.2   EFFECTIVE RAINFALL

The part of the rainfall that is available for infiltration through the soil surface is defined as effective rainfall. It is estimated in the model in three ways: (i) as a percentage of total rainfall; (ii) calculated according to the FAO-56 [7] procedure; and (iii) Taken to be equal to total rainfall. For the present study, the effective rainfall was taken to be 20% of total rainfall [31].

13.3.7.1.3   PLANT WATER UPTAKE IN THE PRESENCE OF SALINE WATER

The actual water uptake rate (S, d^{-1}) formula for the SALTMED model is given below:

$$S(z,t) = [\frac{S_{max}(t)}{1+(\frac{a(t)h+\pi}{\pi_{50}(t)})^3}]\lambda(z,t) \tag{8}$$

where: S (z, t)= water uptake mm/day; l(z) = 5/3L for $z{\leq}0.2$L; l(z) = 25/12L × (1−z/L) for 0.2L £ z£L; l(z) = 0.0 for $z > L$; $S_{max}(t)$ is the maximum potential root water uptake at the time t; z is the vertical depth taken positive downwards; l(z, t) is the depth and time dependent fraction of total root mass; L is the maximum rooting depth; h is the matric pressure head; and p is the osmotic pressure head. The p$_{50}(t)$ is time-dependent value of the osmotic pressure at which $S_{max}(t)$ is reduced by 50%, and $a(t)$ is a weighing coefficient that accounts for the differential response of a crop to matric and solute pressure. The coefficient $a(t)$ equals p$_{50}(t)$/$h_{50}(t)$, where $h_{50}(t)$ is the matric pressure at which $S_{max}(t)$ is reduced by 50%. These p$_{50}$ and h_{50} values were taken for tomato crop from FAO-48 [87].

13.3.7.1.4   MAXIMUM WATER UPTAKE

The maximum water uptake $S_{max}(t)$ is calculated as:

$$S_{max}(t) = ETo(t)*Kcb(t) \qquad (9)$$

13.3.7.1.5   ROOTING DEPTH

The rooting depth follows same pattern as the crop coefficient, Kc. It can be described by following equation [77]:

Root depth(t) = {[Root depth$_{min}$ + (root depth$_{max}$ _ root depth$_{min}$)] × Kc(t)}/Kc$_{max}$ (10)

The maximum root depth was obtained from the experiment.

13.3.7.2   RELATIVE AND ACTUAL CROP YIELD

There is a unique and strong relationship between water uptake and biomass production. Hence, the final yield, the relative crop yield (RY) is estimated as the sum of the actual water uptake over the season divided by the sum of the maximum water uptake under no stress condition as:

$$RY = \frac{\Sigma S(x,z,t)}{\Sigma S_{max}(x,z,t)} \qquad (11)$$

The actual yield AY was obtained by:

$$AY = RY*Y_{max} \qquad (12)$$

where: Y_{max} is the maximum yield obtainable in a given region under optimum and stress-free condition. We used the maximum yield obtained under fresh water treatment during simulation.

13.3.7.3   WATER AND SOLUTE FLOW

The water flow in soils is described mathematically by Richard's equation. It is a partial nonlinear differential equation in time and space. It is based on Darcy's law and mass continuity.

13.3.7.3.1   DARCY'S LAW

$$q = -K(h)\frac{\partial H}{\partial Z} \qquad (13)$$

where: q is the water flux; $K(h)$ is the hydraulic conductivity as a function of soil water pressure head; Z is the vertical coordinate directed downwards with its origin at soil surface; and H is the hydraulic head, which is the sum of the gravity head, z, and the pressure head, y:

$$H = \psi + z \tag{14}$$

13.3.7.3.2 RICHARD'S EQUATION

The vertical transient-state flow water in a stable and uniform segment of the root zone can be described by a modified Richard's equation as:

$$\frac{\partial \theta}{\partial t} = -\frac{\partial}{\partial z}\left[K(\theta)\frac{\partial(\psi + z)}{\partial z} \right] - S_w \tag{15}$$

where: q is volume wetness; t is the time; z is the depth; $K(q)$ is the hydraulic conductivity (a function of wetness); y is the matrix suction head; and S_w is the sink term representing extraction by plant roots. The movement of solute in the soil system, its rate and direction depends greatly on the path of water movement, but it is also determined by diffusion and hydrodynamic dispersion. By the combination of the diffusion, the dispersion and the convection, the overall flux of solute J can be obtained according to Hillel [42] as:

$$J = -(D_h + D_s)\frac{\partial \theta}{\partial x} + \bar{v}\theta C \tag{16}$$

where: C is the concentration of solute in the flowing water; $\bar{v}$ is the average velocity of the flow; D_s is the solute diffusion in soil which decreases due to the fact that the liquid phase occupies only a fraction of soil volume, and also due to the tortuous nature of the path. It can, therefore, be expressed according to the following equation, where D_0 is the diffusion coefficient:

$$D_s = D_0 \theta \xi \tag{17}$$

$$\xi = \frac{\theta^{7/3}}{\theta_s^{2}} \tag{18}$$

where: x is the tortuosity, an empirical factor smaller than unity, which can be expected to decrease with decreasing q [103]. The convection flux generally causes hydrodynamic dispersion too, an effect that depends on the microscopic non-uniformity of flow velocity in the various pores. Thus a sharp boundary between two miscible solutions becomes increasingly diffuse about the mean position of the front. For such a case, the diffusion coefficient has been found by Bresler [19] to depend linearly on the average flow velocity $\bar{v}$, as follows:

$$D_h = \alpha v \tag{19}$$

where: a is an empirical coefficient.

13.3.7.3.3 MASS TRANSFER EQUATION

If one takes the continuity equation into consideration, one-dimensional transient movement of a noninteracting solute in soil can be expressed as:

$$\frac{\partial(\theta C)}{\partial t} = \frac{\partial}{\partial Z}(D_a \frac{\partial C}{\partial Z}) - \frac{\partial(qC)}{\partial Z} - S_s \tag{20}$$

where: C is the concentration of the solute in the soil solution; q is the convective flux of the solution; D_a is a combined diffusion and dispersion coefficient; and S_s is a sink term for the solute representing root adsorption/uptake. Under irrigation from a trickle line source, the water and solute transport can be viewed as two-dimensional flow and simulated by a 'plane flow' model involving the Cartesian coordinates x and z, as emitters are close enough and equally spaced in tomato crop. Also their wetting pattern overlaps. For a stable, isotropic and homogeneous porous medium, the two dimensional flow of water in the soil can be described according to Bresler [19] as:

$$\frac{\partial \theta}{\partial t} = \frac{\partial}{\partial x}[K(\theta)\frac{\partial \psi}{\partial x}] + \frac{\partial}{\partial z}[K(\theta)\frac{\partial(\psi + z)}{\partial z}] \tag{21}$$

where: x is the horizontal coordinate; z is the vertical ordinate (considered to be positive downward); and K(q) is the hydraulic conductivity of the soil. Considering isotropic and homogeneous porous media with principal axes of dispersion oriented parallel and perpendicular to the mean direction of flow, the hydrodynamic dispersion coefficient D_{ij} can be defined as follows:

$$D_{ij} = \lambda_T |V| \delta_{ij} + (\lambda_L - \lambda_T)V_iV_j/|V| + D_s(\theta) \tag{22}$$

where: l_L is the longitudinal dispersivity of the medium; λ_T is the transversal dispersivity of the medium; δ_{ij} is Kronecker delta (i.e., $\delta_{ij} = 1$ if $i = j$ and $\delta_{ij} = 0$ if $i \neq 0$); V_i and V_j are the i-th and j-th components of the average interstitial flow velocity V, respectively; and $D_s(q)$ is the soil diffusion coefficient as defined by Eq. (18). Interstitial flow velocity V is given as:

$$V = \sqrt{v_x^2 + v_y^2} \tag{23}$$

If one considers only flow in the two dimensions and substituting Dij in Eq. (20), the salt flow equation becomes:

$$\frac{\partial(C\theta)}{\partial t} = \frac{\partial}{\partial x}(D_{xx}\frac{\partial C}{\partial x} + D_{xz}\frac{\partial C}{\partial z} - q_xC) + \frac{\partial}{\partial z}(D_{zz}\frac{\partial C}{\partial z} + D_{zx}\frac{\partial C}{\partial x} - q_zC) \tag{24}$$

The Eq. (21) and solute flow Eq. (24) were solved numerically using the initial and boundary conditions.

Boundary and Initial conditions for two dimensional water flow:
W = domain $[0£\ x\ £\ X;\ 0£\ z\ ≤Z]$:

$$q(x, z,0) = \theta n\ (x, z) \text{ for t=0} \tag{25a}$$

$$\frac{\partial \theta}{\partial x} = 0; x = 0, x = X; 0 \le z \le Z; 0 \le t \le T \tag{25b}$$

$$\frac{\partial \theta}{\partial z} = 0; 0 \le x \le X; 0 \le z \le Z; 0 \le t \le T \tag{25c}$$

$$q = \theta s;\ 0£\ x\ £\ r(t);\ z = 0;\ 0£\ t\ £\ T \tag{25d}$$

$$E - K(\theta_s)\frac{\partial H}{\partial z} = 0\ \ r(t)£\ x\ £\ X;\ z=0;\ 0\ £\ t\ £\ T \tag{25e}$$

$$\int_0^{\rho(t)}\left[E - K(\theta_s)\frac{\partial H}{\partial z}\right]dx = \frac{1}{2}f(t)\ ;\ z=0;\ 0 < t\ £\ T \tag{25f}$$

where: $f(t)$ = dripper discharge per unit length of the strip; T = end time of infiltration; θn = initial soil water content; θs = saturated water content; $r(t)$ = the length of the ponded or water saturated area; and E = evaporation.

Boundary and Initial conditions for solute flow c(x, z, t) under plane flow conditions from trickle source:

$$\frac{\partial c}{\partial x} = 0;\ \text{ at } x = 0 \text{ and } x = X \text{ for } t\ ³\ 0 \tag{26a}$$

$$\frac{\partial c}{\partial z} = 0;\ \text{ at } z = Z \text{ for } t\ ³\ 0 \tag{26b}$$

$$-\left[D_{xz}(x,0,t)\frac{\partial c}{\partial x} + D_{zz}(x,0,t)\frac{\partial c}{\partial z}\right] + q_z(x,0,t)c(x,0,t) = 0\ \text{ for } X³\ x > [r(t)] \tag{26c}$$

$$-\left[D_{xz}(x,0,t)\frac{\partial c}{\partial x} + D_{zz}(x,0,t)\frac{\partial c}{\partial z}\right] + q_z(x,0,t)c(x,0,t) = q_z(x,0,t)C_0(t)$$

$$\text{for } 0\ £\ x\ £\ [r(t)] \tag{26d}$$

$$c(x, z = 0, t>0)= C_0(t) \text{ for } 0 \text{ £ } x \text{ £ } [r(t)] \tag{26e}$$

$$\int_0^{\rho(t)} \left\{ [-D_{xz}(x,0,t)\frac{\partial c}{\partial x} + D_{zz}(x,0,t)\frac{\partial c}{\partial z}] + q_z(x,o,t)c(x,0,t) \right\} dx = \frac{1}{2} f(t)C_0(t) \tag{26f}$$

and finally the initial conditions are: $c(x, z,0) = c_n(x, z)$ in W, for $t = 0$, where: $q_z(x,0, t)$ is specific downward water flux at soil surface as given by Darcy's law; C_0 solute concentration at the inlet of trickling water; $C_n(x, z)$ is predetermined initial soil solution concentration; $D_{xz}(x,0, t)$, $D_{zz}(x,0, t)$ are hydrodynamic dispersion coefficients at the soil surface; and the sum of diffusion and mechanical dispersion coefficients is given by D_{ij} (the Eq. (22)). For solving the water and solute transport equations two soil water relations namely the soil moisture–water potential relation and the soil water potential–hydraulic conductivity relation were used. They were taken according to Van Genuchten [109] as:

$$q(h) = \theta r - [(\theta s-\theta r)/(1 - |\alpha h|^n)^m]^2 \tag{27}$$

$$K(h) = KsKr(q) - KsSe^{1/2}[1-(1-Se^{1/m})^m]^2 \tag{28}$$

where: θr and θs denote the residual and the saturated moisture contents, respectively; Ks and Kr are the saturated and relative hydraulic conductivities, respectively; a and n are the shape parameters, $m = 1-1/n$; Se is effective saturation or normalized volumetric soil water content; and a and n are empirical parameters. The soil water potential and hydraulic conductivity as functions of effective saturation are given by van Dam et al. [28] as:

$$Se = (q-\theta r)/(\theta s-\theta r) \tag{29}$$

$$h(Se) = [(S^{-1/m}-1)^{1/n}] /\alpha$$

$$K (Se) =KsSe^{1/2} [1-(1-S_e^{1/m})^m]^2$$

The values of θr, θs, q, Ks, water content at field capacity, and wilting point were obtained from soil field studies at the site. The values of bubbling pressure and l pore size distribution index for sandy loam soil were obtained from the model's database. The constants n and m were given as $n = 1+ 1$ and $m = 1/n$.

13.3.7.4　MODEL INPUT PARAMETERS

1. *Plant characteristics for each growth stage* include the crop coefficient, Kc, Kcb, root depth and lateral expansion, crop height and maximum/potential

final yield observed in the region under optimum conditions. Most of the data on Kc, Kcb and root depth was obtained from FAO 48 [87] and through experiment.

2. *Soil characteristics* include depth of each soil horizon, saturated hydraulic conductivity, saturated soil moisture content, salt diffusion coefficient, longitudinal and transversal dispersion coefficient, initial soil moisture and salinity profiles, tabulated data of soil moisture versus soil water potential and soil moisture versus hydraulic conductivity. The data on saturated hydraulic conductivity, saturated soil moisture content, initial soil moisture and salinity profile relation between soil moisture and soil matric potential, soil moisture and hydraulic conductivity obtained from the experiment.

3. *Meteorological data* include daily values of temperature (maximum and minimum), relative humidity, net radiation, wind speed, and daily rainfall. It was obtained from meteorological observatory at CIPHET, Abohar and Regional fruit research station, Abohar.

4. *Water management data* include the date and amount of irrigation water applied and the salinity level of each applied irrigation, wetting fraction and frequency of irrigation. These data were collected during the experiment period.

5. *Model parameters* include the number of compartments to be fixed in both vertical and horizontal direction in soil domain, tortuosity parameters, uptake parameters, diffusion parameters, position of plant relative to irrigation source and maximum time step for calculation.

13.3.7.5   MODEL OUTPUT PARAMETERS

The model gave output in the form of text and graphical files. These include horizontal and vertical distribution of soil moisture, soil salinity profile, crop transpiration, bare soil evaporation, leaching requirements, irrigation amounts, Kc, Kcb, root depth and final yield.

13.3.7.6   CALIBRATION AND VALIDATION OF SALTMED MODEL

The SALTMED model is calibrated below 100% fresh water treatment for final yield, because yield depends on soil moisture and salinity in the soil profile. The calibrated model was then used to predict final yield for different ratios of fresh saline water and also for different discharge rates of emitters. The predicted yield was then compared with the observed yield and the model was validated.

13.3.8   STATISTICAL ANALYSIS

The growth and quality parameter data were subjected to statistical analysis using factorial randomized block design with analysis of variance (ANOVA) techniques. For the factorial RBD, irrigation levels, discharge rates and salinity treatments were considered as three factors. The individual factor and their interaction were tested at 5% level of significance of difference.

13.3.9   TESTING OF GOODNESS OF FIT OF MODEL

The model was used for simulating the soil moisture and salt distribution profile in 15 treatment combinations for irrigation level of 0.8 in tomato crop. The criteria are described below for testing the goodness of fit of model.

$$RE = 100 \times [O_i - P_i]/[O_i] \tag{30}$$

$$ARE = \frac{\sum_{i=1}^{n} |RE|}{N} \tag{31}$$

$$RMSE = \sqrt{\frac{\sum (P_i - O_i)^2}{N}} \tag{32}$$

where: RE = relative error; O_i = observed value; and P_i = simulated value; ARE = Average Relative error; N = number of pairs of observed and simulated values; and RMSE   =Root mean square error. To get a relative idea of deviation in observed and simulated values within the profile of one treatment, the term relative error is more appropriate and average relative error was used to compare the relative error among treatments. The relative error and average relative error are also called ab-solute percent deviation and absolute average percent deviation. Relative root mean square error is defined below:

$$RRMSE = \frac{RMSE * 100}{O_{mean}} \tag{33}$$

where: RRMSE = Relative root mean square error; and O_{mean} = Observed mean. To get a relative idea of deviation in values, the term RRMSE is more appropriate than RMSE as it gives the percentage error between observed and predicted values. RRMSE approaching zero indicates a perfect fit.

13.4   RESULTS AND DISCUSSION

This research study consisted of five levels of saline irrigation water, which was prepared by mixing saline (tube well) water and fresh (canal) water with five different ratios 100% fresh (T1), 75% Fresh and 25% saline (T2), 50% fresh and 50% saline (T3), 25% fresh and 75% saline (T4) and 100% saline (T5). Fresh water was applied to all treatment plots during first 45 days to avoid stress to plant in early stage because of harsh cold winter in this region. The mixed water was applied through drip irrigation with three types of inline drippers with three discharge rates (1.2, 2.4, and 4.2 lph). The various observations recorded during the experimentation were analyzed and the results are presented and discussed in this section.

13.4.1   TOMATO GROWTH PARAMETERS

13.4.1.1   PLANT HEIGHT

The Figs. 13.9–13.11 show plant heights for 30, 60, 90 and 120 days after transplanting (DAT). The results showed that plant height was affected by different treatments. The plant height varied from 60.8 to 87 cm under different treatments at 120 DAT. The plant height was affected by different salinity treatments except at 90 DAT. There was no significant difference in plant height among T1 to T4 up to 30 DAT, because fresh water was applied to all plots at early stage up to 45 DAT. The maximum height was observed in plots irrigated with fresh water T1 followed by T2, T3, T4 and T5. The significant difference was found in plant height at irrigation levels I1 and I3 on 30 DAT. The average maximum height of 77.8 cm was found at 120 DAT under treatment T1 followed by treatment T2, T3, T4 and T5. The average minimum height was 64 cm in T5 treatment at 120 DAT. The highest salinity levels reduced the plant height compared to fresh water treatment (Fig. 13.9). These findings are in close proximity with Malash et al. [60]. The plant height was reduced by 17.45% in T5 compared to T1 after 120 DAT.

The effects of three discharge rates on plant height at different DAT are presented in Fig. 13.10. The plant height was not significantly different up to 60 DAT in Q1, Q2 and Q3. The significant differences were found after 60 DAT, when irrigation was applied at different discharge rates. The maximum height was observed at discharge rate of emitter Q3 followed by Q2 and Q1 in all DAT.

The effects of different irrigation levels on plant height at different DAT are presented in Fig. 13.11. The maximum average height was in irrigation level I2 followed by I3 and I1 at 90 DAT. There was no significant difference in interaction IxT except 60 DAT. The above results were in line with Sharda [99] and Agrawal et al. [3].

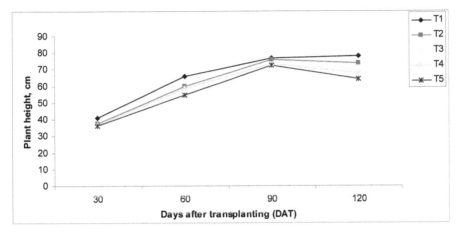

FIGURE 13.9   Effects of five ratios of saline water on plant height of tomato crop.

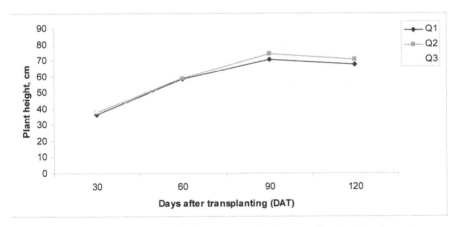

FIGURE 13.10   Effects of three discharge rates of emitter on plant height of tomato crop.

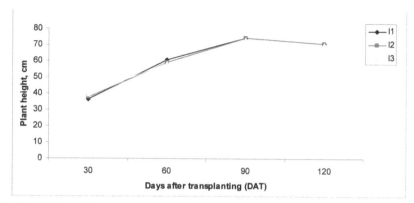

FIGURE 13.11   Effects of three irrigation levels on plant height of tomato crop.

13.4.1.2   PERCENT GROUND COVER

The data recorded for average percent ground cover (PGC) under different salinity ratios, discharge rates and irrigation levels for 30, 60, 90 and 120 DAT are presented in Table 13.5. There was no significant difference found in ground cover under different salinity ratios up to 60 DAT. After that it tends to decrease with increasing salinity levels.

It was observed from the Table 13.5 that the PGC increased up to 90 DAT, thereafter it tends to decrease at slower rate. The PGC goes on decreasing because fruits are at mature stage. The PGC showed significant difference among the salinity ratios up to 30 DAT. The PGC was found maximum at 0.8 and 1.0 irrigation levels compared to 0.6 (I1) irrigation level. At 30 DAT, there was no significant difference in plant cover among discharge rates of emitter (Q1, Q2 and Q3). Due to low evapotranspiration from the crop during winter, no variation was found in PGC. After 60 DAT, the significant difference in PGC were found among irrigation levels, discharge rates and salinity treatments. The PGC decreased with increasing the saline water ratios. The maximum PGC of 49.5% was found in treatment I3Q1T2 after 60 DAT. The PGC was maximum in T1 followed by T2, T3, T4 and T5 treatment. Among the irrigation levels, PGC was maximum at I3 level followed by I2 and I1. There was significant difference between I1 and I3. After 90 DAT, the same trend was observed. This implies that PGC decreased with increasing salinity ratios. It also increased with increasing discharge rates and irrigation levels. The PGC was maximum of 93.67% in I2Q2T1. After 90 DAT, the PGC was found decreasing in all treatment. This is because of increasing salinity levels in the soil due to application of saline water continuously causes osmotic stress and thus reduced water uptake and growth. Also, during reproductive stage of crop, the shredding of leaves occur. These findings are in agreement with those reported by Sharda [99] and Cetin and Demet [20]. The PGC after 120 DAT was maximum at Q2 followed by Q3 and Q1.

The significant difference was found between irrigation levels, discharge rates and salinity ratios. It was maximum of 79.33% at I2Q3T1.

TABLE 13.5 Effects of Irrigation Levels, Discharge Rates, Fresh and Saline Water Ratios (Salinity Levels) on Percent Ground Cover at Different DAT

Irrigation levels (I)	Discharge rate (Q)	Fresh and saline water ratio (T)	Percent ground cover			
	lph	F:S	30	60	90	120
0.6 = I1	1.2 = Q1	100: 0 = T1	12.27	43.33	91.00	74.33
		75:25 = T2	12.57	43.00	88.67	72.33
		50:50 = T3	11.73	41.17	88.17	70.33
		25:75 = T4	13.33	40.67	89.00	68.00
		0: 100 = T5	10.80	38.50	87.33	67.67
	I1Q1	**Mean**	**12.1**	**41.3**	**88.8**	**70.5**
	2.4 = Q2	100: 0	12.50	45.85	91.33	78.33
		75:25	12.90	44.33	90.67	77.67
		50:50	12.27	40.50	89.00	75.00
		25:75	13.53	39.83	90.00	71.67
		0: 100	12.13	39.67	88.00	71.33
	I1Q2	**Mean**	**12.7**	**42.0**	**89.8**	**74.8**
	4.2 = Q3	100: 0	12.93	46.50	93.00	80.67
		75:25	12.78	45.67	92.67	77.00
		50:50	11.67	44.00	91.33	75.33
		25:75	12.00	42.17	90.00	71.00
		0: 100	**13.03**	40.83	89.67	72.67
	I1Q3	**Mean**	**12.5**	**43.8**	**91.3**	**75.3**
	I1	**Mean**	**12.4**	**42.4**	**90.0**	**73.6**

TABLE 13.5   *(Continued)*

0.8 = I2	1.2	100: 0	14.67	44.67	91.33	76.67
		75:25	13.60	46.50	90.00	72.67
		50:50	12.07	42.00	89.00	70.67
		25:75	13.87	41.00	87.33	68.00
		0: 100	12.43	40.00	87.33	69.33
	I2Q1	**Mean**	**13.3**	**42.8**	**89.0**	**71.5**
	2.4	100: 0	15.00	45.83	93.67	79.00
		75:25	13.33	45.33	93.33	78.33
		50:50	11.77	45.33	92.00	75.33
		25:75	13.83	44.17	91.33	74.33
		0: 100	13.00	39.67	89.17	73.33
	I2Q2	**Mean**	**13.4**	**44.1**	**91.9**	**76.1**
	4.2	100: 0	14.50	47.83	93.33	79.33
		75:25	14.33	46.17	92.00	78.00
		50:50	12.00	46.67	91.50	75.67
		25:75	12.67	42.33	88.00	72.67
		0: 100	13.33	41.67	89.67	73.67
	I2Q3	**Mean**	**13.4**	**44.9**	**90.9**	**75.9**
	I2	**Mean**	**13.4**	**43.9**	**90.6**	**74.5**

TABLE 13.5   *(Continued)*

1 = I3	1.2	100: 0	13.67	46.33	90.33	74.33
		75:25	13.83	49.50	89.33	72.00
		50:50	11.33	43.33	88.67	68.00
		25:75	13.10	41.67	88.00	65.67
		0: 100	11.50	41.00	87.00	68.00
	I3Q1	**Mean**	**12.7**	**44.4**	**88.7**	**69.6**
	2.4	100: 0	14.17	47.00	93.00	77.00
		75:25	11.67	46.67	92.33	75.67
		50:50	12.33	43.83	91.00	74.33
		25:75	12.80	44.67	90.33	72.67
		0: 100	11.87	41.67	87.33	70.00
	I3Q2	**Mean**	**12.6**	**44.8**	**90.8**	**73.9**
	4.2	100: 0	14.00	47.67	91.67	77.67
		75:25	12.40	45.60	91.67	75.00
		50:50	11.00	48.83	88.00	74.00
		25:75	13.67	43.93	88.33	71.00
		0: 100	12.93	41.73	89.00	70.00
	I3Q3	**Mean**	**12.8**	**45.6**	**89.7**	**73.5**
	I3	**Mean**	**12.7**	**44.9**	**89.7**	**72.4**
Average			**12.8**	**43.7**	**90.1**	**73.5**
LSD (T)			0.65	1.00	1.68	2.14
LSD (Q)			NS	0.77	1.30	1.66
LSD (I)			0.50	0.77	NS	1.66
LSD (IxT)			NS	NS	NS	NS
LSD (QxT)			1.73	NS	NS	NS
LSD (IxQxT)			NS	NS	NS	NS

13.4.1.3 DRY MATTER OF TOMATO PLANTS

The dry matter accumulation of tomato plants under different salinity ratios of irrigation along with three discharge rates and three irrigation levels at 60, 90, 120 and 150 DAT is shown in Figs. 13.12–13.14. The dry matter of tomato increased with increase in DAT up to 120 DAT. After that it goes on decreasing. The maximum dry matter accumulation was observed in I3Q2T2 in 120 DAT. After 90 DAT, the plant dry weight decreased with increase in salinity levels of irrigation water under all discharge rates of emitters. After 90 DAT, discharge rate of emitters showed significant effect on dry matter. Dry matter decreased with decreasing discharge rate. But the same trend was not observed after 120 and 150 DAT. The maximum dry matter was observed in T1 and minimum in T5 except on 120 DAT. On 120 DAT, minimum dry matter was observed in T3 treatment. After harvest of tomato crop, it was found that the dry matter decreased with increasing salinity levels, increasing discharge rate and increasing irrigation levels. These results are in close proximity with Malash et al. [60] and Reina Sanchez et al. [84]. The dry matter was maximum in I1Q1T1 and minimum in I3Q2T5 treatment after 150 DAT. Saline water reduced water uptake, transpiration and net CO_2 assimilation, which in turn reduce the growth and transport of nutrients into the plants [90]. Reduction in leaf area index (LAI) reduces light interception and thus reduces dry matter [4, 56].

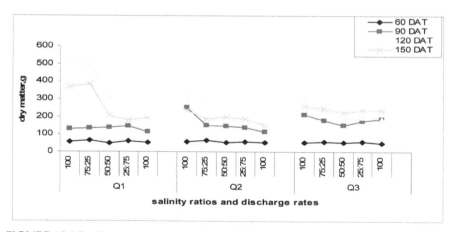

FIGURE 13.12   Dry matter accumulation at different salinity ratios and discharge rate of emitters at I1.

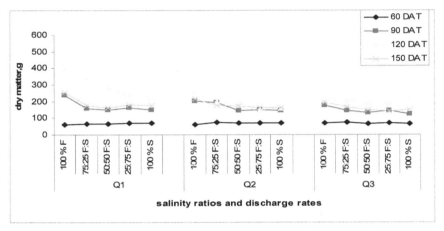

FIGURE 13.13   Dry matter accumulation at different salinity ratios and discharge rate of emitters at I2.

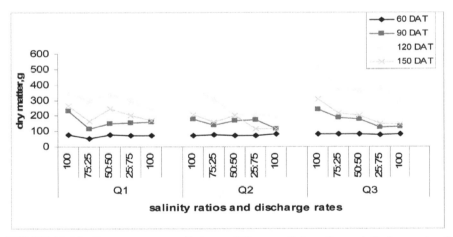

FIGURE 13.14   Dry matter accumulation at different salinity ratios and discharge rate of emitters at I3.

13.4.1.4   PLANT YIELD

Plant yield data under different salinity levels, irrigation levels and discharge rate of emitters were shown in Figs. 13.15–13.17. It was found that the plant yield decreased with increase in salinity levels of irrigation water significantly. The maximum plant yield of 3.91 kg/plant was in I3Q2T1 followed by I3Q1T1. The minimum plant yield of 1.60 kg/plant was in I1Q3T5 treatment. The irrigation levels I2 and I3 showed significant increase in plant yield over I1. The plant yield increased with increasing irrigation level from I1 to I3, and irrigation levels I2 and I3 were found as par with each other. It is also clear from data analysis that the plant yield significantly decreased with increase in discharge rates. The plant yield was maximum under Q1 discharge rate followed by Q2 and Q3 in each salinity treatment T1 to T5. These findings are in close proximity with Badr and Taalab [14], who reported that the maximum yield was obtained with surface drip irrigation applied at 2 lph and lowest yield was obtained at 8 lph with surface drip irrigation in saline water. The average plant yield decreased by 11.51, 25.84, 32.67, and 44.38% in T2, T3, T4 and T5, respectively. The interaction of irrigation levels and discharge rates showed significant effect on plant yield. The yield reduction in saline water mainly was caused due to reduction in fruit size and weight. These findings are agreement with Reina-Sanchez, et al. [84], Malash [60] and Romero-Aranda et al. [91, 92] (Figs. 13.18–13.20).

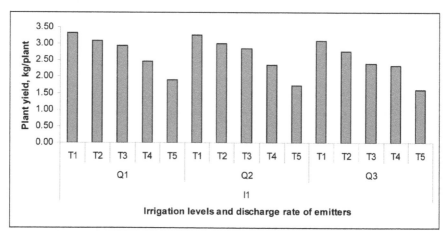

FIGURE 13.15 Plant yield at different salinity ratios and discharge rates at irrigation level I1.

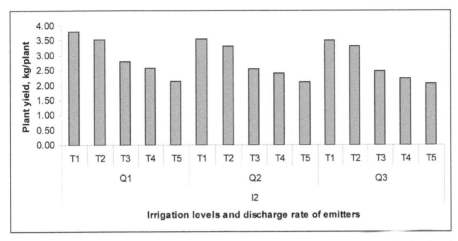

FIGURE 13.16 Plant yield at different salinity ratios and discharge rates at irrigation level I2.

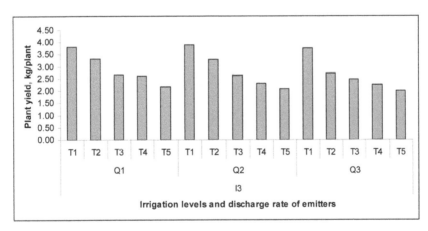

FIGURE 13.17 Plant yield at different salinity ratios and discharge rates at irrigation level I3.

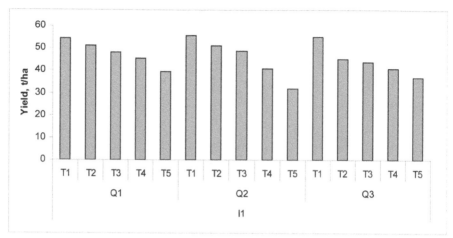

FIGURE 13.18 Total yield at different salinity ratios and discharge rates at irrigation level I1.

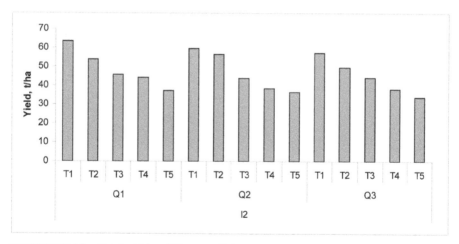

FIGURE 13.19 Total yield at different salinity ratios and discharge rates at irrigation level I2.

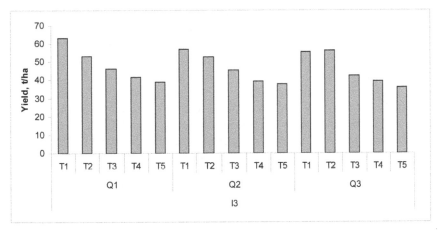

FIGURE 13.20   Total yield at different salinity ratios and discharge rates at irrigation level I3.

13.4.1.5   TOTAL YIELD

The total yield (t/ha) of tomato is presented in Figs. 13.21–13.23. It is clear from these figures that the total yield decreased with increase in salinity levels of irrigation water. As compared with T1, total yield decreased by 10.20% in T2, 21.44% in T3, 29.21% in T4 and 36.83% in T5. The maximum yield of 63.41 t/ha and minimum of 31.74 t/ha was in I2Q1T1and I1Q3T5 treatments, respectively, which confirms the results with Malash et al. [62]. The yield increased with increase in irrigation levels and decreased with increase in discharge rates in all salinity treatments, T1 to T5. The irrigation levels were not significant. But the discharge rate was significant with each other. The Q1 gave more yield compared to Q2 and Q3 discharge rates. The salinity treatments T1 to T5 were significant to each other. The interaction IxQxT was significant which shows I2Q1T1 gave maximum yield of 63.41 t/ha. In T2 treatment, I2xQ2 gave maximum yield 56.49 t/ha, which was significantly different with I2Q1T1. The best yield of 39.47 t/ha was observed T5 treatment in I1xQ1, which was significantly different with other salinity treatments. In fresh water treatment T1 with Ec of 0.38 dS/m, the yield increased with increase in IW/CPE ratios from 0.6 to 0.8. The IW/CPE ratio of 0.8 followed by 1.0 and 0.6 was found best. The IW/CPE ratio of 0.8 and 1.0 was found at par with each other. The irrigation level of 0.8 with 1.2 lph discharge rate was best for T1 treatment, which gave maximum yield of 63.41 t/ha. In treatment T1, the yield decreased with increase in discharge rate of emitters. The discharge rate of 1.2 lph was found best followed by 2.4 and 4.2 lph. The discharge rate of 1.2 and 2.4 lph were found at par with each other. In treatment T2 with Ec of 6.3 dS/m, the IW/CPE ratio of 0.8 and discharge rate of 2.4 lph was found best. The discharge rates of 1.2 and 2.4 lph were significantly different. Also, IW/CPE ratio of 0.6, 0.8 and 1.0 were significantly different to each other. The maximum yield under this treatment was reduced by 10.9% compared to best treatment under fresh water whereas the quality parameters TSS, acidity and ascorbic acid improved compared to T1. In treatment T3 with EC of 9.1 dS/m. The yield was found maximum under irrigation level 0.6 and discharge rate of 2.4 lph. The maximum yield under this treatment was reduced by 23.38% compared to T1 with better quality of tomato compared to T1. In treatment T4 with Ec of 14.5 dS/m, the yield was maximum under irrigation level 0.6 and discharge rate 1.2 lph and which was reduced by 28.62% compared to best yield under T1. The irrigation levels were found at par with each other. The discharge rate 1.2 lph and 2.4 lph were significantly different with each other. The discharge rate 2.4 lph and 4.2 lph were at par with each other. In 100% saline water treatment T5 with EC of 19.5 dS/m, the yield was maximum under I1xQ1 with reduction of 37.75% compared to fresh water treatment. The statistically no significant differences were found in three irrigation levels. The discharge rate of 1.2 and 2.4 lph were found as par with each other and were significantly different with 4.2 lph. The above findings are in line with Sandra Krauss et al. [95].

13.4.1.6   WATER USE EFFICIENCY (WUE)

Table 13.6 indicates the WUE for different salinity ratios treatments, discharge rates and irrigation levels presented in Table 13.6. It can be observed from the data in his table that the WUE was decreased with increase in salinity ratios/salinity levels of irrigation water. There was significant difference among the salinity treatments (T1, T2, T3, T4 and T5). The average maximum WUE of 1.22 t/ha-cm was found in T1 followed by 1.10 t/ha-cm, 0.97 t/ha-cm. 0.88 t/ha-cm and 0.774 t/ha-cm found under treatment T2, T3, T4 and T5, respectively. The irrigation levels and discharge rate of emitters have shown the significant effects on WUE of tomato crop. The WUE decreased with increasing irrigation levels from I1 to I3. Also, the average WUE decreased with increase in discharge rate of emitters under given treatment combination. The interaction IxT and IxQxT had shown the significant effect on WUE. In IxT interaction, I1T1 has shown maximum of 1.54 t/ha-cm and minimum of 0.58 t/ha-cm in I3T5 treatment. In IxQxT interaction, the I1Q2T1 has given maximum WUE of 1.56 t/ha-cm and minimum of 0.56 t/ha-cm in I3Q3T5. The interaction WUE under I1Q1T1, I1Q2T1 and I1Q3T1 were as par with each other.

TABLE 13.6   Effects of Irrigation Levels, Discharge Rate and Fresh and Saline Water Ratio (Salinity Levels) on WUE of Tomato

Irrigation levels (I)	Discharge rate (Q), lph	Fresh and saline water ratios (T) F: S	WUE (t/ha-cm)
0.6	1.2	100: 0	1.53
		75:25	1.44
		50:50	1.36
		25:75	1.27
		0: 100	1.11
	I1Q1	Mean	1.34
	2.4	100: 0	1.56
		75:25	1.44
		50:50	1.37
		25:75	1.15
		0: 100	0.89
	I1Q2	Mean	1.28
	4.2	100: 0	1.55
		75:25	1.27
		50:50	1.23
		25:75	1.15
		0: 100	1.04
	I1Q3	Mean	1.25
	I1	Mean	1.29

TABLE 13.6 *(Continued)*

0.8	1.2	100: 0	1.28
		75:25	1.09
		50:50	0.93
		25:75	0.89
		0: 100	0.75
	I2Q1	**Mean**	0.99
	2.4	100: 0	1.20
		75:25	1.14
		50:50	0.88
		25:75	0.77
		0: 100	0.74
	I2Q2	**Mean**	0.95
	4.2	100: 0	1.15
		75:25	1.00
		50:50	0.89
		25:75	0.77
		0: 100	0.68
	I2Q3	**Mean**	0.90
	I2	**Mean**	0.94
1	1.2	100: 0	0.98
		75:25	0.83
		50:50	0.72
		25:75	0.65
		0: 100	0.61
	I3Q1	**Mean**	0.76
	2.4	100: 0	0.89
		75:25	0.82
		50:50	0.71
		25:75	0.61
		0: 100	0.59
	I3Q2	**Mean**	0.72

TABLE 13.6   *(Continued)*

	4.2	100: 0	0.86
		75:25	0.88
		50:50	0.66
		25:75	0.61
		0: 100	0.56
	I3Q3	**Mean**	0.71
	I3	**Mean**	0.73
	Grand	**Mean**	0.99
LSD (T)			0.031
LSD (Q)			0.024
LSD (I)			0.024
LSD (IxT)			0.054
LSD (QxT)			NS
LSD (IxQxT)			0.094

13.4.2   QUALITY PARAMETERS OF TOMATO

The attributes of quality of tomato fruits such as TSS, acidity and ascorbic acid were studied and given as under.

13.4.2.1   TOTAL SOLUBLE SOLIDS (TSS)

Table 13.7 shows effects of irrigation levels, discharge rate of emitters and ratios of fresh and saline water on TSS of tomato fruits. It can be observed in Table 13.7 that the TSS of tomato fruits significantly different with T1, T2, T3 and TSS of fruits were found as par in T4 and T5 treatments. Reduced water uptake in plants irrigated by saline water led to increases in solute concentrations (particularly sugars) and hence increased TSS contents. TSS increased with increase in salinity of irrigation water, that is, T1 to T5 under all Q1, Q2, and Q3 and for all irrigation levels I1, I2 and I3. These results confirms with the findings of Yurtseven et al. [115], and Malash et al. [60, 62]. The average value of TSS was maximum in Q1 followed by Q2 and Q3. Q1 was significantly different from Q3. Q1 and Q2 treatment and Q2 and Q3 was par with each other. In irrigation levels, average TSS was maximum in I1 followed by I2 and I3. These results are in close proximity with Hanson et al. [39, 40]. The irrigation level I1 was significantly different with I3. The I2 and I3 are observed as par with each other. The maximum TSS of 6.9 was in I1Q1T5 treatment followed by I1Q2T1. The minimum TSS was in tomato irrigated with 100% fresh water in all irrigation levels. Reduced water uptake in plants irrigated by saline water led to increase in solute concentrations (particularly sugars) and hence increased TSS contents with increase in salinity of water.

13.4.2.2 ACIDITY

The effects of irrigation levels, discharge rate of emitters and fresh and saline water ratios on acidity of tomato fruits are shown in Table 13.8. It can be observed that the acidity of tomato fruits increased with increase in saline water ratio. The similar trend was reported by Magan et al. [59]. There was significant difference in acidity of tomato among T1, T3, T4 and T5. T1 and T2 treatments were par with each other. Among the irrigation levels and discharge rate of emitters, there was no significant difference in acidity of tomato fruits. But average acidity was increased with increase in irrigation levels and decreased with increasing discharge rate of emitters. The average acidity was maximum in 0.818% at I2Q1T5 and minimum of 0.689% in I3Q3T1, which confirms the findings of Mitchell et al. [67].

TABLE 13.7 Effects of Irrigation Levels, Discharge Rate and Fresh and Saline Water Ratios (Salinity Levels) on Average TSS (°Brix) of Tomato Fruits

Irrigation levels (I)	Discharge rate (Q), lph	Fresh and saline water ratios (T)	TSS, °Brix
0.6	1.2	100:0	5.8
		75:25	6.1
		50:50	6.2
		25:75	6.5
		0:100	6.9
	I1Q1	**Mean**	**6.3**
	2.4	100:0	5.7
		75:25	6.0
		50:50	6.2
		25:75	6.3
		0:100	6.4

TABLE 13.7 *(Continued)*

		I1Q2	Mean	**6.1**
		4.2	100:0	5.5
			75:25	5.8
			50:50	6.2
			25:75	6.3
			0:100	6.4
		I1Q3	**Mean**	**6.0**
		I1	**Mean**	**6.1**
0.8		1.2	100:0	5.5
			75:25	5.9
			50:50	6.1
			25:75	6.4
			0:100	6.6
		I2Q1	**Mean**	**6.1**
		2.4	100:0	5.3
			75:25	5.9
			50:50	6.2
			25:75	6.4
			0:100	6.5
		I2Q2	**Mean**	**6.0**
		4.2	100:0	5.0
			75:25	5.6
			50:50	5.9
			25:75	6.1
			0:100	6.2
		I2Q3	**Mean**	**5.8**

	I2	Mean	6.0
1	1.2	100:0	5.4
		75:25	5.9
		50:50	6.2
		25:75	6.3
		0:100	6.4
	I3Q1	Mean	6.1
	2.4	100:0	5.3
		75:25	5.8
		50:50	5.9
		25:75	6.0
		0:100	6.1
	I3Q2	Mean	5.8
	4.2	100:0	5.2
		75:25	5.6
		50:50	5.8
		25:75	5.9
		0:100	5.9
	I3Q3	Mean	5.7
	I3	Mean	5.9
	Grand	Mean	6.0
LSD (T)			0.27
LSD (Q)			0.22
LSD (I)			0.22
LSD (IxT)			NS
LSD (QxT)			NS
LSD (IxQxT)			NS

TABLE 13.8 Effects of irrigation levels, discharge rate and fresh and saline water ratios (salinity levels) on average acidity of tomato fruits.

Irrigation levels (I)	Discharge rate (Q), lph	Fresh and saline water ratios (T) F:S	Acidity, %
0.6	1.2	100:0	0.695
		75:25	0.708
		50:50	0.735
		25:75	0.789
		0:100	0.801
	I1Q1	**Mean**	**0.746**
	2.4	100:0	0.674
		75:25	0.710
		50:50	0.727
		25:75	0.741
		0:100	0.785
	I1Q2	**Mean**	**0.727**
	4.2	100:0	0.651
		75:25	0.668
		50:50	0.697
		25:75	0.725
		0:100	0.745
	I1Q3	**Mean**	**0.697**
	I1	**Mean**	**0.723**

0.8	1.2	100:0	0.684
		75:25	0.733
		50:50	0.755
		25:75	0.783
		0:100	0.818
	I2Q1	**Mean**	**0.755**
	2.4	100:0	0.668
		75:25	0.698
		50:50	0.752
		25:75	0.778
		0:100	0.795
	I2Q2	**Mean**	**0.738**
	4.2	100:0	0.658
		75:25	0.687
		50:50	0.720
		25:75	0.769
		0:100	0.797
	I2Q3	**Mean**	**0.726**
	I2	**Mean**	**0.740**
1	1.2	100:0	0.715
		75:25	0.742
		50:50	0.758
		25:75	0.769
		0:100	0.812
	I3Q1	**Mean**	**0.759**
	2.4	100:0	0.683
		75:25	0.723
		50:50	0.738
		25:75	0.744
		0:100	0.754
	I3Q2	**Mean**	**0.728**

	4.2	100:0	0.689
		75:25	0.705
		50:50	0.726
		25:75	0.775
		0:100	0.797
	I3Q3	**Mean**	**0.739**
	I3	**Mean**	**0.742**
	Grand	**Mean**	**0.735**
LSD (T)			0.051
LSD (Q)			NS
LSD (I)			NS
LSD (IxT)			NS
LSD (QxT)			NS
LSD (IxQxT)			NS

TABLE 13.9   Effects of Irrigation Levels, Discharge Rate and Fresh and Saline Water Ratios (Salinity Levels) on Ascorbic Acid of Tomato Fruits

Irrigation levels (I)	Discharge rate (Q), lph	Fresh and saline water ratios (T)	Ascorbic acid, mg/100 g
0.6	1.2	100:0	29.23
		75:25	32.76
		50:50	34.98
		25:75	35.08
		0:100	36.80
	I1Q1	**Mean**	**33.8**

	2.4	100:0	30.18
		75:25	31.33
		50:50	35.10
		25:75	38.04
		0:100	40.04
	I1Q2	**Mean**	**34.9**
	4.2	100:0	31.55
		75:25	31.85
		50:50	32.63
		25:75	38.35
		0:100	40.06
	I1Q3	**Mean**	**34.9**
	Average		**34.5**
0.8	1.2	100:0	31.48
		75:25	33.23
		50:50	34.54
		25:75	37.86
		0:100	40.68
	I2Q1	**Mean**	**35.6**
	2.4	100:0	32.36
		75:25	32.76
		50:50	33.15
		25:75	35.18
		0:100	37.93
	I2Q2	**Mean**	**34.3**
	4.2	100:0	24.78
		75:25	30.21
		50:50	32.13
		25:75	35.20
		0:100	37.44

	I2Q3	Mean	32.0
	I2	Mean	33.1
I	1.2	100:0	30.78
		75:25	37.16
		50:50	37.41
		25:75	38.85
		0:100	38.98
	I3Q1	Mean	36.6
	2.4	100:0	32.89
		75:25	35.80
		50:50	37.23
		25:75	38.65
		0:100	40.89
	I3Q2	Mean	37.1
	4.2	100:0	28.88
		75:25	35.06
		50:50	36.75
		25:75	37.44
		0:100	39.36
	I3Q3	Mean	35.5
	I3	Mean	36.4
	Grand	Mean	34.7
LSD (T)			3.19
LSD (Q)			NS
LSD (I)			NS
LSD (IxT)			NS
LSD (QxT)			NS
LSD (IxQxT)			NS

13.4.2.3 ASCORBIC ACID

The effects of salinity levels, irrigation levels and discharge rate of emitters on ascorbic acid of tomato fruits are presented in Table 13.9. It is clear in Table 13.9 that the ascorbic acid increased with increase in salinity levels of irrigated water significantly. The similar trend was reported by Sandra et al. [95]. The average value of ascorbic acid was minimum of 30.23 mg/100 g under treatment T1 and maximum of 39.87 mg/100 g under T5 treatment. The values of T1 and T2 were par with each other. Also T3 andT4 were par with each other. It was observed that there was no significant effect in ascorbic acid when applying irrigation at different discharge rates and with different irrigation levels. It can be concluded that the ascorbic acid decreased with increase in discharge rate of emitters. The maximum ascorbic acid was found when irrigated with discharge rate of Q1 followed by Q2 and Q3.

13.4.2.4 PH OF TOMATO FRUITS

It can be observed from Table 13.10 and Fig. 13.21 that the pH decreased with increase in salinity levels, which confirm the finding of Yurtseven et al. [115]. The pH also increased with increasing the discharge rates. The maximum and minimum pH were 4.29 and 3.89 in I2Q1T1 and I3Q3T5 treatment, respectively. The pH below 4.5 is desirable to prevent proliferation microorganisms in processed products [36].

TABLE 13.10 pH of Tomato Fruit Under Different Salinity Ratios (Salinity Levels) With Different Discharge Rate and Irrigation Levels

Treatment	I1				I2				I3			
	Q1	Q2	Q3	Mean	Q1	Q2	Q3	Mean	Q1	Q2	Q3	Mean
T1	4.13	4.17	4.19	4.16	4.06	4.2	4.29	4.18	4.09	4.12	4.1	4.10
T2	4.06	4.05	4.09	4.07	4.12	4.21	4.08	4.14	4.12	4.23	4.12	4.16
T3	4.06	4.1	4.09	4.08	4.08	4.1	4.02	4.07	4.04	4.03	4.02	4.03
T4	4	4.01	3.97	3.99	4.01	4	4.03	4.01	4.08	4.04	4.05	4.06
T5	3.98	3.95	3.92	3.95	3.96	3.97	3.92	3.95	3.95	3.91	3.89	3.92
Mean	4.05	4.06	4.05	4.05	4.05	4.1	4.07	4.07	4.06	4.07	4.04	4.05

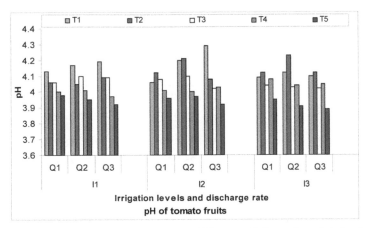

FIGURE 13.21 pH of tomato fruits under different salinity ratios of irrigation water, discharge rates of emitters and irrigation levels.

13.4.2.5 FIRMNESS OF TOMATO FRUITS

The firmness of tomato fruits was determined with the help of textural analyzer. Table 13.11 shows data of firmness of tomato fruits. It can be observed that the firmness of fruits increased with increase in salinity treatment up to 50:50 ratio. After that it again decreased. The firmness of 808 g was maximum in T3 treatment and minimum of 314.29 and 334.70 in T1 and T2 treatments, respectively. These results are in line with Sharaf and Hobson [98]. The firmness in T3 treatment significantly differs from T1, T2, T4 and T5. The firmness in T1 and T2 treatment was par with each other. The firmness of fruits decreased when discharge rate irrigation increased from Q1 to Q3. The overall firmness was better under I1Q1T3 and lowest firmness was found in I3Q3T5.

TABLE 13.11 Effects of Irrigation Levels, Discharge Rate and Fresh and Saline Water Ratios (Salinity Levels) on Average Firmness of Tomato Fruits

Irrigation levels (I)	Discharge rate (Q), lph	Fresh and saline water ratios (T) F:S	Firmness (g)
0.6	1.2	100:0	349.64
		75:25	253.84
		50:50	986.80
		25:75	539.41
		0:100	470.58
	I1Q1	**Mean**	**520.053**
	2.4	100:0	318.59
		75:25	335.98
		50:50	996.42
		25:75	421.82
		0:100	364.65
	I1Q2	**Mean**	**487.493**
	4.2	100:0	247.00
		75:25	274.66
		50:50	645.38
		25:75	326.32
		0:100	376.34
	I1Q3	**Mean**	**373.940**
	I1	**Mean**	**460.495**
0.8	1.2	100:0	225.64
		75:25	475.25
		50:50	811.36
		25:75	440.69
		0:100	288.51
	I2Q1	**Mean**	**448.291**

	2.4	100:0	371.00
		75:25	374.96
		50:50	709.13
		25:75	480.77
		0:100	375.29
	I2Q2	**Mean**	**462.228**
	4.2	100:0	423.10
		75:25	357.51
		50:50	754.78
		25:75	644.73
		0:100	345.68
	I2Q3	**Mean**	**505.158**
	I2	**Mean**	**471.892**
1	1.2	100:0	305.39
		75:25	374.63
		50:50	688.95
		25:75	564.46
		0:100	452.33
	I3Q1	**Mean**	**477.149**
	2.4	100:0	278.98
		75:25	219.55
		50:50	826.59
		25:75	414.63
		0:100	434.86
	I3Q2	**Mean**	**434.923**
	4.2	100:0	309.31
		75:25	346.00
		50:50	852.62
		25:75	453.12
		0:100	351.50
	I3Q3	**Mean**	**462.509**

	I3	Mean	458.193
	Grand	Mean	463.527
LSD (T)			55.98
LSD (Q)			NS
LSD (I)			NS
LSD (IxQ)			75.11
LSD (QxT)			96.96
LSD (IxQxT)			167.94

13.4.3   SIMULATION OF YIELD, MOISTURE AND SALT DISTRIBUTION

SALTMED model developed by Ragab [77] was used for simulating the soil moisture and salt distribution in the soil profile irrigated with saline water for tomato crop. The input data used for the model were climate data, irrigation data, crop parameter and soil parameter. The climate data at the site included: maximum and minimum temperature, relative humidity, sunshine hours, and wind speed and pan evaporation data. Sunshine hours and radiation data were obtained from observatory of Regional fruit research station, Abohar. Following assumptions were made to simulate the model:

1. the soil is isotropic and homogeneous porous medium and Darcy's law is applicable in both saturated and unsaturated zones.
2. the soil has initially uniform water and salt content.
3. the rooting depth follows the same pattern as the crop coefficient.

The model is calibrated for soil moisture, soil salinity and yield. It is decided first to calibrate the model against the final yield as yield depends on soil moisture and salinity in the soil profile. The model was calibrated at 100% fresh water treatments. The following data were used in the calibration:

1. Meteorological data were obtained from the Regional fruit research station, Abohar and also from the mini observatory at CIPHET, Abohar.
2. The irrigation data showing measurement of average flow rate of emitters during each irrigation, duration of each irrigation and salinity of applied water during each irrigation. The data were collected from the field experiment.
3. Crop parameters: maximum plant height, rooting depth, length of each growth stage, planting date, harvesting date were based on the direct field measurements. The crop coefficient values such as Kc, Kcb and Fc were based on FAO 56 [7] paper for tomato crop. The $\eta50$ and p50 parameters were obtained from FAO 48 [87], Rhoades [86].

4. Soil parameters data such as saturated and unsaturated hydraulic conduc-
    tivity were obtained from the field and laboratory experiments. Initial soil
    moisture and salinity profile were based on the data collected in the field
    before start of experiment.
5. Model parameters such as diffusion coefficient, tortuosities were taken from
    model data base. Plant position (Dripper position) was selected as per the
    experiment layout, which is fixed near to plant. The numbers of cells in
    vertical and horizontal directions were taken according to Ragab [77]. The
    calculations were performed over 20 cells in horizontal and 40 cells in ver-
    tical direction. Thus, each cell is 2.5×2.5 cm² (i.e., field domain is 100×50
    cm²).

During calibration, the model requires daily values of plant parameters such
as crop coefficient kc and kcb and rooting depth etc. These parameters were not
available on daily basis. But, the model generates the daily values by interpolation
between the values given for each growth stage. The interpolated calibrated crop
coefficient kc and kcb and rooting depth are shown in Fig. 13.22. Once all input
data of meteorology, irrigation, crop, soil and model have been entered, the calibration process starts. The process of calibration was carried out: First using initial
measured values of crop and soil parameters; and then by gradually changing the
values of crop parameters one at a time until calibrated yield equal or near equal to
observed one. The calibration was carried out for fresh water treatment T1I2Q1 as it
gives the maximum yield. The initial crop and soil parameters taken for calibration
are shown in Table 13.12. The Fig. 13.23 shows the calibration against crop yield
under drip irrigation for the tomato crop under T1Q1 treatment at experimental site.

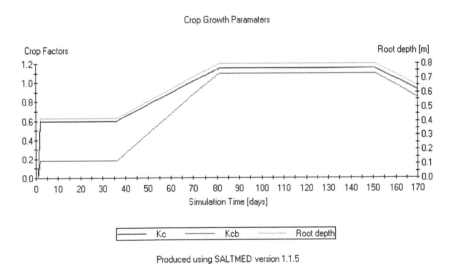

FIGURE 13.22   Calibrated crop parameters for tomato crop.

TABLE 13.12   Initial Crop and Soil Parameters

Parameters	Crop and soil parameters	Values
Crop	Kc (crop coefficient) Initial	0.6
	Mid	1.2
	End	0.8
	Kcb (crop transpiration coefficient) Initial	0.18
	Mid	1.1
	End	0.7
	Maximum crop height, cm	80
	Maximum depth of crop, cm	80
Soil	Saturated moisture content, m^3/m^3	0.41
	Pore size distribution index	0.332

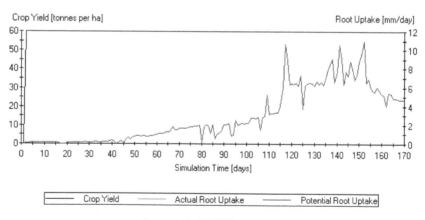

Predicted Crop and Root Uptake

FIGURE 13.23 Crop yield obtained by calibration with 100% fresh water with drip irrigation.

In Fig. 13.22, the center line represents the ratio of actual to potential water uptake. On the harvest day, cumulative ratio of actual to potential water uptake is used as the reduction factor. When this reduction factor if multiplied by maximum yield obtained for this tomato crop, one can obtain the actual yield (Section 13.3 in this chapter). The calibrated yield for drip irrigation under discharge rate of emitter Q1 (1.2 lph) was observed as 60 t/ha. These values were identical to the maximum observed yield, that is, 63.41 t/ha for same treatment with the deviation of 5.38%. It was seen from the Fig. 13.23 that the simulated yield was not subjected to water stress, as actual and calibrated yield were found nearly equal to observed one.

13.4.3.1 VALIDATION OF THE MODEL FOR CROP YIELD

With the calibrated crop and soil parameters, the model predicted the crop yield under different treatment of salinity ratios (T1 to T5) and the discharge rates (Q1, Q2 and Q3) over the season for three irrigation levels (I1, I2, I3). The predicted and observed yield data are presented in Tables 13.13–13.15. It can be seen from the Table 13.13 that at irrigation level I1, the relative error was minimum of 1.36% in the treatment T1 and maximum of 23.62% was in treatment T5 (irrigation with 100% saline water). The positive difference in deviation in T1, T2, T4 and T5 means that observed yield is more than the predicted yield. In the observed yield, the yield decreased with increase in rate of discharge of emitters within the treatments except in T1Q2, T4Q2 and T5Q2. The simulated yield decreased with increase in discharge rate of emitters from 1.2 lph to 4.2 lph in each treatment. The average relative error among all salinity treatments and discharge rates under irrigation level I1 was 10.86%.

In irrigation level I2 (Table 13.14), the minimum and maximum relative error of −0.82 and 17.32% was in T1Q2 and T4Q1 treatment. The observed and simulated yield was decreased with increase in salinity treatments. The average relative error was 8.79% considering all discharge rates and salinity treatments.

The relative error in observed and simulated yield under drip irrigation with different mixing ratios of saline and fresh water and discharge rates of emitters in I3 are presented in Table 13.15. It was found that the predicted and observed yield was decreased with increase in salinity of irrigation water. In irrigation level I3, the minimum and maximum relative error of 1.56 and 14.90 was for T2Q2 and T5Q3, respectively. The average relative error under I3 level was 8.12% considering all T-Q combinations.

The performance of the model tested by regression analysis for crop yield between the observed and simulated yield for all irrigation and discharge rate levels. The good agreement was found with correlation coefficient r^2 of 0.896 and is shown in Fig. 13.24. The graphical representation of the actual and potential water uptake (actual yield obtained by the model for treatment T2Q1 in irrigation level I1 and level I3) are presented in Figs. 13.25 and 13.26 that indicates predicted yield of 48 t/ha in I1 and 55 t/ha in I3, respectively. The observed and predicted yield under drip irrigation with different ratios of salinity of water and discharge rates at irrigation level I1, I2 and I3 are presented in Figs. 13.27 to 13.29.

TABLE 13.13   Relative Error in Observed and Simulated Yield Under Drip Irrigation with Different Mixing Ratios of Saline and Fresh Water and Discharge Rates of Emitters in I1

Treatment	Discharge rate	Simulated yield	Observed yield	Relative error
	Q1	53.7	54.44	1.36
	Q2	53.23	55.53	4.14
T1	Q3	53	55.03	3.69
	Q1	47.76	51.2	6.72
	Q2	46.85	51	8.14
T2	Q3	44.3	45.12	1.82
	Q1	43.2	48.20	10.37
	Q2	42.53	48.59	12.47
T3	Q3	40.58	43.66	7.05
	Q1	35.56	45.27	21.45
	Q2	33.32	40.73	18.19
T4	Q3	31.78	40.84	22.18
	Q1	32.23	39.47	18.34
	Q2	30.67	31.74	3.37
T5	Q3	28.2	36.92	23.62
			ARE	10.86
			RMSE	5.60
			RRMSE	12.23

TABLE 13.14   Relative Error in Observed and Simulated Yield Under Drip Irrigation With Different Mixing Ratios of Saline and Fresh Water and Discharge Rates of Emitters in I2

Treatment	Discharge rate	Simulated yield	Observed yield	Relative error
	Q1	60	63.41	5.38
	Q2	60	59.51	−0.82
T1	Q3	59.72	55.27	−8.05

	Q1	51.57	54.01	4.52
	Q2	48.05	56.5	14.96
T2	Q3	46.52	49.72	6.44
	Q1	43.86	45.93	4.51
	Q2	44.83	43.74	−2.49
T3	Q3	42.54	44.05	3.43
	Q1	36.48	44.14	17.35
	Q2	33.41	38.41	13.02
T4	Q3	32.17	38.14	15.65
	Q1	33.57	37.23	9.83
	Q2	31.27	36.59	14.54
T5	Q3	30.2	33.89	10.89
			ARE	**8.79**
			RMSE	**4.48**
			RRMSE	**9.60**

TABLE 13.15 Relative Error in Observed and Simulated Yield Under Drip Irrigation with Different Mixing Ratios of Saline and Fresh Water and Discharge Rates of Emitters in I3

Treatment	Discharge rate	Predicted yield	Observed yield	Relative error
	Q1	58.78	62.94	6.61
	Q2	60	56.86	−5.52
T1	Q3	60	55.41	−8.28
	Q1	54.75	52.63	−4.03
	Q2	52.5	53.33	1.56
T2	Q3	51.1	52.82	3.26

	Q1	50.48	46.33	−8.96
	Q2	47.52	45.45	−4.55
T3	Q3	45.31	42.32	−7.07
	Q1	38.78	41.7	7.00
	Q2	35.84	39.38	8.99
T4	Q3	34.23	39.33	12.97
	Q1	33.69	39	13.62
	Q2	32.29	37.76	14.49
T5	Q3	30.45	35.78	14.90
			ARE	**8.12**
			RMSE	**3.83**
			RRMSE	**8.20**

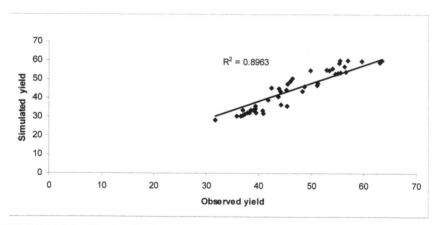

FIGURE 13.24   Simulated and observed yield (t/ha) under different treatments.

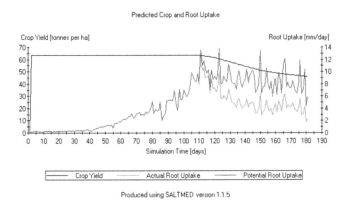

FIGURE 13.25   Crop yield predicted by the model for treatment T2Q1 in irrigation level I1.

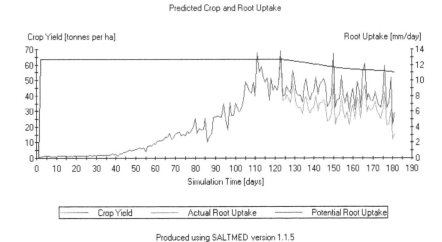

FIGURE 13.26   Crop yield predicted by the model for treatment T2Q1 in irrigation level I3.

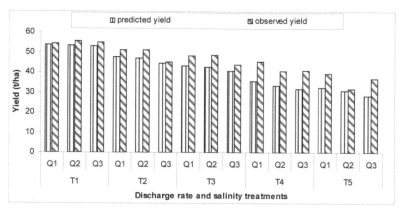

FIGURE 13.27 Simulated and observed yield under drip irrigation with different salinity ratios at different discharge rates of emitters at irrigation level I1.

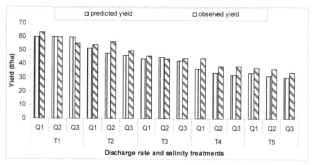

FIGURE 13.28 Simulated and observed yield under drip irrigation with different salinity ratios at different discharge rates of emitters at irrigation level I2.

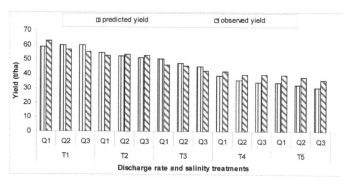

FIGURE 13.29 Simulated and observed yield under drip irrigation with different salinity ratios at different discharge rates of emitters at irrigation level I3.

13.4.4 SIMULATION OF THE SOIL MOISTURE DISTRIBUTION IN SOIL PROFILE

The model runs for simulating the soil moisture and salt distribution with initial moisture content and with initial salinity in the soil profile which was measured layer wise in the field on first day of start of model. The model run for treatment T1Q1 as the same treatment was used for calibrating the yield. The model gave the soil moisture and salinity on daily basis in the grid form within the domain. The model runs from first day to 65 days. On 65th day the observed and simulated values of soil moisture were compared. At each grid point, the relative error between the observed and simulated was calculated.

In TiQ1, it was observed that the relative error was minimum 2.91% at 30 cm depth and 30 cm away from drippers. The maximum relative error was 18.87% at 60 cm depth and 60 cm away from dripper. The overall average relative error was 7.38% during calibration (Table 13.16).

TABLE 13.16 Simulated and Observed Values of Soil Moisture During Calibration in T1Q1 Treatment

Depth cm	P 0 cm	O 0 cm	RE, %	P 10 cm	O 10 cm	RE, %	P 20 cm	O 20 cm	RE, %	P 30 cm	O 30 cm	RE, %
10	0.251	0.245	−2.657	0.250	0.239	−4.471	0.247	0.237	−4.177	0.242	0.235	−3.284
20	0.249	0.240	−3.411	0.248	0.230	−7.796	0.245	0.223	−9.735	0.240	0.222	−8.049
30	0.242	0.233	−3.777	0.241	0.232	−3.881	0.239	0.232	−3.019	0.237	0.230	−2.913
40	0.236	0.284	16.886	0.236	0.276	14.608	0.235	0.276	14.861	0.235	0.214	−9.876
60	0.323	0.345	6.360	0.323	0.345	6.357	0.323	0.332	2.668	0.324	0.399	18.868

P: predicted O: observed

13.4.5 VALIDATION FOR SOIL MOISTURE DISTRIBUTION

The calibrated parameter was used to generate the data of the soil moisture and salinity distribution. The data of soil moisture and EC in the soil profile observed depth wise on 65th day was used as initial value in each treatment T-Q combinations for predicting the soil moisture and EC on 7th April 2009 by running the model for one month considering start of simulation date as 65th (6th March 2009). The observed and simulated values of soil moisture in soil profile, relative error and average relative error in T-Q combinations treatments for irrigation level I2 were determined during validation as on 7th April 2009.

13.4.5.1   SOIL MOISTURE DISTRIBUTION UNDER IRRIGATION WITH FRESH WATER

It can be observed from Fig. 13.30 that the observed soil moisture (m³/m³) decreased with increase in the depth up to 30 cm depth and then it increased with the depth. The soil moisture decreased along the lateral distance from the emitter. The soil moisture in soil profile varied from 0.213 to 0.424. When irrigation is applied at Q1 discharge rate, the soil moisture of 0.22 was observed near the emitter source up to lateral distance of 20 cm and vertical distance of about 20 cm. When water was applied at discharge Q2, the moisture was spread laterally more as compared to Q1. The soil moisture was 0.23 up to 10 cm depth. In Q3, moisture content was spread laterally more near the surface as compared to Q1 and Q2. The average soil moisture of 0.25 was at the surface near to emitter source and along lateral distance of 25 cm. A close match was found between the observed and predicted values of moisture content in the soil profile in the upper 20–40 cm layer compared to lower profiles. In the lower profiles, moisture content increased because the water table at the experimental site was at 120 cm depth. The capillary action causes to increase the moisture up to depth of 60 cm from the surface.

13.4.5.2   SOIL MOISTURE DISTRIBUTION UNDER 75:25 (FRESH:SALINE WATER, T2) TREATMENT

It was observed from Fig. 13.31 that in treatment T2 for discharge rates Q1, Q2 and Q3, the soil moisture decreased with depth vertically up to 30 cm depth and then it increased with depth. The soil moisture was decreased with increase in lateral distance. In T2 treatment, the soil moisture varied from 0.212 to 0.42. The minimum relative error of -1.052% was in Q3 at 30 cm depth below the emitter. The moisture accumulation was more in Q1 as compared to Q2 and Q3 at lower depth. The soil moisture at 30 cm depth and 10 cm distance was 0.23 in Q1. The discharge rate has its effect on moisture spreading. The low discharge rate emitters reduce formation of the saturated zone below the emitters. The moisture at depth of 60 cm was increased due to capillary action. The higher discharge rate emitter spreads water on the surface having shallow depth. A close match was found between the observed and predicted values of soil moisture in soil profile up to 40 cm depth with minimum deviation of 1.3% in Q1 and maximum deviation of 10.04% in Q2.

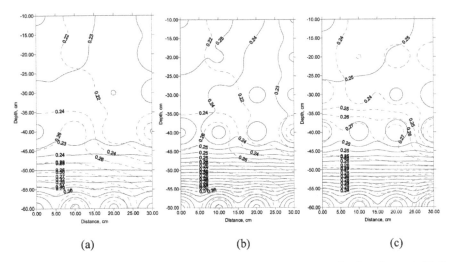

FIGURE 13.30   Observed (dotted) and predicted (solid) contours of soil moisture (m³/m³) distribution profile under: (a) Q1, (b) Q2, and (c) Q3 for treatment T1 on 97th days after transplanting (DAT).

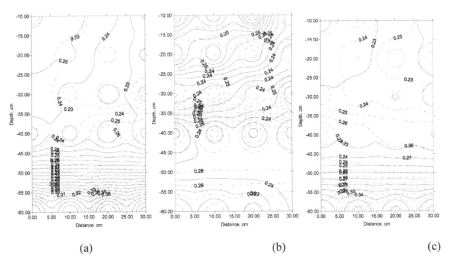

FIGURE 13.31   Observed (dotted) and predicted (solid) contours of moisture distribution profile under: (a) Q1, (b) Q2, and (c) Q3 for treatment T2 on 97th DAT.

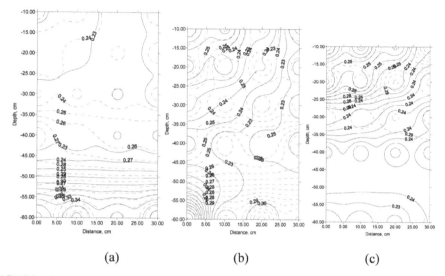

FIGURE 13.32   Observed (dotted) and predicted (solid) contours of moisture distribution profile under: (a) Q1, (b) Q2, and (c) Q3 discharge rates, for treatment T3 on 97th DAT.

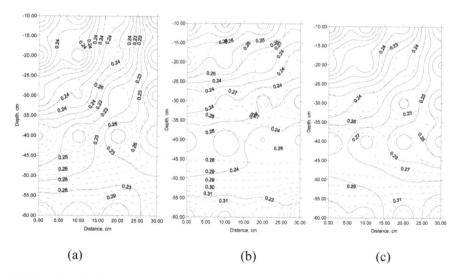

FIGURE 13.33   Observed (dotted) and predicted (solid) contours of moisture distribution profile under: (a) Q1, (b) Q2, and (c) Q3 discharge rates for treatment T4 on 97th DAT.

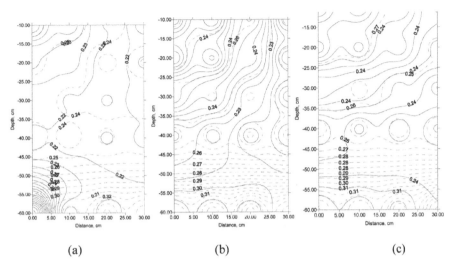

(a)   (b)   (c)

FIGURE 13.34 Observed (dotted) and predicted (solid) contours of moisture distribution profile under: (a) Q1, (b) Q2, and (c) Q3 discharge rates for treatment T5 on 97th DAT.

13.4.5.3 SOIL MOISTURE DISTRIBUTION AT 50:50 RATIO (FRESH:SALINE WATER) TREATMENT

From Fig. 13.32, it was observed that the soil moisture varied from 0.212 and 0.377. The maximum soil moisture was 0.28 in Q3 discharge rate near emitter source and up to 10 cm depth compared to Q1 and Q2. In all discharge rates, the soil moisture was decreased with increase in the depth in soil profile up to 30 cm depth and also along the lateral distance. The average moisture content in soil profile was more in Q1 compared to Q2 and Q3. The irrigation water spread laterally more in Q3 as compared to Q1 and Q2. The minimum deviation of 0.535% in soil moisture was in Q3 at 30 cm depth below the emitter (at 0 cm lateral distance). The maximum deviation of 23.95% was in Q2 at 30 cm distance and at 60 cm depth.

13.4.5.4 SOIL MOISTURE DISTRIBUTION AT 25:75 (FRESH:SALINE WATER) RATIO

It was observed from Table 13.16 and Fig. 13.25 that the soil moisture decreased vertically up to of 40 cm depth in Q1 and Q2. In T4 treatment, soil moisture varied from 0.211 to 0.344. Then, it was increased with increasing the depth. The soil moisture was decreased along the lateral distance. The moisture was higher at 0–10 cm depth and 45–60 cm depth. The minimum deviation in soil moisture was 1.607% in Q2 at 10 cm distance and at 40 cm depth. The maximum deviation of 19.53% was at 40 cm depth below the emitter. A close match was found between observed and predicted soil moisture in Q1 compared to Q2 and Q3.

13.4.5.5   SOIL MOISTURE DISTRIBUTION AT IRRIGATION WITH 100% SALINE WATER

It was observed from Fig. 13.34 that the soil moisture varied from 0.213 to 0.351. The soil moisture was maximum of 0.274 in Q3 near the emitter source up to 10 cm depth. The soil moisture was decreased with increase in depth and along lateral distance. The minimum deviation of 1.187% was at Q1 compared to Q2 and Q3. At lower depth, moisture was observed more due to capillary action of ground water. The maximum moisture was 0.27 in Q3 discharge rate along 10 cm distance and at depth of 10 cm.

These results are in close proximity with those of Badr and Taalab [13], Ah Koon et al. [4] and Hanson et al. [39, 40].

13.4.6   SALINITY DISTRIBUTION IN SOIL PROFILE

The contour maps of observed salinity in soil profile during crop period over the season at 30 90 and 150 DAT for three discharge rates and for five salinity ratios of irrigation water are depicted in Figs. 13.35–13.43. It was observed that the salinity increased laterally with increase in lateral distance and also with increase in depth below soil surface in all five salinity ratios. At low discharge rates, salts moved away from the emitter and also below the root zone. Thus, salts concentrate at the bottom of the root zone in T1 treatment.

In treatment T1, with Q1 discharge rate (Fig. 13.35), the salts concentrated at bottom at 75 cm depth and 30 cm away from emitter source after 30 DAT. The salt accumulation in the root zone profile increased with increase in duration of crop. After 90 DAT, the EC was increased along the lateral distance. It was increased to 1.2 and 3.5 dS/m after 90 and 150 DAT at 45–60 cm depth, respectively. The same trend was found in T2 and T3. In T1Q3 after 90 DAT, the salt concentration was about 2–3 dS/m at 10–20 cm depth and 20 cm away from emitters At 15–30 cm depth, more water moves laterally than vertically, the salts were accumulated at the top within 15 cm and below 60 cm. After 15 cm depth, the EC decreased with increase in depth up to 60 cm, and then it goes on increasing with depth. The lower discharge rate keep salts away from emitter at deeper depth compared to higher discharge rates. At higher salinity ratios, higher discharge rate spreads water on the surface and thus increase in accumulation of salts on surface due to high evaporation rates. Also salt accumulation increased in the soil profile with increase in salinity ratio and increase in duration of crop in each treatment. The results confirm the findings of Hanson et al. [39].

Therefore in surface drip irrigation, it can be concluded that soil salinity increased throughout the growing season in all treatments. Saline water irrigation produced three salinity zones: **upper salinity zone** near soil surface with high concentration; a **wide intermediate zone** where salinity levels are low near the dripper at the surface; and **lower zone** where the salinity levels increases with depth and with distance from the water source. However, the decreasing the discharge rates of applied water resulted in lower average salinity profiles and led to reduction in soil salinity at the end of the growing season compared to higher discharge rates. The higher discharge rate developed higher salt concentration in the root zone near the soil surface due to shallow wetted depth promote salt accumulation at the soil surface due to salt build up by evaporation components. This effect showed in the yield of crop, because lower discharge rate increased the yield as compared to higher discharge rates. The above results are in line with Badr and Taalab [13].

13.4.7 VALIDATION OF MODEL FOR EC DISTRIBUTION

The model validated for the EC distribution data along with simulated and observed values, relative error are plotted in Fig. 13.44. It can be observed from Fig. 13.36 that in T1Q1 treatment, the simulated EC and observed EC was decreased with increase in depth up to 45 cm and then it increased up to 75 cm depth and up to 20 cm lateral distance. The maximum EC in the soil profile was 2.17 dS/m at 30 cm lateral distance and 75 cm depth. The minimum EC was observed at 45 cm depth and 10 cm away from the emitter. With discharge rate Q2, the maximum EC was observed at 75 cm depth below the emitter. The EC goes on increasing with increasing the lateral distance on the surface. The salts were accumulated at 30 cm depth and up to 20 cm lateral distance. With Q3 discharge rate, maximum EC was 30 cm depth below the emitter (0 cm lateral distance) because at Q3 water spreads laterally more than vertically, the soil moisture with discharge rate Q3 remains at shallow depth.

In T2 treatment (Fig. 13.45) with discharge rate Q1, the maximum EC was in the soil profile at 75 cm below the emitter. The salinity was increased with increase in distance laterally up to 20 cm distance. With Q2 discharge rate, the maximum EC of 7.8 dS/m was at 75 cm depth and 30 cm away from the emitter source. In Q3 discharge rate, the maximum EC of 5.98 dS/m was at 75 cm depth and 20 cm distance from emitter.

In T3 treatment (Fig. 13.46) with discharge rate Q1, the maximum EC of 11.65 dS/m was at 75 cm depth. The EC observed values were decreased with increase in the lateral distance. With Q2 discharge rate, the maximum EC of 11.43 was at 75 cm depth below the emitter. The EC was increased with increasing the depth in soil profile. With Q3 discharge rate, the maximum EC of 7.93 dS/m was at 75 cm depth at 20 cm distance.

In T4 treatment (Fig. 13.47) with discharge rate Q1, it was observed that the EC was decreased with increase in depth up to 45 cm depth and again it increased with depth up to 75 cm. The maximum EC of 12.34 dS/m was at 20 cm distance and 75 cm depth. At Q2 discharge rate, the same trend was observed as Q1 and it was maximum of 12.62 dS/m at 75 cm depth below the emitter. With increase in the discharge rate Q3, EC increased along the lateral distance up to 20 cm and again it decreased up to 30 cm. The maximum EC of 12.33 dS/m was at 75 cm depth, as the salts are concentrated at 75 cm depth.

In T5 treatment (Fig. 13.48) with discharge rate Q1, it was observed that EC decreased with increase in depth up to 45 cm depth and again it increased with depth up to 75 cm. The maximum of EC of 22.25 dS/m was at 20 cm distance and 75 cm depth. At Q2 discharge rate, the same trend was observed as Q1 and maximum observed EC of soil 22.35 dS/m was at 30 cm distance and 75 cm depth. With increasing discharge rate Q3, the EC increased along the lateral distance at depth of 75 cm.

It can be concluded that the salts were concentrated at top of soil surface due to high evaporative demand. In between, salt concentration was decreased with increase in soil depth. At bottom, salts increased with increase in depth. These findings are in close proximity with those of Hanson et al. [38] and Badr and Taalab [13].

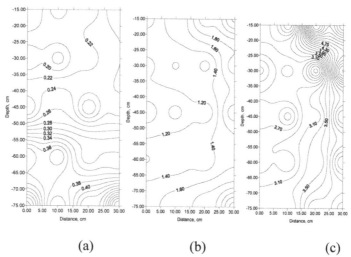

(a)   (b)   (c)

FIGURE 13.35   Observed salinity distribution in soil profile during crop period in T1Q1 treatment: (a) 30 DAT, (b) 90 DAT, and (c) 150 DAT.

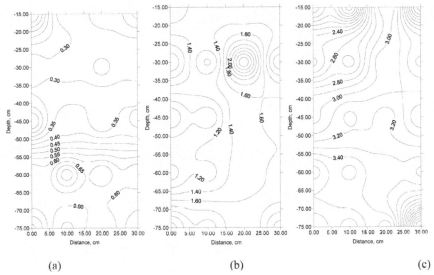

(a)   (b)   (c)

FIGURE 13.36   Observed Salinity distribution in soil profile during crop period in T1Q2 treatment: (a) 30 DAT, (b) 90 DAT, and (c) 150 DAT.

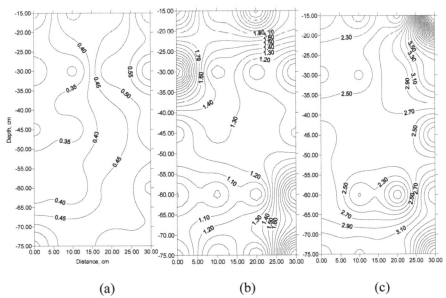

(a)   (b)   (c)

FIGURE 13.37   Observed Salinity distribution in soil profile during crop period in T1Q3 treatment: (a) 30 DAT, (b) 90 DAT, and (c) 150 DAT.

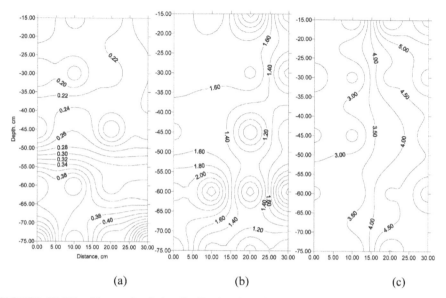

FIGURE 13.38   Observed salinity distribution in soil profile during crop period in T2Q1 treatment: (a) 30 DAT, (b) 90 DAT, and (c) 150 DAT.

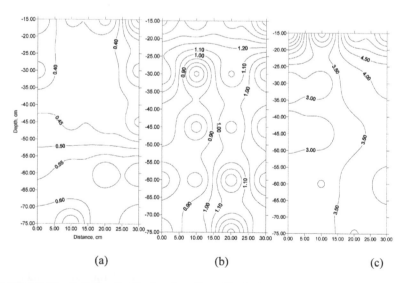

FIGURE 13.39   Salinity distribution in soil profile during crop period in T2Q2 treatment after: (a) 30 DAT, (b) 90 DAT, and (c) 150 DAT.

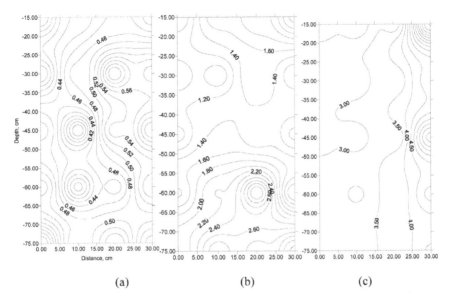

FIGURE 13.40 Salinity distribution in soil profile during crop period in T2Q3 treatment after: (a) 30 DAT, (b) 90 DAT, and (c) 150 DAT.

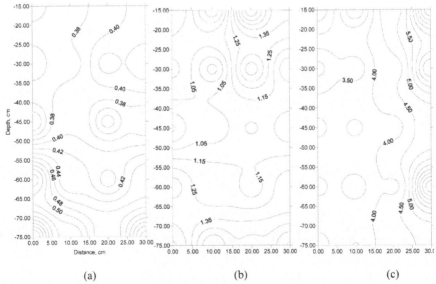

FIGURE 13.41 Salinity distribution in soil profile during crop period in T3Q1 treatment: (a) 30 DAT, (b) 90 DAT, and (c) 150 DAT.

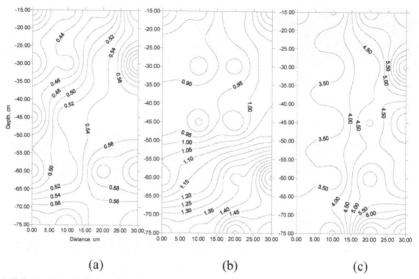

(a)   (b)   (c)

FIGURE 13.42   Salinity distribution in soil profile during crop period in T3Q2 treatment after: (a) 30 DAT, (b) 90 DAT, and (c) 150 DAT.

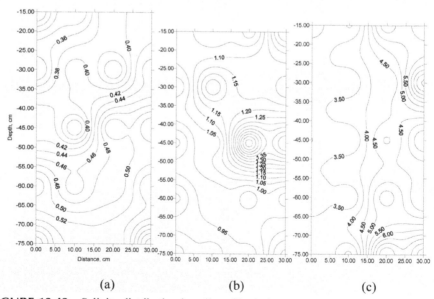

(a)   (b)   (c)

FIGURE 13.43   Salinity distribution in soil profile during crop period in T3Q3 treatment after: (a) 30 DAT (b) 90 DAT (c) 150 DAT.

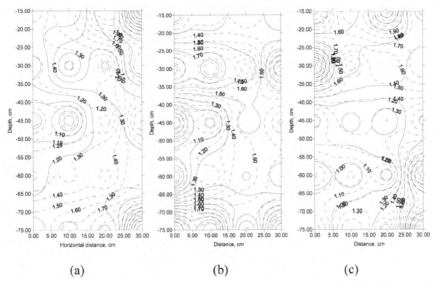

(a)   (b)   (c)

FIGURE 13.44   Observed (dotted) and predicted (thick) contours of soil salinity profile under: (a) Q1, (b) Q2, and (c) Q3 discharge rates on 97th day for treatment T1.

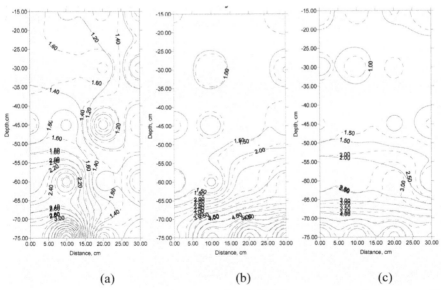

(a)   (b)   (c)

FIGURE 13.45   Observed (dotted) and predicted (thick) contours of soil salinity profile under: (a) Q1, (b) Q2, and (c) Q3 discharge rates on 97th day for treatment T2.

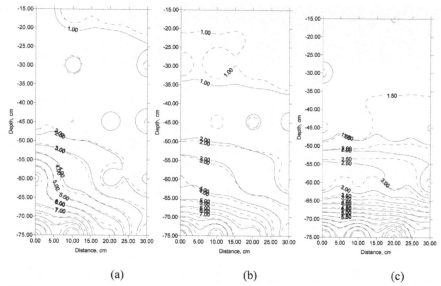

(a)   (b)   (c)

FIGURE 13.46   Observed (dotted) and predicted (thick) contours of soil salinity profile under: (a) Q1, (b) Q2, and (c) Q3 discharge rates on 97th day for treatment T3.

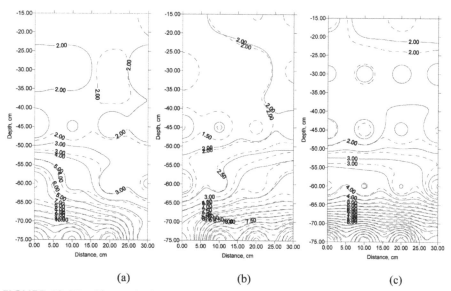

(a)   (b)   (c)

FIGURE 13.47   Observed (dotted) and predicted (thick) contours of soil salinity profile under: (a) Q1, (b) Q2, and (c) Q3 discharge rates on 97th day for treatment T4.

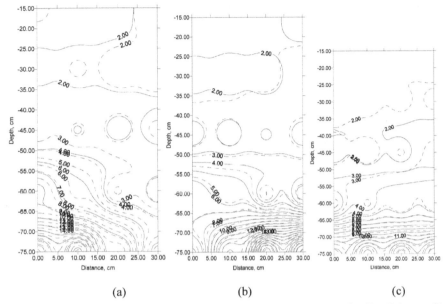

(a) (b) (c)

FIGURE 13.48 Observed (dotted) and predicted (thick) contours of soil salinity profile under: (a) Q1, (b) Q2, and (c) Q3 discharge rates on 97th day for treatment T5.

13.4.8 TESTING OF GOODNESS OF FIT OF MODEL

To evaluate the model performance, the average relative error between the observed and simulated values of soil moisture and salinity were calculated and these are given in Tables 13.17 and 13.18, respectively. The performance of the model was also tested by regression analysis between the observed and simulated moisture and salinity in the profile (Figs. 13.49 and 13.50), respectively. The average relative error between simulated and observed values of soil moisture in soil profile varied from 7.57 to 14.64, RMSE varied from 0.0269 to 0.0517, and RRMSE varied from 10.22 to 18.68 with $R^2 = 0.735$. The average relative error between simulated and observed values of salinity (EC) in soil profile varied from 12.68 to 21.22, RMSE varied from 0.248 to 1.24 and RRMSE varied from 7.65 to 27.01% good agreement with $R^2 = 0.928$.

TABLE 13.17 Average Relative Error, RMSE and RRMSE Between the Simulated and Observed Values of Moisture Content in the Soil Profile During Validation

Treatments	Average relative error	RMSE	RRMSE
T1Q1	10.63	0.0313	12.63
T1Q2	13.50	0.0390	15.65
T1Q3	9.69	0.0269	10.22
T2Q1	7.57	0.0287	10.91
T2Q2	7.63	0.0288	11.17
T2Q3	11.04	0.0424	15.92
T3Q1	7.67	0.0296	11.15
T3Q2	9.21	0.036	14.15
T3Q3	14.15	0.0517	17.94
T4Q1	12.06	0.0479	18.68
T4Q2	13.05	0.044	16.26
T4Q3	14.64	0.0512	18.04
T5Q1	9.84	0.0475	18.38
T5Q2	11.71	0.0485	17.96
T5Q3	13.05	0.0493	17.72
Overall	11.03	0.040	15.12

TABLE 13.18 Average Relative Error, RMSE and RRMSE Between the Simulated and Observed Values of Salinity (Ec) in the Soil Profile During Validation

Treatments	Average relative error	RMSE	RRMSE
T1Q1	17.19	0.354	24.83
T1Q2	16.02	0.248	18.48
T1Q3	18.92	0.317	20.97
T2Q1	17.88	0.349	8.28
T2Q2	17.48	0.424	10.34
T2Q3	18.09	0.511	12.49
T3Q1	16.75	0.453	10.56

T3Q2	17.91	0.478	11.84
T3Q3	14.20	0.323	7.65
T4Q1	17.93	1.17	25.43
T4Q2	18.35	0.727	16.02
T4Q3	12.68	0.493	10.91
T5Q1	15.21	0.642	12.87
T5Q2	17.67	1.05	22.52
T5Q3	21.22	1.24	27.01
Overall	**17.16**	**0.585**	**16.01**

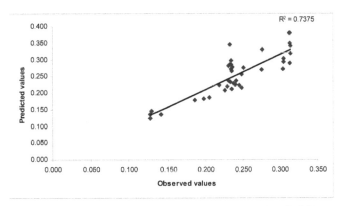

FIGUR 13.49 Observed and simulated values of soil moisture for all different salinity and discharge rate treatments.

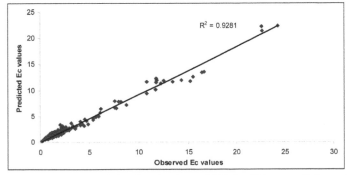

FIGURE 13.50 Observed and simulated values of EC for different salinity and discharge rate treatments.

13.4.9 INTERPRETATION OF STATISTICAL ANALYSIS OF YIELD DATA

The observed yield data was statistically analyzed. The ANOVA are presented in Appendix B. It was found that there was significant difference in discharge rate of emitters on yield of crop. Also, the interaction IxQxT was found significant. The significant difference was found between the each salinity of irrigation water treatments. The best IxQxT in each salinity treatments was taken for comparing the yield of crop along with the quality parameters of the tomato and these are presented in Table 13.20.

TABLE 13.19 Yield and Quality Parameters of Tomato With Best Combinations of Discharge Rates and Irrigation Levels Under Different Salinity of Water Treatments

Best Treatment	Yield	WUE	Decrease in yield	TSS	Acidity	Ascorbic acid
	t/ha	t/ha-cm	%	°Brix	%	mg/100 g
T1I2Q1 (0.38 ds/m)	63.41	1.56	-	5.5	0.684	31.48
T2I2Q2 (6.30 dS/m)	56.49	1.14	10.91	5.9	0.698	32.76
T3I1Q2 (9.1 dS/m)	48.56	1.37	23.38	6.2	0.737	35.10
T4I1Q1 (14.7 dS/m)	45.26	1.27	28.62	6.5	0.789	35.08
T5I1Q1 (19.5 dS/m)	39.47	1.11	37.75	6.9	0.801	36.80

Best treatment: T1I2Q1, (0.38 ds/m)

It was observed from Table 13.19 that the best treatment among all salinity treatments was T1I2Q1, which gave maximum yield. The WUE under best treatment in each salinity shows that the maximum WUE was in T1 followed by T3, T4, T2 and T5.

When fresh water is available, the yield got maximum at irrigation level of 0.8 and applying irrigation water with 1.2 discharge rate with WUE of 1.56 t/ha-cm. The yield is increased with increase in irrigation level from 0.6 to 1.0. The irrigation level 0.8 was found best followed by 1.0 and 0.6. The yield decreased with increase in salinity treatments. The quality parameters were increased with increase in application of saline water. When 100% saline irrigation was applied, there was about 37.75% yield reduction compared to fresh water but it improved the quality of tomato fruits. When available irrigation water at EC of 6.3 dS/m, the best yield can be obtained with irrigation level 0.8 and 2.4 discharge rate of emitters with 10.91% reduction in yield but reducing the WUE to 26%. The TSS, acidity and ascorbic acid increased in treatment T2 compared to T1. When the irrigation water at 9.1 salinity was available, I1Q2 combination gave the best yield with reduction in yield of 23.38% with significant increase in quality parameters compared to T1 reducing the WUE of 12.17% compared to T1. When the salinity ratios in irrigation water was at 14.7 dS/m, the I1Q1 combination was best for getting maximum yield with the given salinity treatment. Therefore, considering the objective in our mind about yield reduction, WUE and quality, the discharge rate and irrigation levels can be decided to get maximum yield from availability of different quality of saline water.

13.5 CONCLUSIONS

Based upon this research study, the following specific conclusions are drawn:

1. The average relative error between simulated and observed values of soil moisture in soil profile varies from 7.57 to 14.64%, RMSE varies from 0.0269 to 0.0517, and RRMSE varies from 10.22 to 18.68% with $R^2 = 0.735$. The overall average of relative error, RMSE and RRMSE were 11.03%, 0.04 and 15.12%, respectively.

2. The average relative error between simulated and observed values of salinity (EC) in soil profile varies from 12.68 to 21.22, RMSE varies from 0.248 to 1.24 and RRMSE varies from 7.65 to 27.01% with good agreement with $R^2 = 0.928$. The overall average of relative error, RMSE and RRMSE were 17.16%, 0.585 and 16.01%, respectively.

3. The relative error for crop yield varies from 1.36 to 23.62, 0.82 to 17.32 and 1.56 to 14.90 at irrigation levels I1, I2 and I3, respectively with $R^2 = 0.893$. The average relative error for crop yield was 10.86, 8.79 and 8.12% in different salinity and discharge rate treatments at irrigation levels I1, I2 and I3, respectively. The RRMSE for crop yield was 12.23, 9.60 and 8.20 at irrigation level of I1, I2 and I3, respectively.

4. In 100% fresh water treatment with EC of 0.38 dS/m, the yield was maximum compared to saline water treatment. The maximum yield of 63.41 t/ha was in fresh water treatment at irrigation level 0.8 and discharge rate of 1.2 lph.

5. If available water is of 6.3 dS/m (with 75:25% of Fresh:Saline), then IW/CPE ratio of 0.8 and discharge rate of 2.4 lph gave best result. The maximum yield under this treatment was reduced by 10.91% but gave better

quality parameters (TSS, acidity and ascorbic acid) compared to 100% fresh water treatment.

6. In 50:50% (Fresh:Saline) water with EC of 9.1 dS/m, the yield was maximum at irrigation level 0.6 and discharge rate 2.4 lph. The maximum yield under this treatment was reduced by 23.38% but quality of tomato was improved compared to T1 and T2.

7. In treatment 25:75% (F:S) T4 with EC of 14.5 dS/m, the yield was maximum at irrigation level 0.6 and discharge rate 1.2 lph; was reduced by 28.62% compared to 100% fresh water treatment.

8. In 100% saline water treatment (T5) with EC of 19.5 dS/m, the yield was maximum under irrigation level 0.6 and discharge rate of 1.2 lph or 2.4 lph, with reduction of 37.75% compared to fresh water treatment.

9. The average plant yield was significantly different among all salinity treatments. The plant yield was reduced by 44.52% under 100% saline water treatments as compared to 100% fresh water treatment. The plant yield was reduced by 11.27% when irrigated with 25% saline water as compared to 100% fresh water.

10. The discharge rate of 1.2 lph gave maximum yield when 100% fresh water used for irrigation compared to 4.2 lph. The 1.2 and 2.4 lph discharge rate were at par and it can be used for irrigating tomato crop with 100% fresh water.

11. A lower rate of water application of 1.2 lph reduces the salinity in the soil profile as compared to higher discharge rates of 2.4 and 4.2 lph, because the salts accumulate on the periphery of water front outside the root zone of crop with 1.2 lph discharge rates.

12. TSS was maximum when irrigated with 100% saline water with irrigation level of 0.6 and discharge rate of 1.2 lph and was minimum under fresh water. The TSS of tomato fruit was increased with decrease in irrigation levels and discharge rates.

13. The average acidity of tomato fruits increased with increase in salinity treatment and irrigation levels and decreased with increase in discharge rates. An average acidity of 0.818% was maximum when irrigated with 100% saline water at irrigation level of 0.8 with discharge rate of 1.2 lph and minimum was 0.689% when irrigated with fresh water at irrigation level of 1.0 and discharge rate of 4.2 lph.

14. The average value of ascorbic acid was minimum of 30.23 mg/100 g and was maximum of 39.87 mg/100 g under fresh water and 100% saline water treatments, respectively. No significant effects were found in ascorbic acid when irrigation was applied at three different discharge rates.

15. The WUE in tomato crop decreased with increase in salinity ratios/salinity levels of irrigation water. The maximum and minimum WUE of 1.22 t/ha-cm and 0.714 t/ha-cm were under treatment T1 and T5, respectively. The WUE

decreased with increasing irrigation levels from I1 to I3. In 100% fresh water treatment, maximum WUE of 1.56 t/ha-cm was found when irrigated at irrigation level I1 with discharge rate of 2.4 lph.

It is recommended that if fresh water is available for irrigation, the maximum yield can be obtained by applying water with discharge rate of 1.2 lph at IW/CPE = 0.8. To get better quality of tomato with reduction in yield of 11% compared to fresh water, it is recommended that the mixing of 75% fresh and 25% saline water with EC of 6.3 dS/m can be applied with discharge rate of 2.4 lph and IW/CPE = 0.8. The increase in salinity ratio increased the quality of tomato but reduced the yield. Therefore, keeping in mind the availability of both fresh and saline water, these values of yield reduction and quality of tomato, the discharge rate and IW/CPE ratio are recommended.

13.6 SUMMARY

Drip irrigation is one of the efficient methods of irrigation, which is especially useful in water scarcity areas with poor quality of irrigation water. The increased use of drip irrigation improves the sustainability of irrigation system around the world. The frequent application of water through drip system results in maintenance of high matric potential and results in better tolerance of crop to irrigation with saline water. When saline water is applied through drip irrigation, many management practices need to be modified such as salinity levels, irrigation method and water management practices.

Using low quality irrigation water with drip irrigation, prevents possible damage to the foliage compared to sprinkler irrigation and because of salt accumulated at the wetting front, soil salinity in the root zone is found near to the initial salinity of the irrigation water when irrigation water is managed properly. Mathematical models and laboratory experiment that describe water and solute transport in the soil are available. However, the soil salinity transport studies at field level are limited. Also, the water application rate is one of the factors, which determines the soil moisture and salt distribution around the dripper and related root distribution and plant water uptake patterns. The discharge rate of emitters as well as salinity levels in irrigation water has influence on the distribution of soil moisture and salt in the root zone. The characteristics of soil moisture and salt distribution influence the growth and yield of crop substantially.

In southwestern Punjab, canal water is mostly used for the irrigation. If canal water is not sufficient, the tube-well water is used to meet the water requirement of crop. The quality of subsurface ground water is not very good in this region. In order to study effects of different salinity levels of irrigation water on yield and quality of tomato crop, the experiment was conducted at research farm of Central Institute of Post-Harvest Engineering and Technology (CIPHET), Abohar – India. The five levels of saline water were used for irrigation with drip system. The saline water used for irrigation was prepared by mixing saline (tube-well) and fresh (canal) water in five different ratios: 100% fresh (T1), 75% Fresh+25% saline (T2), 50% Fresh+50% saline (T3), 25% fresh+75% saline (T4), and 100% saline water (T5). The mixed water was applied with three different discharge rates of inline emitters of 1.2 lph, 2.4 lph and 4.2 lph. The average salinity of irrigation water (ECw) was measured under different salinity ratios and it was found to be 0.38, 6.30, 9.1, 14.7, 19.5 dS/m in T1, T2, T3, T4 and T5, respectively. The soil moisture and salinity distribution influenced by different discharge rates of emitters and salinity of irrigation water during the experimental period were studied.

Tomato hybrid seedlings were transplanted in the field. The seedlings were covered with polyethylene by low tunnel during winter. The cultural practices as recommended by PAU, Ludhiana were followed during experimentation. The crop was irrigated with five different salinity of mixed water. Drip irrigation considered was a line source as the wetting pattern of tomato plants overlap over each other. The planar two dimensional flow models were used for simulation. The plant growth parameters were plant height, percent ground cover, dry matter, plant yield, total yield under different salinity levels and rate of water application that were recorded during the growing period. Also, the quality parameters such as TSS, acidity, pH, ascorbic acid, firmness were studied. The WUE under different salinity and discharge rate of emitters were studied after harvest of crop.

Models can be useful tools in agricultural water management. They cannot only help in irrigation scheduling and crop water requirement calculations but also can be used to predict yields and soil salinization. Most of the models were earlier used for the study are single process oriented. Ragab [77] developed SALTMED model which uses well established water and solute transport, evapotranspiration and crop water uptake equations. The evapotranspiration has been calculated using Penman-Monteith equation. The actual water uptake in the presence of saline water is determined. The salt and water transport were determined by well-established Richard's equation for two-dimensional flows and dispersion convection equation, respectively. The water and solute flow equations were solved numerically using finite difference explicit scheme in the model itself. SALTMED model was used in the study for simulating salt and moisture distribution and crop yield. The necessary data required for simulating and validating salt and moisture distribution in tomato crop was collected through the experiment. The model regime uses climate data files, irrigation data files, crop data and soil data files. The crop parameters data such as maximum height, root depth, the values of Kc at different growth periods were collected from literature and the actually conducting experiment. Soil parameters data such as saturated hydraulic conductivity and unsaturated hydraulic conductivity were based on field measurement and moisture retention curves based on laboratory measurements. The initial soil moisture and salinity in the soil profile were observed in the field. The model runs initially with given parameters to get values of simulated soil moisture and salinity in the soil profile on daily basis.

The model was calibrated against the final yield under fresh water treatment T1 and discharge rate Q1 as it gave maximum yield. The soil moisture and salt distribution data were collected during each growth stage for each treatment combinations in the field. The soil moisture data was calibrated for 65th day for the T1Q1 treatment, which gave the higher yield. It was validated at middle of growth stages at 97th day for soil moisture and salt distribution, respectively. The model was tested for goodness of fit using relative error (RE), average relative error (ARE), root mean square error (RMSE) and relative root mean square error (RRMSE).

KEYWORDS

- acidity
- arid region
- ascorbic acid
- average relative error, ARE
- canal water
- Central Institute of Post-Harvest Engineering and Technology, CI-PHET
- days after transplanting, DAT
- discharge rate
- dry matter
- emitter
- firmness
- fresh water
- ground cover
- ground water
- growing period
- irrigated agriculture
- irrigation
- moisture wetting pattern
- permeability
- plant growth parameter
- plant height
- plant yield
- rainfall
- relative error, RE

- relative root mean square error, RRMSE
- root mean square error, RMSE
- saline water
- salt distribution
- SALTMED model
- saturated hydraulic conductivity
- semiarid region
- sodic water
- soil moisture
- soil profile
- tomato
- total yield
- unsaturated hydraulic conductivity
- water application
- water scarcity
- water use efficiency, WUE

REFERENCES

1. Abu-Awwad, A. M. (2001). Influence of different water quantities and qualities on lemon trees and soil salt distribution at the Jordan Valley. *Agri. Water Manage.*, *52*, 53–71.
2. Agrawal, M. C., Khanna, S. S. (1983). Efficient soil and water management in Haryana. Bulletin Haryana Agricultural University, Hisar, India. pp. 118.
3. Agrawal, P. N., Purohit, R. C., Singh, J. (2003). Effect of water salinity on tomato under drip irrigation. Proceedings of 37th ISAE Annual Convention and symposium held at Udaipur, on January 29–31.
4. Ah-Koon, P. D., Gregory, P. J., Bell, J. P. (1990). Influence of drip irrigation emission rate on distribution and drainage of water beneath a sugar cane and a fallow plot. *Agri. Water Mangt.*, *17*, 267–282.
5. Ainechee, G. S., Boroomand, N., Behzad, M. (2009). Simulation of Soil Wetting Pattern under Point Source Trickle Irrigation. *J. App. Sci.*, *9*, 1170–1174.
6. Alarcon, J. J., Bolarin, M. S., Sanchez-Blanco, M. J., Torrecillas, A. (1994). Growth, yield and water relations of normal fruital and cherry tomato cultivars irrigated with saline water. *J. Hortic. Sci.*, *69(2)*, 283–288.
7. Allen, R. G., Pereira, L. S., Raes, D., Smith, M. (1998). *Crop Evapotranspiration*, Irrigation and Drainage paper No. 56. FAO, Rome, Italy, 300 pp.
8. Amor, F., Martinez, V., Cerda, A. and del-Amor, F. M. (1999). Salinity duration and concentration affect fruit yield and quality, and growth and mineral composition of melon plants grown in perlite. *Hort. Sci.*, *34(7)*, 1234–1237.

9. Amorin, A., Fernandes, P. D., Gheyi, H. R., Azevedo, N. C., Amorim, J. R., Azevedo, N. C. (2002). Effect of irrigation water salinity and its mode of application on garlic growth and production. *Pesquisa Agropecuaria Brasileira, 37*, 167–176.

10. Angelino, G., Ascione, S., Ruggiero, C. (2003). Growth and water relations of sun-cured tobacco irrigated with saline water. Beitrage zur Tabakforschung International, *20(6)*, 394–401.

11. Anonymous, (2008). Package of practices for cultivation of vegetables. pp. 21–27. Punjab Agricultural University, Ludhiana.

12. Assouline, S. S., Cohen, D. M., Hardod, T., Rosner, M. (2002). Micro drip irrigation for field crops: effects on yield, water uptake and drainage in sweet corn. *Soil Sci. Soc. Am. J., 66*, 228–233.

13. Badr, M. A., Hussein, S. D. (2008). Yield and fruit quality of drip-irrigated cantaloupe under salt stress conditions in an arid environment. *Australian J. Basic Applied Sci., 2(1)*, 141–148.

14. Badr, M. A., Taalab, A. S. (2007). Effect of drip irrigation and discharge rate on water and solute dynamics in sandy soil and tomato yield. *Australian J. Basic Applied Sci., 1(4)*, 545–552.

15. Beltrao, J., Faria, J., Miguel, G., Chaves, P., Trindade, D., Ferreira-Jones, H. G. (2000). Cabbage yield response to salinity of trickle irrigation water. *Acta Hort., 537*, 641–645.

16. Ben-Asher, J., van, D., Feddes, R. A., Jhorar, R. K. (2006). Irrigation of grapevines under saline water II: Mathematical simulation of grapevine growth and yield. *Agric. Water Mangt., 83, 22–29.*

17. Bielorai, H., Shalhevet, J., Levy, Y. (1978). Grapefruit response to variable salinity in irrigation water and soil. *Irrig. Sci., 1*, 61–70.

18. Brar, J. S., Singh, B. (1993). Underground irrigation water quality in south-western Punjab. *J. Res. Punjab Agric. Univ.* 30(1–2):15–23.

19. Bresler, E. (1975). Two-dimensional transport of solute during nonsteady infiltration for a trickle source. *Soil Sci. Soc. Am. Proc. 39*, 604–613.

20. Cetin, Oner and Demet, U. (2008). The effect of drip line spacing, irrigation regime and planting geometries of tomato on yield, irrigation WUE and net return. *Agric. Water Mangt., 95*, 949–958.

21. Chauhan, S. K., Chauhan, C. P., Minhas, P. S. (2007). Effect of cyclic use and blending of alkali and good quality waters on soil properties, yield and quality of potato, sunflower and Sesbania. *Irrig. Sci., 26*, 81–89.

22. Choudhary, O. P., Ghuman, B. S., Josan, A. S., Bajwa, M. S. (2006). Effect of alternating irrigation with sodic and nonsodic waters on soil properties and sunflower yield. *Agric. Water Mangt., 85*, 151–156.

23. Claire, M., Cote, K. L., Bristow, P. B., Charles, W., Freeman, J., Cook, C., Eter, J. T. (2003). Analysis of soil wetting and solute transport in subsurface trickle irrigation. *Irrig. Sci. 22*, 143–156.

24. Coelha, F. E., Or, D. (1997). Application of analytical solutions for flow from point sources to drip irrigation management. *J. Am. Soc. Soil Sci., 61*, 1331–1341.

25. Cote, C. M., Bristow, K. L., Charles W., Cook, C., Thorburn, P. J. (2003). Analysis of soil wetting and solute transport in subsurface trickle irrigation. (Special issue: Micro-irrigation: advances in system design and management). *Irrig. Sci. 22*, 143–156.

26. Cuarter, J., Rafael, F. M. (1999). Tomato and salinity. *Scientia Hort., 78*, 83–125.

27. Cucci, G., Cantore, V., Boari, F., Caro, A., Ferreira, M. I. (1999). Water salinity and influence of SAR on yield and quality parameters in tomato. *Acta Hort., 537*, 663–70.

28. Dam, van, J. C., Stricker, J. N. M., Droogers, P. (1994). Inverse method to determine soil hydraulic functions from multistep outflow experiments. *Soil Sci. Soc. Am. J., 58*, 647–652.

29. Dasberg, S., Or, D. (1999). Drip Irrigation. Springer–Verlag, Berlin, 162 pp.

30. Dehghanisanij, H., Yamamoto, T., Inoue, M., Akbari, M. (2007). Water flow and solute transport under drip irrigation in sand dune field. *J. Applied Sci., 7(20)*, 2997–3005.

31. Dunage, V. S., Balakrishnan, P., Patil, M. G. (2009). WUE and economics of tomato using drip irrigation under net house conditions. *Karnataka J. Agric. Sci., 22(1)*, 133–136.

32. Fares, A., Parsons, L. R., Wheaton, T. A., Morgan, K. T., Simunek, J., Genuchten, M. T. (2001). *Proceedings of the Florida State Horticultural Society, 114,* 22–24.
33. Feng, G. L., Mairi, A., Letey, J. (2003). Evaluation of the model for irrigation management under saline conditions: II salt distribution and rooting pattern effects. *Soil. Sci. Soc. Am. J., 67,* 77–80.
34. Franco, J. A., Perez-Saura, P. J., Fernandez, J. A., Parra, M., Garcia, A. L. (1999). Effect of two irrigation rates on yield, incidence of blossom-end rot, mineral content and free amino acid levels in tomato cultivated under drip irrigation using saline water. *J. Hort. Sci. and Biotech., 74(4),* 430–435.
35. Gawad, G., Abdel, A. A., Gaihbe, A., Kadouri, F. (2005). The effects of saline irrigation water management and salt tolerant tomato varieties on sustainable production of tomato in Syria (1999–2002). *Agric. Water Mangt., 78,* 39–53.
36. Giordano, L. B., Silva, J. B. C., Barbosa, V. (2000). Tomato for industrial processing. EMBRAPA, CNPH, Brasilia, pages 36–59.
37. Golabi, M., Naseri, A. A., Kashkuli, H. A. (2009). Evaluation of SALTMED model performance in irrigation and drainage of sugarcane farms in Khuzestan province of Iran. *J. Food Agric. Environ., 7(2),* 874–880.
38. Hanson, B., Don, M. (2007). The effect of drip line placement on yield and quality of drip-irrigated processing tomatoes. *Irrig. Drainage Syst., 21,* 109–118.
39. Hanson, B., Hutmacher, R. B., May, D. M. (2006). Drip irrigation of tomato and cotton under shallow saline ground water conditions. *Irrig. Drain. Syst. 20,* 155–175.
40. Hanson, B., May, D. (2004). Effect of subsurface drip irrigation on processing tomato yield, water table depth, soil salinity and profitability. *Agric. Water Mangt., 68,* 1–17.
41. Harbi, A. L., Wahb, A. R., Allah, M. A., Omran, A. M., 2009. Effects of salinity and irrigation management on growth and yield of tomato grown under greenhouse conditions. *Acta Hort., 807(1),* 201–206.
42. Hillel, D. (1977). *Computer Simulation of Soil Water Dynamics: A Compendium of Recent Work.* IDRC, Ottawa, Canada, 214 pp.
43. Huang, Q., Sheng Xiu, S. Y., Huang, Q., Song, Y. D. (2003). The movement of water and salt in sandy land after irrigated with saline water. *Acta Pedologica Sinica. 40(4),* 547–553.
44. Incalcaterra, G., Curatolo, G., Iapichino, G., Pardossi, A., Tognoni, F. (2003). Influence of the volume and salinity of irrigation water on winter melon (*Cucumis melo inodorus Naud*) grown under plastic tunnel. *Acta Hort. 609,* 423–427.
45. Jain, A. K., Kumar, R. (2007). Water management issues – Punjab, North-West India. Indo-US Workshop on Innovative E-technologies for Distance Education and Extension/Outreach for Efficient Water Management, March 5–9, ICRISAT, Patancheru/Hyderabad, Andhra Pradesh, India.
46. Jinguan, W., Renduo, Z., Shengxiang, G. (1999). Modeling salt water movement with water uptake by roots. *Plant and Soil, 215,* 7–15.
47. Kadam, J. R., Bhingardeve, S. D., Walke, V. N. (2007). Effect of saline water and fertigation on the yield contributing parameters of eggplant. International *J. Agric. Sci., 3(1),* 162–164.
48. Kadam, J. R., Patel, K. B. (2001). Effect of saline water through drip irrigation system on yield and quality of tomato. *J. Mah. Agric. Univ., 26(1),* 8–9.
49. Karlberg, L., Per-Erik, J., David, G. (2007). Model-based evaluation of low-cost drip-irrigation systems and management strategies using saline water. *Irrig. Sci., 25,* 387–399.
50. Karlberg, L., Rockström, J., Annandale, G., Martin, S. (2007). Low-cost drip irrigation: A suitable technology for southern Africa, an example with tomatoes using saline irrigation water. *Agric. Water Mangt., 89(1/2):*59–70.
51. Khroda, G. (1996). Strain, social and environmental consequences and water management in the most stressed water systems in Africa. In: Rached, E., Rathgeber, E., D. B. Brooks (Eds.),

Water Management in Africa and the Middle East – Challenges and opportunities, IDRC, Canada. pp. 295.

52. Khumoetsile, M., Dani, O. (2000). Root zone solute dynamics under drip irrigation: A review. *Plant and Soil*, 163–190.

53. Lei, T., Xiao, J., Mao, J., Wang, J., Liu, Z., Zhang, J., 2003a. Effect of drip irrigation with saline water on WUE and quality of watermelons. *Water Resources Mangt., 17*, 395–408.

54. Lei, T., Xiao, J., Mao, J., Wang, J., Liu, Z., Zhang, J., 2003b. Experimental investigation into effects of drip irrigation with saline groundwater on WUE and quality of honeydew melons in Hetao Region, Inner Mongolia. *Trans. of Chinese Soc. of Agri. Eng., 19(2)*, 80–84.

55. Li, J., Zhang. J., Rao, M. (2005). Modeling of water flow and nitrate transport under surface drip fertigation. *Trans. of the ASAE, 48(2)*, 627–637.

56. Li, J., Stanghellini, C. (2001). Analysis of the effect of EC and potential transpiration on vegctative growth of tomato. *Sci. Hortic., 89(1)*, 9–21.

57. Lubana, P. P. S., Narda, N. K. (2001). Modeling Soil water dynamics under trickle emitters: A Review. *J. Agric. Eng. Res., 78(3)*, 217–232.

58. LuDian, Q., Wang, Q., Wang, W. Y., Shao, M. A. (2002). Factors affecting soil water movement and solute transport for film drip irrigation. *Acta Pedologica Sinica, 39(6)*, 794–801.

59. Magan, J. J., Gallardo, M., Thompson, R. B., Lorenzo, P. (2008). Effect of salinity on fruit yield and quality of tomato grown in soil-less culture in greenhouse in Mediterranean climate conditions. *Agric. Water Mangt., 95*, 1041–1055.

60. Malash, N. M., Flowers, T. J., Ragab, R. (2008). Effect of irrigation methods, management and salinity of irrigation water on tomato yield, soil moisture and salinity distribution. *Irrig. Sci., 26*, 313–323.

61. Malash, N. M., Flowers, T. J., Ragab, R. (2005). Effect of irrigation systems and water management practices using saline and nonsaline water on tomato production. *Agric Water Mangt., 78*, 25–38.

62. Malash, N. M., Ghaibeh, A., Abdelkarim, G., Yeo, A., Flowers, T. J., Ragab, R., Cuartero, J., Aksoy, U., Anac, D. (2002). Effect of irrigation water salinity on yield and fruit quality of tomato. *Acta Hort. 573*, 423–434.

63. Mane, M. S., Ayare, B. L., Magar, S. S. (2008). *Principles of Drip Irrigation System*. pp. 191. Jain Brothers Publishers, New Delhi.

64. Meiri, A., Plantz, Z. (1985). Crop production and management under saline conditions. *Plant Soil, 89*, 253–271.

65. Minhas, P. S. (1996). Review – saline water management for irrigation in India. *Agri. Water Management, 30*, 1–24.

66. Minhas, P. S., Bajwa, M. S. (2001). Use and management of poor quality waters for the rice–wheat based production system. *J. Crop Prod. 4*, 273–305.

67. Mitchell, J. P., Shenna, C., Grattan, S. R., May, D. M. (1991). Tomato fruit yield and quality under water deficit and salinity. *J. Am. Soc. Horti. Sci., 116*, 215–221.

68. Nagaz, K., Mohamed, M. M., Netij, B. M. (2007). Soil salinity and yield of drip irrigated potato under different irrigation regimes with saline water in arid conditions of southern Tunisia. *J. Agric. 6(2)*, 324–327.

69. Olympios, C. M., Karapanos, I. C., Lionoudakis, K., Apidianakis, I., Pardossi, A., Serra, G., Tognoni, F. (2003). The growth, yield and quality of greenhouse tomatoes in relation to salinity applied at different stages of plant growth. *Acta Hort., 609*, 313–320.

70. Oron, G., DeMalach, Y., Gillerman, L., David, I., Lurie, S. (2002). Effect of Water Salinity and irrigation technology on yield and quality of Pear. Biosystems Engineering, *81(2)*, 237–247.

71. Oster, J. D., Jaiwardhane, N. S. (1998). Agricultural management of sodic soils in summer. Naidu, M. E. (Eds.) Sodic Soils: Distribution, Processes, Management and Environmental Consequences. Oxford University. 125–147.

72. Ouda, A. H. (2006). Predicting the effect of water and salinity stresses on wheat yield and water needs. *J. Applied Sci. Res.*, *2(10)*, 746–750.

73. Pascale, S., Barbieri, G., Pascale, S., Ferreira, M. I., Jones, H. G. (2000). Yield and quality of carrot as affected by soil salinity from long-term irrigation with saline water. *Acta Hort.*, *537(2)*, 621–628.

74. Pascale, S., Ruggiero, C., Barbieri, G., Pascale, S., Ferreira, M. I., Jones, H. G. (2000). Effects of irrigating pepper (*Capsicum annuum* L.) plants with saline water on plant growth, water use efficiency, and marketable yield. *Acta Hort.*, *537(2)*, 687–695.

75. Philip, J. R. (1997). Effect of root water extraction on wetted regions from continuous irrigation sources. *Irrigation Sci.*, *17*, 127–135.

76. Postel, S., Polak, P., Gonzales, F., Keller, J. (2001). Drip irrigation for small farmers: A new initiative to alleviate hunger and poverty. *Water Int.*, *26(1)*, 3–13.

77. Ragab, R. (2002). A holistic generic integrated approach for irrigation, crop and field management: the SALTMED model. *Environmental Modeling and Software*, *17*, 345–361.

78. Ragab, R., Malash, N., Gawad, G., Abdel, A. A., Ghaibah, G., 2005a. A holistic generic integrated approach for irrigation, crop and field management, I: The SALTMED model and its calibration using field data from Egypt and Syria. *Agric. Water Management*, *78*, 67–88.

79. Ragab, R., Malash, N., Gawad, G., Abdel, A. A., Ghaibah, G., 2005b. A holistic generic integrated approach for irrigation and field management, II: The SALTMED model validation using field data of five growing seasons from Egypt and Syria. *Agric. Water Management*, *78*, 89–107.

80. Raine, S. R., Meyer, W. S., Rassam, D. W., Hutson, J. L., Cook, F. J. (2007). Soil – water and solute movement under precision irrigation: knowledge gaps for managing sustainable root zones. *Irrig. Sci. 26*, 91–100.

81. Ranganna, B. (1986). *Manual of analysis of fruit and vegetable products*. Tata McGraw-Hill Publisher, New Delhi.

82. Rawls, W. J., Brakensiek, D. L. (1989). Estimation of soil water retention and hydraulic properties. In: *Unsaturated Flow in Hydrological Modeling by Morel-Seytoux*, H. (Ed.). NATO ASI Series 275. Kluwer Academic, London, pages 275–300.

83. Razuri, R. L., Linares, D. J., Rosale, J. G., Romero, C. E., Hernandez J. D., 2005. Validation of the model LEACM to predict the salinity in a soil of the Valley de Quibor, in the onion crop under drip irrigation. *Revista Forestal Latinoamericana*, *38*, 97–118.

84. Reina-Sanchez, A., Romero-Aranda, R., Cuartero, J. (2005). Plant water uptake and WUE of greenhouse tomato cultivars irrigated with saline water. *Agric. Water Mangt.*, *78*, 54–66.

85. Restuccia, G., Marchese, M., Mauromicale, G., Restuccia, A., Battaglia, M., Malfa, G., Lipari, V., Noto, G., Leonardi, C. (2003). Yield and fruit quality of tomato grown in greenhouse with saline irrigation water. *Acta Hort.*, *614(2)*, 699–704.

86. Rhoades, J. D., Bingham, F. T., Latey, J., Hoffman, G. J., Dedrick, A. R., Pinter, P. J., Peproyl, J. A. (1989). Use of Saline Drainage Water for Irrigation: Imperial Valley Study. *Agric. Water Mangt.*, *16*, 25–36.

87. Rhoades, J. D., Kandiah, A., Mashak, A. M. (1992). *The use of saline waters for crop production*. Irrigation and Drainage Paper 48. FAO, Rome Italy, pages 150.

88. Roberts, T., Lazarovitch, N., Warrick, A. W., Thompson, T. L. (2009). Modeling salt accumulation with subsurface drip irrigation using HYDRUS-2D. *J. Am. Soil. Sci. Soc. 73(1)*, 233–240.

89. Robinson, R. R. (2007). Drip irrigation water and salt flow model for table grapes in Coachella Valley, California. *Irrig. Drainage Syst.*, *21*, 79–95.

90. Romero-Aranda, R., Sorai, T., Cuartero, J. (2001). Tomato plant–water uptake and plant–water relationships under saline growth conditions. *Plant Sci.*, *160*, 265–272.

91. Romero-Aranda, R., Sorai, T., Cuartero, J. (2000). Tomato plant-water uptake and plant-water relationships under saline growth conditions. *Plant Sci.*, *160*, 265–272.
92. Romero-Aranda, R., Sorai, T., Cuartero, J. (2002). Greenhouse mist improves yield of tomato plants grown under saline conditions. *J. Am. Soc. Hort. Sci.*, *127*, 644–648.
93. Rui M A., Machado, Maria R G and Oliveira (2005) Tomato root distribution, yield and fruit quality under different subsurface drip irrigation regimes and depths. Irrig Sci., *24*, 15–24.
94. Samiha, A., El Fetouh, O. (2007). Predicting the effect of water and salinity stresses on yield and water consumption of wheat. *International J. Natural and Eng. Sci.*, *1*, 45–50.
95. Sandra, K., Wilfried, H. S., Johanna, G., Markus, W. (2006). The Influence of different electrical conductivity values in a simplified recirculating soilless system on inner and outer fruit quality characteristics of tomato. *J. Agric. Food Chem.*, *54(2)*, 441–448.
96. Shalhevet, J. (1994). Using water of marginal quality for crop production: major issues. *Agric. Water Mangt.*, *25*, 233–269.
97. Shannon, M. C., Grieve, C. M. (2000). Options for using low-quality water for vegetable crops. *Hort. Sci., 35(6)*, 1058–1062.
98. Sharaf, A. R., Hobson, G. E. (1986). Effect of salinity on the yield and quality of normal and nonripening mutant tomatoes. *Acta Hortic.*, *190*, 175–181.
99. Sharda, R. (2008). *Development of Soil Water Dynamics Model For Subsurface Drip Irrigation.* PhD dissertation. Punjab Agricultural University, Ludhiana, India.
100. Sharma, B. R. (2001). Availability, status of development and opportunities for augmentation of ground water policy initiative. CSSRI, Karnal, India. November, 6–7. pp. 1–18.
101. Sharma, B. R., Minhas, P. S. (2005). Strategies for managing saline/alkali waters for sustainable agricultural production in South Asia. *Agric. Water Mangt. 78*, 136–151.
102. Sharma, Y., Tiwari, C., Verma, B. L., Singhania, R. A., Sharma, Y. (2003). Effect of mixing of saline and canal water on yield of groundnut and wheat. *Crop Research, HAU, Hisar, 26(2)*, 249–253.
103. Simunek, J., Suarez, D. L. (1994). Two-dimensional transport model for variably saturated porous media with major chemistry. *Water Resources Res.*, *30*, 1115–1133.
104. Singh, C. S., Gupta, S. K., Sewa, R. (1996). Assessment and management of poor quality water for crop production, A simulation model (SWAM). *Agric. Water Mangt.*, *30*, 25–40.
105. Singh, D. K., Rajput, T. B. S., Singh, D. K., Sikarwar, H. S., Sahoo, R. N., Ahmad, T. (2006). Simulation of soil wetting pattern with subsurface drip irrigation from line source. Agric. Water Managt., *83*, 130–34.
106. Singh, P. P. S. (1995). Development of a soil water dynamics model for determining wetting pattern of point source trickle emitters. Ph.D. Thesis, 136 pp. *Punjab Agric. Univ., Ludhiana*, 141004, India.
107. Singh, P., Sanwal. S. (2003). Efficient use of saline drainage effluent with high frequency drip irrigation. International Commission on Irrigation and Drainage (ICID), New Delhi – India. 152 pages.
108. Singh, S. D., Gupta, J. P., Singh, P. (1978). Water economy and saline water use by drip irrigation. *Agronomy J. 70*, 948–951.
109. Van Genuchten, Th, M. (1980). A closed-form equation for predicting the hydraulic conductivity of unsaturated soils. *Soi. Sci. Soc. Am. J., 44*, 892–898.
110. Wallender, W. W., Tanji, K. K., Clark, B., Hill, R. W., Stegman, E. C., Gilley, J R., Lord, J. M., Wang, D., Shannon, M. C. (2000). Salt distribution and plant uptake under drip and sprinkler irrigation with saline water. National Irrigation Symposium Proceedings of the 4th Decennial Symposium, Phoenix, Arizona, USA, 612–617.
111. Wallender, W. W., Tanji, K. K., Clark, B., Hill, R. W., Stegman, E. C., Gilley, J. R., Lord, J. M., Robinson, R. R. (2007). Drip irrigation water and salt flow model for table grapes in Coachella Valley, California. *Irrig. Drainage Syst.*, *21*, 79–95.

112. Wang, D., Kang, Y. H., Wang, S. (2007). Distribution characteristics of different salt ions in soil under drip irrigation with saline water. *Trans of Chinese Soc. of Agric. Eng., 23(2),* 83–87.

113. Wang, D., Shannon, M. C. (2000). Salt distribution and plant uptake under drip and sprinkler irrigation with saline water. National Irrigation Symposium. Proceedings of the 4th Decennial Symposium, Phoenix, Arizona, USA. pp. 612–617.

114. Yazar, A., Gencel, B., Sezen, M. S. (2003). Corn yield response to saline irrigation water applied with a trickle system. *J. Food Agric. And Envir., 1(2),* 198–202.

115. Yurtseven, E., Kesmez, G. D., Unlukara, A. (2005). The effect of water salinity and potassium levels on yield, fruit quality and water consumption of native central anatolian tomato species (*Lycopercicon esculantum*). *Agric. Water Managt., 78,* 128–135.

116. Zhou, Q., Shaozhong, K., Lu, Z., Li, F. (2007). Comparison of APRI and Hydrus-2D models to simulate soil water dynamics in a vineyard under alternate partial root zone drip irrigation. *Plant Soil, 291,* 211–223.

APPENDIX I: PHOTOS OF DRIP IRRIGATED TOMATO

CHAPTER 14

WATER USE EFFICIENCY FOR SWEET PEPPERS

A. S. LODHI, ARUN KAUSHAL, and KAMAL G. SINGH

14.1 INTRODUCTION

Sweet pepper (*Capsicum annuum* L. var. *grossum*) or also called bell pepper is one of the most popular and high value vegetable crop grown throughout the world [9]. Sweet pepper is more sensitive to environment, especially soil moisture and temperature. Soil moisture is one of the predominant factors influencing sweet pepper productivity [6] and drip irrigation is the best alternative. Hanson et al. [4] indicated that the consumed water in the drip method is 40% less than that of the furrow method. Ngouajio et al. [11] showed that drip irrigation reduced water consumption by 20% compared to the furrow irrigation.

This study has been planned with the objective to determine the impact of irrigation regimes on growth, yield and WUE of sweet pepper.

14.2 MATERIALS AND METHODS

A Field experiment was conducted at the Research Farm of the Department of Soil and Water Engineering, PAU, Ludhiana, India. Ludhiana is situated at latitude of 30°54'N and longitude of 75°48'E and at a mean height of 247 meters above sea level. This place is characterized by very hot and dry summer (April to June) followed by a hot and humid monsoon period and cold winters during December to January. The average rainfall of the area is 600 mm most of which is received during the monsoon season. Average minimum and maximum temperature in the region varies from 3°C in winter to 43°C in summer season, respectively.

Mechanical properties of experimental field were determined by standard methods and are given in Table 14.1.

---

*In this chapter, the currency is expressed in Indian Rupees (1.00 US$ = Rs. 60.93; 1.00 Rs. = 0.02 US$).

TABLE 14.1   Soil Characteristics at Experimental Site

Soil depth (cm)	% of			Texture
	Sand	Silt	Clay	
0–15	70.75	18.46	10.79	Sandy loam
15–30	70.17	18.08	11.75	Sandy loam

Table 14.2 shows the initial levels of nitrogen (N), phosphorous (P), potassium (K), organic carbon, pH and electrical conductivity of the soil that were determined by standard methods and as reported by the Department of Soils, PAU, Ludhiana.

Water samples were analyzed for carbonates, bicarbonate, chloride, Ca-Mg, residual sodium carbonate and Electrical Conductivity (EC) etc. by titration method to check if the water was fit for irrigation purpose [3]. It was found that all the parameters of irrigation water were within the safe limits (Table 14.3).

TABLE 14.2   Initial Fertility Status of Soil

Soil depth (cm)	Fertility (Kg/ha)			Organic C (%)	pH	EC mmho/cm
	N	P	K			
0–15	Low	30.5	480	0.27	8.9	0.21
15–30	Low	31.75	322.5	0.12	8.8	0.16

TABLE 14.3   Irrigation Water Quality Parameters

Carbonate (meq/L)	Bicarbonate (meq/L)	Chloride (meq/L)	Calcium-Magnesium (meq/L)	Residual Sodium Carbonate (meq/L)	EC micro-mhos/cm
Nil	7.4	1.4	7.0	0.4	914

The experimental area was tilled twice with a disc harrow followed by a cultivator and planking. The soil at the experiment field was sandy loam having pH of 8.9. The soil was low in organic carbon and available nitrogen, medium in phosphorous and high in potash. As per the recommendation of Punjab Agricultural University, Ludhiana in its "Package of Practices for Vegetable Crops," [1] the farm yard manure at the rate of 55t/ha was added to the field, one month before the field preparation so that it could thoroughly mix in the soil and get decomposed by the time of sowing of crop.

Sweet pepper seeds were planted in nursery trays in polyhouse in October 2008 and the seedlings were transplanted in November 2008. In paired sowing, 60 cm wide beds were raised, row-to-row spacing between paired rows was 45 cm and row space between pairs was 75 cm but plant-to-plant spacing was 30 cm. Irrigation was applied as per treatments. In the single furrow, the row-to-row spacing was 60 cm and plant-to-plant spacing was 30 cm.

A field plot measuring approximately 550.8 m² (54 m × 10. 2 m) was prepared and the experiment was laid out in split plot design keeping five irrigation treatments in main plots and three different heights of low tunnel in sub plots and replicated three times. The three different low tunnel height treatments were 45 cm, 60 cm and 75 cm.

Out of the five treatments, three were drip irrigated and two were furrow irrigated. After the installation of drip irrigation system, it was tested for design discharge, uniformity of emitters and for clogging problem. At a pressure of 1 kg/cm² the average discharge per emitter was measured and the Christiansen uniformity coefficient was found as 98.4%:

$$E_{uc} = \left(1 - \frac{\Delta \bar{q}}{\bar{q}}\right) \tag{1}$$

where: E_{uc} = Christiansen uniformity coefficient; $\Delta \bar{q}$ = mean absolute deviation of the emitter flow from the mean value (L/min); and $\bar{q}$ = average discharge (L/min). In drip irrigation treatments, water was applied for three different irrigation levels at IW/CPE ratio of 0.60, 0.75 and 0.90. Drip irrigation was applied after 10 mm cumulative pan evaporation. Total volume of water applied per plant was calculated as:

$$V_d = \frac{Ac \times CPED \times (IW/CPE) \times Aw}{Euc} \tag{2}$$

where: V_d = volume of water applied per plant in drip irrigation system (liter); Ac= Cropped area (m²) which is calculated by row-to-row spacing (m) × plant-to-plant spacing (m); CPED = The desired cumulative pan evaporation (mm) after which the drip irrigation was applied; IW/CPE = Ratio of irrigation water to cumulative pan evaporation; Aw = Fractional wetted area which was taken as 75% (075); E_{uc}= Christiansen uniformity coefficient. The time of irrigation for operating drip system per application was calculated as:

$$T(drip) = \frac{Np \times V_d}{Ne \times qe} \tag{3}$$

where: $T(drip)$ = drip irrigation time (hrs.); N_p = Number of plants per lateral; V_d = volume of water applied per plant in drip irrigation system (liter); Ne = number of emitters in one lateral; and qe = average emitter discharge (L/h). In furrow irrigation treatment, water was applied using siphon tubes having discharge of one liter/sec for both furrow irrigation with paired and single row planting. Irrigation water was applied after 30 mm cumulative pan evaporation for both furrow-irrigated treatments. The furrow irrigation time was calculated as:

$$T\ (furrow) = \frac{d \times w \times l}{q \times 360}$$

(4)

where: $T(furrow)$ = furrow irrigation time (hrs.); d = depth of water to be applied (cm); w = furrow spacing (m); l = furrow length (m); and q = discharge from siphon tube (L/sec). The irrigation time was calculated by Eqs. (3) and (4) for drip and furrow irrigation, respectively (Table 14.4).

TABLE 14.4   Irrigation Time For Different Irrigation Treatments

Irrigation treatment	Water applied after CPE (mm)	Time of water application
I1	10.0	49 min. 23 sec.
I2	10.0	1 hr. 2 min.
I3	10.0	1 hr. 14 min.
I4	30.0	3 min. 4 sec.
I5	30.0	6 min. 7 sec.

Five plants were tagged randomly in each subplot to measure height, leaf area index (LAI), dry matter accumulation (DMA), days to flowering, days to fruit initiation and days to fruit maturity. Cumulative number of fruits harvested from each subplot was found by counting the fruits in various picking stages till the end of crop season. Five fruits were selected from each subplot and measurements of length and girth at the maximum point and average fruit length and girth in cm was calculated. The fruits from each subplot were picked at green mature stage and weighed at each picking. The weight of all the pickings were added and yield per plant was worked out and subsequently the yield per hectare was calculated.

The data were analyzed using ANOVA. For the split plot design, irrigation treatments were considered as main plots and different low tunnel heights as subplot. The significance of differences was tested at 5% level.

14.3   RESULTS AND DISCUSSION

14.3.1   GROWTH PARAMETERS

Among all the irrigation treatments, drip irrigation with I3 ratio gave the highest plant height and LAI throughout the season followed by I2, I1, I4 and I5. This may be due to better moisture distribution in drip irrigation than the conventional irrigation [5]. While drip irrigation with 0.75 IW/CPE ratio gave the highest DMA throughout the season followed by drip irrigation at 0.90 IW/CPE ratio, drip irrigation at 0.60 IW/CPE ratio, furrow irrigation paired row planting and furrow irrigation single row planting (Table 14.5).

14.3.2   PHENOLOGY

Among all the irrigation treatments, drip irrigation with I1 gave the minimum number of days to flowering, fruiting and fruit maturity followed by I2, I3, I4 and I5. This may be due to deficient irrigation water in I1 treatment. The results are in line with that of Khan et al. [7], who reported that the plants under any kind of stress conditions tends to have a shortened life span and try to complete the life cycle in hasten which causes the lowest days to flowering, fruiting.

14.3.3   YIELD AND YIELD ATTRIBUTES

Among all the irrigation treatments, I2 treatment gave the highest number of fruits per plant, fruit length and girth, mean sweet pepper yield followed by I3, I1, I4 and I5. Moisture at optimum level enhances the cell metabolism resulting in an increase in released energy, which induces growth [12]. Best drip irrigated treatment (i.e., drip irrigation, IW/CPE= 0.75) gave an increase of 30.67% over the furrow irrigated paired row planting and an increase of 33.74% over furrow irrigated single row planting.

TABLE 14.5   Effects of Irrigation Regimes on Growth and Yield of Sweet Pepper

Treatments	Plant height (cm)	Leaf area index	Dry matter accumulation (gm)
Growth parameters			
I1	56.93	4.12	90.55
I2	58.83	4.17	94.50
I3	60.78	4.26	92.55
I4	56.26	4.05	88.60
I5	55.08	3.98	87.73
CD (5%)	1.33	0.15	0.63
Phenology			
Treatments	Days to flowering	Days to fruit initiation	Days to fruit maturity
I1	82.60	89.15	114.37
I2	87.35	94.46	119.20
I3	93.42	100.28	124.20
I4	95.20	101.60	124.57
I5	95.97	102.88	125.37
CD (5%)	1.50	1.92	2.19

Yield and yield Attributes			
Treatments	Number of fruits per plant	Average fruit size	Yield (100 kg/ ha)
		Fruit length (cm)	Fruit girth (cm)	
I1	9.56	5.78	18.17	222.46
I2	10.07	6.38	20.41	288.11
I3	9.70	5.83	18.98	276.95
I4	9.48	5.65	17.99	220.48
I5	9.38	5.53	17.88	215.42
CD (5%)	0.12	0.14	0.39	5.00

14.3.4 IRRIGATION WATER SAVING UNDER DRIP IRRIGATION

Quantity of water applied under different irrigation treatments is presented in Table 13.6. The highest amount of irrigation water was 78 cm in furrow irrigation and lowest water was applied in drip irrigation IW/CPE = 0.60 (I1). The percentage saving was 39.72% for irrigation treatment I2 (drip irrigation, IW/CPE=0.75) over the conventional furrow irrigated treatments. All the drip irrigated treatments saved considerable amount of water over furrow irrigation [8, 10].

TABLE 13.6 Comparison of Irrigation Water Use in Different Irrigation Treatments

Irrigation treatment	Pre- irrigation depth (cm)	Depth of water per irrigation (cm)	Total number of irrigations	Total depth of irrigation (cm)	Percentage saving over furrow irrigation
I1	3	0.45	77	38.21	51.01%
I2	3	0.57	77	47.01	39.72%
I3	3	0.68	77	55.82	28.44%
I4 and I5	3	3.00	25	78.00	—

14.3.5 WATER USE EFFICIENCY (WUE)

WUE was calculated by dividing pepper yield with water used and the results are given in Table 14.7. The data given in the table clearly revealed that WUE in I2 gave the highest value followed by I3, I1, I4 and I5. Among the treatments with good yield and less irrigation water used gave higher WUE due to optimum moisture present in the soil. The results are in line with that of Antony and Singandhupe [2].

TABLE 14.7   Comparison of WUE (q/ha/cm) With Different Irrigation Regimes

Treatments	I1	I2	I3	I4	I5	CD	(5%)
WUE 100 kg/(ha-cm)	5.82	6.12	4.96	2.82	2.76	**0.09**	

14.4   CONCLUSIONS

The highest plant height and leaf area index were observed in I3 treatments while the highest DMA, highest number of fruit per plant, maximum fruit length, fruit girth and highest sweet pepper yield were observed in I2 treatments. The lowest number of days to flowering, fruit initiation and fruit maturity were observed in I1 treatment. Drip irrigated treatments gave better yield than furrow irrigation. Best drip irrigated treatment (I2) gave an increase of 30.67% over I4 treatment and an increase of 33.74% over I5 treatment. The percentage of water saving for drip irrigation treatments I1, I2 and I3 were 51.01%, 39.72% and 28.73%, respectively over the furrow irrigation treatment. The WUE was highest I2 treatment among the all irrigation treatments.

14.5   SUMMARY

Field experiment was conducted in the Department of Soil and Water Engineering, PAU, Ludhiana. Drip and furrow irrigation methods were used to study effects on growth and yields of the sweet pepper using split plot design. The irrigation treatments were: drip irrigation with IW/CPE ratio of 0.60(I1), drip irrigation with IW/CPE ratio of 0.75(I2), drip irrigation with IW/CPE ratio of 0.90(I3), furrow irrigation with paired row planting (I4) and furrow irrigation with single row planting (I5). The dry matter accumulation, number of fruits per plant, fruit size, total yield and WUE were observed highest in I2 treatments. The highest plant height and leaf area index were observed in I3 treatment. The minimum number of days to flowering, fruit initiation and fruit maturity were observed in I1 treatment. There were significant effects of irrigation on leaf area index, plant height, dry matter accumulation, and number of days to flowering, fruit initiation, fruit maturity, fruit girth, number of fruits per plant, fruit length, yield and WUE. Best drip irrigated treatment I2 gave an increase of 30.67% over I4 and an increase of 33.74% over I5. The percentage of water saving for drip irrigation treatment I1, I2 and I3 were 51.01%, 39.72% and 28.73%, respectively, over the furrow irrigation treatment. Irrigation treatment I2 gives maximum yield and highest WUE.

KEYWORDS

- days to flowering
- drip irrigation
- dry matter accumulation
- fruit girth
- fruit initiation
- fruit length
- fruit maturity
- growth
- irrigation
- leaf area index
- number of fruits
- plant height
- sweet pepper
- vegetables
- water use efficiency
- yield

REFERENCES

1. Anonymous, (2008). *Package of Practices for Vegetable Crops*. pp. 20–25. Punjab Agricultural University, Ludhiana.
2. Antony, E., Singandhupe, R. B. (2004). Impact of drip and surface irrigation on growth, yield and WUE of capsicum (*Capsicum annuum L.*). *Agricultural Water Management, 65,* 121–132.
3. Das, D. K. (2002). *Introductory Soil Science*. pp. 238. Kalyani Publishers, Delhi, India.
4. Hanson, B. R., Schwankl, L. J., Schulbach, K. F., Pettygrove, G. S. (1997). A comparison of furrow, surface drip, and subsurface drip irrigation on lettuce yield and applied water. *Agricultural Water Management, 33,* 139–157.
5. Hsaio, T. C. (1993). Growth and productivity of crops in relation to water status. *Acta Horticulturae, 331,* 137–48.
6. Kaushal, A., Lodhi, A. S., Singh, K. G. (2011). Economics of growing sweet pepper under low tunnels. *Progressive Agriculture, 11(1),* 67–72.
7. Khan, M. H., Chattha, T. H., Saleem, N. (2005). Influence of different irrigation intervals on growth and yield of bell pepper (*Capsicum annuum grossum*). *Research Journal of Agriculture and Biological Science, 1,* 125–128.
8. Locascio, J. S. (2005). Management of Irrigation for Vegetables: Past, Present, Future. *Hort Technology, 15(3),* 482–485.

9. Lodhi, A. S., Kaushal, A., Singh, K. G. (2013). Effect of irrigation regimes and low tunnel heights on microclimatic parameters in the growing of sweet peppers. *International Journal of Engineering Science Invention, 2(7),* 20–29.

10. Mane, M. S., Ayare, B. L., Magar, S. S. (2006). *Principles of Drip Irrigation System.* pp. 32. Jain Brothers publication, New Delhi, India.

11. Ngouajio, M., Wang, G., Goldy, R. (2007). Withholding of drip irrigation between transplanting and flowering increases the yield of field-grown tomato under plastic mulch. *Agricultural Water Management, 87,* 285–291.

12. Pandey, V., Ahmed, Z., Tewari, H. C., Kumar, N. (2005). Effect of greenhouse models on plant-growth and yield of capsicum in North-west Himalayas. *Indian Journal of Horticulture, 62,* 312–313.

CHAPTER 15

IRRIGATION WATER REQUIREMENTS OF GREEN PEA

KAMAL G. SINGH, MUKESH SIAG, and G. MAHAJAN

15.1 INTRODUCTION

Peas occupy an important place among vegetable crops in the submountainous region of India. Being a cool weather-loving crop, it occupies a place of prominence among winter vegetables in *kandi* area, because the *kandi* area is cooler as compared to other pea-growing regions of Punjab. But, due to various physiographic constraints, existing legal constraints and the present method of utilization, the utilizable water for irrigation is very limited. In the present day context, improvements in irrigation practices are needed to increase crop production and to sustain the productivity level. Therefore, drip irrigation is a most efficient method of irrigation, which optimizes the use of irrigation water by providing it uniformly and directly to the roots of the plants. Drip irrigation can be helpful in *kandi* area where water is scarce and very expensive to pump and the fields have uneven topography. Irrigation with drip may be more efficient in *kandi* area due to precise application of water because evaporation is reduced, runoff is reduced or eliminated, deep percolation is reduced, and irrigation uniformity is improved so it is no longer necessary to "over water" parts of a field to adequately irrigate the more difficult parts. The major drawback of the drip irrigation system is its high initial investment; however, cost can be recovered in a short span if proper water and nutrient management and design principles are followed. Among the various components of the drip irrigation system, the cost of the lateral is a major factor, which influence the total system cost. Any effort made to reduce the length of lateral per unit area will result in reduction of system cost. Many scientists reported that drip irrigation in vegetable crops is very economical by reducing the cost and water use by 50%, when these crops were planted in paired row patterns [1, 3]. With drip irrigation, precise application of nutrients is possible. Fertilizer costs and nitrate losses can be reduced and nutrient applications can be better timed to needs of a crop. The response of pea to the combined effect of

---

*In this chapter, the currency is expressed in Indian Rupees (1.00 US$ = Rs. 60.93; 1.00 Rs. = 0.02 US$).

drip with different levels of irrigation in conjunction with different nitrogen levels and their economic feasibility are not well known. The information on seasonal water requirement of pea crop under drip irrigation is not available.

This study was conducted to evaluate the yield response and economic viability of drip irrigation in combination with different nitrogen levels.

15.2   MATERIALS AND METHODS

15.2.1   EXPERIMENTAL SETUP

The field experiment was conducted at the research farm of Department of Soil and Water Engineering, PAU, Ludhiana during to study the water and nutrient requirements of green pea. The experiment consisted of 11 treatments (Table 15.1): Three levels of irrigation (1.0 Epan, 0.75 Epan and 0.5 Epan); three levels of nitrogen (100, 75 and 50% of recommended dose of N) that were tested against two check basin methods of irrigation at recommended dose of N when the crops were sown in either paired row or single row (conventional method) in randomized block design. In check-basin method (surface flooding), the irrigations were provided on the basis of 1.0 × cumulative pan evaporation (Epan).

The pea cultivar Punjab 88 was sown in the second week of October. In normal sowing, the distance between the rows was 30 cm and plant-to-plant spacing was 10 cm. However, in paired sowing, the row-to-row spacing between paired rows was 30 cm and row-to-row spacing between pairs was also 10 cm but plant-to-plant spacing space was 30 cm. Thus, in paired row sowing total as well as number of rows and plants were same. The recommended fertilizers in pea were 50 kg N and 67.5 kg P_2O_5/ha. Whole of the phosphorus was basal applied (before sowing of crop) in all the treatments. In all the treatments, a basal dose of FYM @ 20t/ha was also applied before sowing. Other cultural operations were same in all treatments.

The drip system consisted of polyethylene laterals of 12 mm in diameter, laid parallel (each lateral served 2 rows of crop). The laterals were provided with on-line emitters of 3 lph capacity at 30 cm apart. The different levels of water supply were maintained by managing the number of holes in each lateral. In drip irrigation system, N was fertigated at 10 days interval in six equal doses of N starting from 30 days after sowing (DAS).

15.2.2   BENEFIT COST ANALYSIS

To calculate the net return of green pea, the cost of different inputs like expenses incurred on preparation of field, plowing, seed, sowing, cost of fertilizers, manure and their application, weeding, crop protection measure and cost of irrigation water, harvesting and selling prices were taken from the Department of Agricultural Economics, PAU, Ludhiana. Market prices were taken to compute cost of drip irrigation system. The subsidy of Rs. 25,000 per hectare was deducted from the fixed cost of drip irrigation system. An additional cost of operation and maintenance of drip irrigation system at Rs. 500 per month was added to the cost of cultivation for drip system. The annual cost of drip irrigation system was divided equally between the three crops that can be taken in a sequence round the year. The seasonal cost of drip irrigation includes depreciation, prevailing bank rate of interest (8%/year/crop). The useful life of main line, sub main, fertilizer tank, valve, filter and pumping unit was considered 20 years, while the cost of inbuilt drip was considered 10 years. The maximum yields obtained under drip and check basin method of irrigation were taken to find out the net returns per hectare and then the net income under both systems of irrigation for pea crop have been compared and analyzed.

15.3 RESULTS AND DISCUSSION

15.3.1 GREEN PEA YIELD

The Table 15.1 revealed that for same quantity of water and nitrogen fertigation a significantly higher pod yield 121.8 per 100 kg/ha was obtained as compared to 113.0 per 100 kg/ha and 96.8 per 100 kg/ha in check basin method of irrigation when the crop was sown in paired rows and in single row respectively. It further revealed that when the quantity of water through drip irrigation was reduced, the pod yield of pea increased significantly as compared to check basin method of irrigation at the same level of nitrogen. A significant increase in pod yield was observed with increase in N application when drip irrigation was based at 0.5 Epan. However, when drip irrigation was based at 0.75 Epan and 1.0 Epan, the pea yield increased significantly at 100% of recommended nitrogen, while statistically same when N applied at 50% and 75% of recommended N, respectively. At recommended nitrogen, when drip irrigation and fertigation was done at 0.5 Epan, pea pod yield was highest (154.3 per 100 kg/ha) and increased by 36.5% and 59.4% over check basin method of irrigation when the crop was sown in paired or single rows, respectively. At recommended nitrogen, further increase in water supply caused reduction in yield.

15.3.2 SAVINGS IN WATER AND NITROGEN

Table 15.1 depicts the amount of water applied to the different irrigation treatments for pea crop. When 100% of recommended nitrogen was applied through drip at low level of irrigation, maximum yield of 154.3 per 100 kg/ha was obtained and water used for this treatment was 13.4 cm along with maximum WUE (1151 kg/(ha-cm)) and nitrogen use efficiency (6.17 100 kg/kg of N). At low level of drip irrigation, 49.6% of water was saved as compared to conventional irrigation method. Drip irrigation at 0.5 Epan caused maximum WUE and nitrogen use efficiency at all the levels of nitrogen as compared to drip irrigation at 0.75 Epan and 1.0 Epan

TABLE 15.1 Effects of Different Treatments on Yield, Water Use Efficiency and Nitrogen Use Efficiency of Green Pea

Treatments	Yield	Irrigation water applied	WUE	Nitrogen use efficiency
	100 kg/ha	cm	kg/(ha-cm)	100 kg/(kg of N)
Drip at 0.5 Epan + 50% recommended N + PR	113.6	13.4	848	4.54
Drip at 0.5 Epan + 75% recommended N + PR	137.4	13.4	1025	5.50
Treatments	Yield	Irrigation water applied	WUE	Nitrogen use efficiency
	100 kg/ha	cm	kg/(ha-cm)	100 kg/(kg of N)
Drip at 0.5 Epan + 100% recommended N + PR	154.3	13.4	1151	6.17
Drip at 0.75 Epan + 50% recommended N + PR	113.9	17.1	666	3.04
Drip at 0.75 Epan + 75% recommended N + PR	119.4	17.1	698	3.18
Drip at 0.75 Epan + 100% recommended N + PR	138.0	17.1	807	3.68
Drip at 1.0 Epan + 50% recommended N + PR	101.7	20.8	489	2.03
Drip at 1.0 Epan + 75% recommended N + PR	111.9	20.8	538	2.24
Drip at 1.0 Epan + 100% recommended N + PR	129.8	20.8	624	2.60
Surface irrigation at 1.0 Epan + 100% recommended N + PR	113.0	27.0	419	2.26

TABLE 15.1   *(Continued)*

Surface irrigation at 1.0 Epan + 100% recommended N + NS	96.8	27.0	359	1.94
LSD (0.05)	14.9	—	—	—
PR = Paired rows, NS= Normal sowing				

TABLE 15.2   Economic Analysis of Drip Irrigation versus Check Basin Method of Irrigation For Green Pea

Parameters	Drip irrigation	Check basin irrigation
	Rs.	Rs.
1. Main line, Sub main, Fertilizer tank, Control valve, Filter, Pumping unit	—	—
a. Fixed cost	60,000	Nil
b. Life (years)	20	Nil
c. Depreciation per crop (taking 3 crops per year: (Bottle gourd-okra-green pea)	1000	Nil
d. Interest (at 8% per year) per crop	534	Nil
e. Total (c + d)	1534	Nil
2. Laterals with inbuilt emitters	—	—
a. Fixed cost	1,66,680	Nil
b. Life (years)	10	Nil
c. Depreciation per crop (taking 3 crops per year: (Bottle gourd-okra-green pea)	5556	Nil
d. Interest (at 8% per year) per crop	1481.6	Nil
e. Total (c + d)	7037.6	Nil
3. Cost of cultivation (Rs./ha)	64,500	62,500
1. Seasonal cost = 1(e) + 2 (e) + 3	73,071	62,500
2. Yield of produce (100 kg/ha) from Table 1	154.3	96.8
3. Selling price (Rs./100 kg)	950	950
4. Income from the produce in Rs. = (5×6)	1,46,585	91,960
5. Net seasonal income, Rs.= (7 – 4)	73,514	29,460

15.3.3 ECONOMIC ANALYSIS

To calculate the net return for drip irrigated green pea, the cost of different inputs and outputs were recommended by Department of Agricultural Economics, PAU, Ludhiana. These values were used for calculating the net income from green pea crop. Table 15.2 depicts the cost of different components for the drip irrigation and net return from the crop. Table 15.2 shows that net seasonal income from green pea was Rs. 73,514 and Rs. 29,640 per ha for drip and check basin method of irrigation, respectively. In green pea, the net increase in income by drip irrigation method was 40.1% more as compared to check basin method of irrigation and further increase in yield is possible where water is scarce by increasing the area under cultivation with the saved amount of water. These results are in conformity with Raina et al. [2], who also reported higher pea yield, water use efficiency (WUE) and benefit-cost ratio under drip irrigation system.

15.4 SUMMARY

The research was undertaken to evaluate the effects of various levels of water and N fertigation on yield and water use efficiency of green pea crop. In this experiment, various combination of three levels of water (Epan × 1.0, Epan × 0.75 and Epan × 0.5) and three levels of N (100, 75 and 50% of recommended N of 45 kg/ha) through drip were compared with check basin method of irrigation at two method of planting (normal sowing NS; and paired sowing, PS). The highest yield (154.3 per 100 kg/ha) was observed when 100% of recommended N was fertigated through drip irrigation at lower level of irrigation (0.5 × Epan). This increase in yield was 36.5% and 59.4% over check-basin method of irrigation, when the crop was sown in paired rows and normal sown, respectively. The study further revealed that the increase in N through fertigation caused increase in yield at all the levels of drip irrigation (0.5 Epan, 0.75 Epan and 1.0 Epan), but the magnitude of increase was highest at lowest level of water supply. Further results revealed that WUE and nitrogen use efficiency also improved significantly with drip irrigation system.

KEYWORDS

- **benefit–cost ratio, BCR**
- **bottle guard**
- **check basin**
- **drip irrigation**

- **fertigation**
- **green pea**
- **nitrogen use efficiency**
- **paired row**
- **plastic mulch**
- **water requirement**
- **water supply**
- **water use efficiency**

REFERENCES

1. Kumar, A., Singh, A. K. (2002). Improving nutrient and water use efficiency through fertigation. *J. Water Management, 10,* 42–48.
2. Raina, J. N., Thakur, B. C., Bhandari, A. R. (1998). Effect of drip irrigation and plastic mulch on yield, water use efficiency and benefit: cost ratio of pea cultivation. *J. Indian Soc. Soil Sci., 46,* 562–567.
3. Tiwari, K. N., Mal, P. K., Singh, R. M., Chattopadhyay, P. (1998). Response of okra to drip irrigation under mulch and no-mulch condition. *Agricultural Water Management, 38,* 91–102.

CHAPTER 16

IRRIGATION SCHEDULING OF CAULIFLOWER

CHETAN SINGLA, KAMAL G. SINGH, and NILESH BIWALKAR

16.1 INTRODUCTION

The cauliflower (*Brassica oleracea* L.) plant belongs to the family cruciferae. Its varieties are very responsive to temperature, photoperiod and irrigation. Early varieties if sown late produce "button" head and late varieties if sown early will go on giving leafy growth and will produce curds very late. It is one of the most important winter vegetables of India. India produces 4.694 million metric tons (Mmt) per year of cauliflower from 0.256 million-ha area with an average productivity of about 18.3 mt/ha. The major cauliflower producing states are Punjab, Bihar, Uttar Pradesh, Orissa, West Bengal, Assam, Haryana and Maharashtra. Advance technology for cauliflowers cultivation is use of Hybrid seeds and drip irrigation. In the plains, it is available from September to May. It is consumed as a vegetable in curries, soups and pickles.

Since greenhouse technology has recently been introduced in Punjab, very limited work has been done for the standardization of different techniques of growing of early vegetable crops in greenhouse. Early cultivars of cauliflower are tropical in nature requiring comparatively higher temperature for reproductive onset and therefore, these are grown during July–August allowing adequate growth. However, high moisture rains coinciding with the period of cropping frequently disturb the growth and development of the crop in open.

Crop production functions represent empirical relationships between the crop yields and the inputs [2]. These functions are the tools in the hands of the planners for the efficient management of cropping systems, in India, for any situation. Arkley [1] found a linear relationship between water transpired and dry matter of the crop

*Authors acknowledge the financial support and other facilities received from AICRP on Application of plastics in agriculture, Punjab Agriculture University, Ludhiana.
**In this chapter, the currency is expressed in Indian Rupees (1.00 US$ = Rs. 60.93; 1.00 Rs. = 0.02 US$).

yield, both in humid and semiarid locations. Arkley [1] also indicated advantages of empirical crop yield water use in making water allocations and management decisions in an intelligent way, particularly in situations of irrigations water shortage for agriculture purpose. According to Ayer [2] and Reddy [3], optimum application of water may be less than the water required, to ensure that water is nonlimiting.

This chapter discusses the research study the response cauliflower to drip irrigation levels and nitrogen fertilizer through the modeling approach. In this study, water and fertilizer use optimization functions were developed for cauliflower production.

16.2 MATERIALS AND METHODS

A field study of fertigation of drip irrigated cauliflower (*Brassica oleracea var. botrytis Linn.*) was undertaken during summer season at the Irrigation Research Farm of the Department of Soil and Water Engineering, PAU, Ludhiana – India. There were three irrigation levels at (irrigation water to cumulative pan evaporation ratio) IW/CPE = 0.5 (W1), 0.75 (W2) and 1.0 (W_3) with three rates of nitrogen application 100% (N_3), 75% (N_2) and 50% (N_1) of recommended dose of Nitrogen. Nine treatments were replicated thrice. The soil at the experimental field was sandy loam with a pH of 8.5. The harvesting of marketable curbs was commenced 80 days after transplanting (DAT) and was completed within four weeks. With this data, multivariant production functions were fitted:

$$Y = F (WN/P) \tag{1}$$

where: Y is the Yield of cauliflower, per 100 kg/ha; W is the irrigation water applied, cm; N is the nitrogen level in kg/ha; P is the quantity of other fixed input; and slash (/) denotes that only W and N are variable inputs. The general function was expressed in the form of:

$$Y = a_c + a_1 W + a_2 N + a_3 N^n + a_4 W^n \tag{2}$$

where: a_c represents the combined effect of all fixed inputs; a_1, a_2, a_3 and a_4 are regression coefficients; and n is the exponent. The Eq. (1) was fitted to data for Y, W and N with exponent values of $n = 1$, 2, 0.0.75 and 0.50 to give linear, quadratic, three halves and square root functions, respectively, using least square multiple regression analysis. The generalized forms of equations were:

$$\text{Linear: } Y = a_c + a_1 W + a_2 N$$

$$\text{Quadratic: } Y = a_c + a_1 W + a_2 N + a_3 N^2 + a_4 W^2$$

Three halves: $Y = a_c + a_1 W + a_2 N + a_3 N^{0.75} + a_4 W^{0.75}$

Square root: $Y = a_c + a_1 W + a_2 N + a_3 N^{0.50} + a_4 W^{0.50}$ (3)

where: Y = predicted yield of cauliflower (100 kg/ha); W = total water use (cm); N = nitrogen level (kg/ha); and a_1 to a_4 = regression coefficients; and a_c = regression constant.

The linear, quadratic, square root and three halves functions were fitted for cauliflower crop and the best fit was selected. Least square approximation technique was used for fitting these functions. For best-fitted functions, yield was determined. During field experimentation, the completely randomized design (CRD) was followed in this study. The data collected for all the dependent parameters were analyzed statistically with computer program CPCSI. The analysis of variance was performed and the critical difference at 5% level of significance was calculated for testing the significance of difference between different treatments of irrigation and fertilizer.

16.3 RESULTS AND DISCUSSION

16.3.1 CURD YIELD

Cauliflower yield as influenced by the application of different doses of nitrogen and irrigation water is presented in Table 16.1. The total cumulative pan evaporation was 534 mm. The amount of water applied at 0.5, 0.75 and 1.0 IW/CPE was 26.70, 40.05 and 53.40 cm, respectively. The amount of nitrogen fertilizer applied at 50, 75 and 100% of recommended dose was 62.50, 93.75 and 125 kg of N/ha. Tables 16.1 and 16.2 reveal that highest yield of cauliflower was recorded under irrigation schedule based on IW/CPE ratio of 0.5 and 100% of recommended dose of nitrogen which was comparable with other drip irrigation treatments. This treatment (IW/CPE = 0.5 and 100% of recommended dose of nitrogen, that is, 125 kg/ha) established its superiority by yielding 120.7 per 100 kg/ha and 105.5 per 100 kg/ha as compared to other eight treatments in fan pad cooled greenhouse and naturally ventilated greenhouse respectively. These findings show that an irrigation schedule based on IW/CPE ratio of 0.5 secured yields as high as that of higher ratio for 0.75 and 1.0. These results depicted that low water and high fertilizer treatment by drip irrigation were superior to other treatments. It might be due to the fact that optimum requirement of water was met by just IW/CPE = 0.5 (less amount of water) rather ratio of 1.0 (high amount of water) and 0.75 (medium level of water).

TABLE 16.1 Effects of Different Treatments on Cauliflower Yield in Fan Pad Cooled Greenhouse

Irrigation levels (IW/CPE)	N fertilizer levels (%)	Yield in greenhouse (100 kg/ha)		
		1st year	2nd year	3rd year
0.50 = I1	50 = N1	92	95.9	93.95
	75 = N2	115	118.2	116.60
	100 = N3	120	121.4	120.70
0.75 = I2	50	94	93.8	93.90
	75	117	115	116.00
	100	120	119.5	119.75
1.00 = I3	50	91	92.32	91.66
	75	103	104.4	103.70
	100	110	110.36	110.18
C.D. at P = 0.05%	N.S.	N.S.	N.S.	N.S.

TABLE 16.2 Effects of Different Treatments on Cauliflower Yield in Naturally Ventilated Greenhouse

Irrigation levels (IW/CPE)	N fertilizer levels (%)	Yield in greenhouse (100 kg/ha)		
		1st year	2nd year	3rd year
0.50	50	86.07	91.70	88.88
	75	100.4	101.36	100.88
	100	105.15	105.95	105.95
0.75	50	87.00	89.00	88.00
	75	93.00	97.36	95.19
	100	106.40	104.40	105.40
1.00	50	79.40	79.76	79.58
	75	75.58	76.54	76.06
	100	91.00	94.10	92.55
C.D. at P = 0.05%	N.S.	N.S.	N.S.	N.S.

16.3.2 EFFECTS OF N ON YIELD

Table 16.1 reveals that, as the quantity of applied nitrogen was increased from 50 to 100% of the recommended dose, the yield goes on increasing. In fan pad cooled greenhouse, the yield increased from 93.95 to 106.6 and 120.7 per 100 kg/ha indicating 13.46 and 28.47% increase at nitrogen levels of 93.75 and 125 kg/ha, respectively, at irrigation with IW/CPE = 0.5. While at irrigation with IW/CPE = 0.75 and 1.0, the relative yield increase was 23.53, 27.52 and 13.13, 20.20% over the nitrogen level of 62.5 kg/ha at the respective nitrogen levels. Similarly in naturally ventilated greenhouse, the yield increase was 13.50, 18.75 and 8.17, 9.77% over the nitrogen level of 62.5 kg/ha, at irrigation levels of IW/CPE = 0.5 and 0.75.

16.3.3 EFFECTS OF IRRIGATION ON YIELD

Tables 16.1 and 16.2 reveals that the yield of cauliflower curd was more with IW/CPE = 0.5 as compared to IW/CPE = 0.75 and 1.0 for all levels of nitrogen fertilizers. Irrigation schedule based on IW/CPE = 0.5 established its superiority by yielding 93.95, 116.6 and 120.7 per 100 kg/ha in fan pad cooled greenhouse and 88.88, 100.88 and 105.5 per 100 kg/ha in naturally ventilated greenhouse.

TABLE 16.3 Water Use Efficiency of Cauliflower in Fan Pad Cooled Greenhouse

Treatments	Water used	Mean yield	WUE
	cm	100 kg/ha	Kg/(ha-cm)
W1N1	26.70	93.95	351.8
W1N2	26.70	116.60	436.70
W1N3	26.70	120.90	452.05
W2N1	40.05	93.90	234.45
W2N2	40.05	116.00	289.63
W2N3	40.05	119.75	299.00
W3N1	53.40	53.40	53.40
W3N2	53.40	53.40	194.19
W3N3	53.40	53.40	206.32

TABLE 16.4 Water Use Efficiency of Cauliflower in Naturally Ventilated Greenhouse

Treatments	Water used (cm)	Mean Yield 100 kg/ha	Water Use Efficiency kg/ (ha-cm)
W1N1	26.70	88.88	332.88
W1N2	26.70	100.88	377.82
W1N3	26.70	105.55	395.31
W2N1	40.05	88.00	219.72
W2N2	40.05	95.19	237.67
W2N3	40.05	105.40	263.17
W3N1	53.40	79.58	149.02
W3N2	53.40	76.06	142.43
W3N3	53.40	92.55	173.31

TABLE 16.5 Values of Constants and Coefficients FOR yield of Cauliflower in Different Mathematical Models in Fan Pad Cooled Greenhouse

Model	a_c	a_1	a_2	a_3	a_4	R^2
Linear	85.034	−0.321	0.362	—	—	0.89
Square root	−115.88	−0.321	−1.897	—	43.05	0.95
Three halves	−48.28	−0.321	−4.133	—	18.55	0.95
Quadratic	34.81	−0.321	1.519	—	−0.0062	0.95

TABLE 16.6   Values of Constants and Coefficients For Yield of Cauliflower in Naturally Ventilated Greenhouse

Model	a_c	a_1	a_2	a_3	a_4	R^2
Linear	91.913	−0.588	0.237	—	—	0.88
Square root	197.80	−0.588	1.425	—	−22.6429	0.91
Three halves	162.027	−0.588	2.600	—	−9.75385	0.91
Quadratic	118.33	−0.588	0.372	—	0.00325	0.91

TABLE 16.7   Actual and Predicated Yield of Cauliflower With Prediction Models For FAN pad Cooled Greenhouse

Treat-ment	Water needed	Nitrogen level	Field yield	Predicted yield (100 kg/ha)				RMSE
				Linear	Square Root	Quad-ratic	Three halves	
	(cm)	(kg/ha)	100 kg/ha					5.89 (Linear)
W2N1	40.05	62.5	93	94.8	93.04	92.77	92.64	4.18 (Square root)
Treat-ment	Water needed	Nitrogen level	Field yield	Predicted yield (100 kg/ha)				RMSE
				Linear	Square Root	Quad-ratic	Three halves	
	(cm)	(kg/ha)	100 kg/ha					5.89 (Linear)
W2N2	40.05	93.75	115.6	106.1	110.26	110.12	110.05	4.27 (Quadratic)
W2N3	40.05	125	119.7	117.4	115.4	115.4	115.5	4.28 (Three halves)

TABLE 16.8 Actual and Predicated Yield of Cauliflower With Prediction Models For Naturally Ventilated Greenhouse

Treat-ment	Water needed	Nitrogen level	Field yield	Predicted yield (100 kg/ha)				RMSE
				Linear	Square root	Quad-ratic	Three halves	
	(cm)	(kg/ha)	100 kg/ha					5.76 (Linear)
W2N1	40.05	62.5	88.0	83.17	84.08	84.25	84.25	5.85 (Square root)
W2N2	40.05	93.75	99.19	90.58	88.38	88.51	88.45	5.72 (Quadratic)
W2N3	40.05	125	105.4	97.98	99.01	99.10	98.96	5.80 (Three halves)

16.3.4   WATER USE EFFICIENCY (WUE) OF CAULIFLOWER

Total water used during the crop period and WUE for different treatments are presented in Tables 16.3 and 16.4. The amount of water used varied, depending upon the different treatments and the system used. The WUE with respect to yield indicated that it was low when irrigations were scheduled at 0.75 and 1.0 IW/CPE. As WUE is a function of crop yield and water used, the decrease in crop yield with increase in water application reduced the WUE under IW/CPE = 0.75 and 1.0. However, at IW/CPE = 0.5, the yield was increased linearly with water application, and WUE. The WUE at this treatment (W1N3) was 452.05 kg/(ha-cm) in fan pad cooled greenhouse and 395.31 in naturally ventilated greenhouse with water saving of 52.43% as compared to conventionally grown cauliflower.

Regression coefficients and constants for all production functions for yield are shown in Tables 16.5 and 16.6. Cauliflower yield showed a declining response to irrigation water applied, showing negative regression coefficients in all the production functions (Tables 16.5 and 16.6). This implies that cauliflower does not need higher amount of water for its optimum yield, as indicated by reduction in yield at IW/CPE = 1.00. It can be observed that quadratic, square root and three halves functions were the best fit to the data, because the regression coefficients for these functions were significant at high value of R^2.

The measured and predicted values of yield for regression models in different treatments are presented in Tables 16.7 and 16.8. Table 16.1 shows that yield was maximum at 100% of nitrogen dose and IW/CPE = 0.5. To evaluate the overall performance of all the models, the root mean square error (RMSE) between the measured and predicted values was calculated:

$$RMSE = \sum [(O - P)^2/N] \tag{4}$$

where: O = observed value (100 kg/ha); P = predicted value (100 kg/ha); and N = total number of observations.

After comparing the linear, square root, three halves and quadratic models, square root model was the best model in fan pad cooled greenhouse. In general, predicted values are in close agreement with the experimental results in all the models; and more closely with the square root model. The value of RMSE varied from 4.18 to 5.89 in fan pad cooled greenhouse and 5.72 to 5.85 in naturally ventilated greenhouse. Square root model (RMSE = 4.18) and quadratic model (RMSE = 5.72) were best-fitted models for fan pad cooled greenhouse and naturally ventilated greenhouse, respectively.

16.4   CONCLUSIONS

The maximum yield was 120.7 and 105.5 per 100 kg/ha of recommended nitrogen dose in fan pad cooled greenhouse and naturally ventilated greenhouse, respectively. For assessing the impact of irrigation levels and nitrogen dosages on crop yield, the square root model was the best prediction model with R^2 =0.956 in fan pad cooled greenhouse (RMSE = 4.18) and quadratic model with R^2 = 0.901 in naturally ventilated greenhouse (RMSE = 5.72). The quadratic models having negative regression coefficients also indicated a declining response to amount of irrigation water. Curd yield was increased with increased rates of nitrogen. WUE was highest with IW/CPE = 0.5 and 100% rate of nitrogen.

16.5   SUMMARY

A field study on fertigation of drip irrigated cauliflower (*Brassica oleracea var. botrytis Linn.*) was conducted with three rates of nitrogen application (100%, 75% and 50%) of recommended dose applied in three splits, that is, 50% of the dose at planting time, 25% at 30 days after planting and remaining 25% at 60 days after planting; and with three levels of irrigation at (irrigation water to cumulative pan evaporation ratio) IW/CPE= 0.5, 0.75 and 1.0 in fan pad cooled greenhouse and naturally ventilated greenhouse, respectively. Various predication models were linear, quadratic, square root and three halves that govern the relationship between irrigation water, nitrogen fertilizer and the yield in respect of off-season cauliflower.

The studies reveal that the yield 120.7 and 105 per 100 kg/ha of early cauliflower were maximum under irrigation schedule based on IW/CPE = 0.5 and 100% of recommended nitrogen dose in both the greenhouses. Square root model was the best prediction model for assessing the impact of irrigation water and nitrogen dose on crop yield, with R^2 = 0.956 in fan pad cooled greenhouse (RMSE = 4.18) and with R^2 = 0.901 in naturally ventilated greenhouse (RMSE = 5.72). The prediction model indicates that cauliflower shows a declining response to amount of irrigation water having negative regression coefficients. The crop response was positive to nitrogen application in square root model, indicating its favorable effect on crop yield at higher dosages.

KEYWORDS

- cauliflower
- crop response
- crop yield
- drip irrigation
- fan pad cooled greenhouse
- irrigation schedule
- linear model
- naturally ventilated greenhouse
- nitrogen level
- prediction functions
- root mean square error
- square root model
- yield

REFERENCES

1. Arkley, R. J. (1963). Relationships between plant growth and transpiration. *Hilgardla, 34*, 559–584.
2. Ayer, H. W., Hoyt, P. G., Cotner, M. L. (1980). Crop water production functions in economic analysis. *Proc. Irrig. Drain. Spec. Conf., Am. Soc. Civ. Eng.*, pages 346–364.
3. Reddy, Y. M., Sankara, G. H. (1977). Crop yield water use production functions. Proceedings of the national Seminar on Technology for Agricultural Development held at Chandigarh by the Institution of Engineers India) 9–11 April.
4. Rao, Siva K. S. V. V. (1987). Irrigated crop management through production functions. *J. Agric. Eng., 24(3)*, 317–322.
5. Rao, V., Raikhelkar, S. V., Sondge, V. D. (1994). Effect of irrigation and fertilization on yield and its components in sesame. *Indian J. Agric. Sci., 64(2)*, 93–100.
6. Shivakumar, H. K., Ramachandrappa, D. K., Nanjappa, H. V., Sreenivas, B. T., Aravinda, M. K. (2001). Production functions for irrigation water and planting methods in sunflower. *Crop Res., 21(2)*, 139–142.
7. Selvaraj, P. K., Krishnamurthy, V. V., Manickasundram, P., Martin, G., Ayyaswamy, M. (1996). Effect of irrigation schedules and nitrogen levels on the yield of turmeric through drip irrigation. *Madras Agric. J., 84(6)*, 347–348.

APPENDIX I: PHOTOS OF DRIP IRRIGATED CAULIFLOWER

PART III
MULCHING AND CROP PERFORMANCE

CHAPTER 17

USE OF MULCHES IN SOIL MOISTURE CONSERVATION: A REVIEW

SUSHANT MEHAN and KAMAL G. SINGH

17.1 INTRODUCTION

Mulch is a protective covering, usually of organic matter such as leaves, straw, or peat, placed around plants to prevent the evaporation of moisture, and the growth of weeds. The word mulch has probably been derived from the German word "*molsch*" means soft to decay, which apparently referred to the gardener's use of straw and leaves as a spread over the ground as mulch [1].

Mulching reduces the deterioration of soil by way of preventing the runoff and soil loss, minimizes the weed infestation and checks the water evaporation. Thus, it facilitates for more retention of soil moisture and helps in control of temperature fluctuations, improves physical, chemical and biological properties of soil, as it adds nutrients to the soil and ultimately enhances the growth and yield of crops. Further, reported that mulching boosts the yield by 50–60% over no mulching under rain-fed situations [2, 24, 43].

17.2 CLASSIFICATION OF MULCHES

Advancement in plastic chemistry has resulted in development of films with optical properties that are ideal for a specific crop in a given location. Horticulturists need to understand the optimum above and below ground environment of a particular crop before the use of plastic mulch. These are two types.

Photo-degradable plastic mulch: This type of plastic mulch film gets destroyed by sun light in a shorter period.

Bio-degradable plastic mulch: This type of plastic mulch film is easily degraded in the soil over a period of time.

---

*In this chapter, the currency is expressed in Indian Rupees (1.00 US$ = Rs. 60.93; 1.00 Rs. = 0.02 US$).

17.2.1 COLOR OF FILM

Soil environment can be managed precisely by a proper selection of plastic mulch composition, color and thickness. Films are available in variety of colors including black, transparent, white, silver, blue red, etc. But the selection of the color of plastic mulch film depends on specific targets. Generally, the following types of plastic mulch films are used in horticultural crops.

1. **Black plastic film**: It helps in conserving moisture, controlling weed and reducing outgoing radiation.
2. **Reflective silver film**: It generally maintains the root-zone temperature cooler.
3. **Transparent film**: It increases the soil temperature and preferably used for solarization.

Apart from the above classification there is another way of classifying Methods in mulching [2, 3]:

1. **Surface mulching:** Mulches are spread on surface to reduce evaporation and increase soil moisture.
2. **Vertical mulching:** It involves opening of trenches of 30 cm. depth and 15 cm width across the slope at vertical interval of 30 cm.
3. **Polythene mulching:** Sheets of plastic are spread on the soil surface between the crop rows or around tree trunks.
4. **Pebble mulching:** Soil is covered with pebbles to prevent transfer of heat from atmosphere.
5. **Dust mulching:** Interculture operation that creates dust to break continuous capillaries, and deep and wide cracks thus reducing evaporation from the exposed soil areas.
6. **Live vegetative barriers** on contour key lines not only serve as effective mulch when cut and spread on ground surface, but also supply nitrogen to the extent of 25 to 30 kg per ha, besides improving soil moisture status.

17.3 EFFECTS OF MULCHES ON SOIL MOISTURE CONSERVATION

Water is essential for growth and development. It is also a major cost in agricultural systems. The success of many agricultural forms relies on conservative and efficient use of water. Moisture retention is undoubtedly the most common reason for which mulch is applied to soil.

Ingman [25] claimed that the use of things made with plastic or plastic components have become a routine part of our daily lives. In a similar way, over the past 50 years world agricultural systems have rapidly adopted the use of many types of plastic products to grow the food we eat because of the productive advantages they afford. Plastic use in agriculture (plasticulture) continues to increase every year in its worldwide use despite the many negative factors of plastic waste disposal and the ever-diminishing supply of petroleum. There is a common lack of awareness regarding what plastic mulch is, and also a lack of applied research of its use in agricultural communities. However, the use of plastic mulch may actually be one of the most significant water conservation practices in modern agriculture: quite possibly surpassing the water savings of drip irrigation. Even though most of the world's use of freshwater is spent for irrigation purposes, little research explores how plastic mulch use as a water conservation practice may influence the current and future status of water resources. He used a multidisciplinary approach to understand why Chinese farmers on the margins of the Gobi desert continue to use plastic mulch, and in particular, how its use may relate to water conservation [25]. Next, the study asks to what extent the plasticization of agriculture may influence the income and standard of living for agricultural communities. He was able to prove the role of plastic mulch in conserving soil moisture.

Mulch is used to protect the soil from direct exposure to the sun, which would evaporate moisture from the soil surface and cause drying of the soil profile. The protective interface established by the mulch stops raindrop splash by absorbing the impact energy of the rain, hence reducing soil surface crust formation. The mulch permits soil surface to prevent runoff allowing a longer infiltration time. These features result in improved water infiltration rates and higher soil moisture. An auxiliary benefit of mulch reducing soil splash is the decreased need for additional cleaning prior to processing of the herb foliage [7]. Organic and inorganic mulches have shown to improve the soil moisture retention. This increased water holding ability enables plants to survive during dry periods. The use of plastic mulch can be improved if under-mulch irrigation is used in combination with soil moisture monitoring. The influence of rainfall events is not as great when plastic mulch is used, necessitating active irrigation management. Under mulch, irrigation of vegetable crops has been shown to improve crop yields more than overhead irrigation systems [16]. Mulch enables the soil moisture levels to maintain for longer periods. In some cases while providing improved moisture conditions within the soil, the mulch changes microclimate so that it uses more water [15, 53], thus negating the initial benefits. Plastic mulch conserved 47.08% of water and increased yield by 47.67% in tomato when compared to nonmulched control [20]. Plastic mulching resulted in 33 to 52% more efficient use of irrigation water in bell pepper compared to bare soil.

The conservation of soil moisture through mulching is one of the important best management practices (BMP). The microclimatic conditions are favorably affected by optimum degree of soil moisture. When soil surface is covered with mulch helps to prevent weed growth, reduce evaporation and increase infiltration of rainwater during growing season.

Different mulching materials helped bell pepper (*C. annuum cv. California Wonder*) to perform better at water deficits from 25–75% and plastic mulch had highest water use efficiency. Treatment receiving mulch recorded significantly higher net returns and benefit – cost ratio (1.80) compared to control as a result of soil water conservation. A 34–50% reduction in soil water evaporation was observed as a result of crop residue mulching. Mulch slows down evaporation and reduces the irrigation requirement [9, 27]. Liu et al. [31 to 34] also reported that mulching improves the ecological environment of the soil and increases soil water contents.

Plastic mulch helps prevent soil water loss during dry years and sheds excessive water away from the crop root zone during periods of excessive rainfall. This can reduce irrigation frequency and amount of water. It may help to reduce the incidence of moisture related physiological disorders such as blossom end rot on tomato, fruit cracking in lime and pomegranate. Research has shown that mulch provides many benefits to crop production through soil and water conservation, enhanced soil biological activity and improved chemical and physical properties of the soil [19]. Cooper [17], Menezes et al. [38], Chung [14] and Aliudin [5] reported that mulches conserved more soil moistures, enhanced vegetative growth and yield of garlic.

Adeoye [4] recorded high moisture content up to a depth of 60 cm in grass-mulched soil together with good infiltration and reduced evaporation. Chen [13] also reported high water content in the top 5 cm of soil (an increase of 4.7% in clayey, 3.1% in loamy and 0.8–1.8% in sandy soil) with polyethylene mulch from sowing to the emergence of groundnut seedlings [44].

According Bhelia [9], increased plant dry weight for mulched plants was due to the capabilities of mulch to maintain soil moisture as well as increased efficiency in water uptake by plants. The growth and leaf yield of mulberry appears to be direct reflection of the soil moisture status.

Orzolek et al. [41] observed that use of polyethylene mulch in the field, increase in the soil temperature especially in early spring, reduced weed problems, and increased moisture conservation, reduction in certain insect pest, higher crop yield and more efficient use of soil nutrients. Sood and Sharma [47] reported similar beneficial effects of mulching through improvement of soil.

The rate of evaporation and water losses in agricultural land depends on a number of factors, from which the irrigation methods are very important. Open ditches and flooded basins experience evaporation losses ranging from 10 to 40% [50]. The use of mulches helps to conserve water, by reducing evaporation from soil surface, cooling soil, controlling weed growth, reducing soil erosion, and compaction [46, 48, 50]. The mulches compose of organic, inorganic (natural) and man-made materials. Organic mulches are readily available, provide nutrients, and do not added chemicals to the environment [45].

Budelman [12] examined the impact of three types of organic mulches on surface soil temperature reduction and moisture conservation on a sandy soil (85% sand) during a 60-day dry period. According to him, the initial impact ranged from a reduction in soil temperature of 5.6–9.8°C and an increase in soil moisture from 4% to 5.6% for different mulches applied in this experiment [35].

Not all types of mulches are appropriate that can be used with any irrigation systems. For instance, natural or organic mulches slow down the water movement in surface irrigation. Drip irrigation systems are often effective when used in conjunction with any type of mulches, particularly with plastic mulches [36]. In contrast, plastic mulches can readily be used in conjunction with surface irrigation systems such as furrow irrigation to save water by reducing evaporation losses.

Plastic mulches may be clear, black or any color type. Short wave rays of sun penetrate in clear plastic and warm up the soil (+10°C) and the long wavelengths reradiating from the soil, that is also trapped, can intensify the situation. In this case, evaporating water causes greenhouse effects that in turn increases early growth in cool season, but also stimulates weed growth under clear plastic. Black plastic is heated due to short wave absorption which in turn warm soil (+6°C) by conduction and increases many crop yield and control weed growth [46].

Plastic mulch seems to be one of the effective ways to conserve water in the soil reservoir to be up-taken gradually by plants. Plastic mulch is spread over the soil surface to trap the heat and accelerate the rate of soil warming and cause early harvest of crop product.

Combination of plastic mulch and drip irrigation was popularized in USA for more than two decades and recently in some parts of Iran, due to "quicker to market" benefit of faster seed germination and plant growth. To determine yield, quality, and water use efficiency (WUE) of muskmelon, six irrigation systems including drip and furrow with mulch on *Cucumis melo L. group Cantalupensis, Caravelle* were evaluated by Leskovar et al. [31]. The results showed that the average water applied for drip systems was 53% lower than that for conventional furrow systems, and WUE was 2.3-folds more. However, the combination of plastic and furrow irrigation has not yet been fully examined in the field to find if there is any effect on soil water conservation and likelihood increase in yield of cucumber and tomato crops. The purpose of this study was to find the effect of plastic mulch use with furrow irrigation on cucumber and tomato yield in the field at flowering and production stages.

Plastic film mulch technology is a cropping system feature for water saving that is used extensively in arid and semiarid areas of northwest China [37]. When evaporative demand is fairly strong, film mulching can greatly reduce soil water evaporation. However, research on water flow beneath and through various open-hole ratios of the perforated film mulches is limited, and questions concerning soil water flow and soil heat transfer for this type of water-saving system remain unanswered. It is, therefore, very important to perform research on soil water evaporation and soil temperature distribution with various open-hole ratios of the perforated plastic mulches. Yi et al. [52] conducted a series of soil water evaporation experiments using different open-hole ratios of perforated plastic mulches. The columns received mulches with various open-hole ratios: 0% (covered with a solid plastic mulch), 1.39%, 2.84%, 7.24%, 30.5%, and 100% (non mulched bare surface). In conjunction with the water movement of evaporation from film open-hole studies, soil temperature distributions were also analyzed. Their measurements indicated that film open-hole mulch had a restraining effect on evaporation and that the restraining effect decreased with the increase in open-hole ratios. Compared with bare soil evaporation, the percentage of evaporation reduction rates for open-hole ratios of 0%, 1.39%, 2.84%, 7.24%, and 30.5% were 69.26%, 33.09%, 22.80%, 20.05%, and 11.82%, respectively. The results showed a linear relationship between cumulative water evaporation and square root of time for the different open-hole ratios of the perforated plastic mulches, and the coefficients of the linear function were significant. On this basis, mathematical relations of relative evaporation rate and evaporation based on hole areas of perforated plastic mulches were analyzed and discussed. These results extend the Gardner evaporation equation to bare soils [22] to include water evaporation from soils covered by various perforated plastic mulches. The resulting equations presented in this study provide an approach for describing evaporation from plastic mulch covered soil. The annual evaporative demand of arid and semiarid regions near Xinjiang in northwest China exceeds 2000 mm, whereas annual precipitation is less than 200 mm. Thus, a great need exists in the area to make efficient use of all agricultural water.

Plastic mulch can also increase soil temperature, and the increase in soil temperature can cause changes in soil water distribution [34]. Thus the coupled transport of water and heat in soils should be researched in order to understand the fundamental features of soil water evaporation with different open-hole ratios of plastic mulch.

Research regarding bare soil water evaporation has been conducted by experimental methods, theoretical analysis, and numerical calculation [6, 10, 11, 21, 22, 23, 29, 39, 40, 42, 45, 48]. The evaporation equation for bare soils developed by Gardner [23] is one of the most popular equations for calculation of soil water evaporation. He found that cumulative evaporation was a linear function of the square root of time. Behzad [8] comparing his study with the previous studies concluded that using clear plastic mulch conjunction with furrow irrigation system had significant effect on soil moisture retention by 74.4%. No significant water conservation was considered under black plastic mulch in his study. Considerable yield increments (60 and 48.7%) and (65.8 and 46.7%) were achieved for tomato and cucumber crops under both clear and black plastic mulches, respectively, at flowering and production stages. A more field study is recommended to understand the effect of furrow irrigation with plastic mulches on water conservation.

Organic vegetable producers in drier, cooler climates such as ours on the front range of Colorado like to use black polyethylene plastic film as mulch on vegetable row crops for multiple reasons. McDonald [27] suggested that when drip irrigation is laid underneath the plastic film, it delivers water and fertilizer to the plants and evaporation is reduced. But, because there is no surface evaporation of water, it is easy to over-irrigate crops. For this reason, a moisture probe should be used to check root zone moisture levels.

In addition to providing water conservation, this synthetic mulch controls weeds and warms the soil, making for an earlier crop. The black plastic mulch can be covered with hay or straw to protect crops from excessive heat later in the summer.

In addition to black plastic film which can only be used one season, black woven landscape cloth is often used [28], which can be reused up to seven years. Organic mulches such as straw, hay, grass clippings, pine needles, and leaves also conserve moisture. These organic mulches add organic matter to the soil after they decompose. However, one needs to pay attention how different organic mulches can change the soil chemistry. Finally, green living mulches, or cover crops, can help to conserve moisture if the right cover crop is used for the right agricultural crop under given soil and climate conditions.

According to a research conducted by Chakraborty [18], it is concluded that mulching is one of the important agronomic practices in conserving the soil moisture and modifying the soil physical environment. Field experiments were conducted in a sandy loam soil to evaluate the soil and plant water status in wheat under synthetic (transparent and black polyethylene) and organic (rice husk) mulches with limited irrigation and compared with adequate irrigation with no mulch (conventional practices by the farmers). Though all the mulch treatments improved the soil moisture status, rice husk was found to be superior in maintaining optimum soil moisture condition for crop use. The residual soil moisture was also minimum, indicating effective utilization of moisture by the crop. The plant water status, as evaluated by relative water content and leaf water potential, was favorable. Specific leaf weight, root length density and dry biomass were also greater in this treatment. Optimum soil and canopy thermal environment of wheat with limited fluctuations were observed, even during dry periods. This produced comparable yield with less water use, enhancing the water use efficiency. Therefore, it may be concluded that under limited irrigation condition, mulching will be beneficial for wheat as it is able to maintain better soil and plant water status, leading to higher grain yield and enhanced water use efficiency.

17.4 CONCLUSIONS

Use of moisture conservation measures is a prime need of the hour. There is an urgent need to evaluate methods to prevent excessive loss of water from soil surface, which can otherwise be used by the crop for its physiological development. Mulching has been advocated as an effective practice for conserving soil moisture. It works as an insulating barrier which checks evaporation from soil surface. Moreover, use of polyethylene mulch has been reported to conserve soil moisture appreciably. Hence, under prevailing drought and water scarcity conditions, conservation of soil moisture and to ensure availability of soil moisture to crop is of greater importance [51]. The black polyethylene mulch also reduces all types of weeds in addition to soil moisture conservation. Therefore, black plastic mulch is more beneficial.

17.5 SUMMARY

India is an agricultural economy, which focuses on various water conservation strategies, due to limited water resources. Moreover, most of these resources are either out of reach for agricultural use or not in proper condition to be used directly. There is considerable loss from soil surface. It has now become the need of an hour to save the water, which has been wasted recklessly. Mulching is a technique that can help to conserve water. Besides, mulching helps in weed management as well. This helps in retaining best soil structure for the crop growth. Mulches prevent surface runoff over soil, prevents water splashing, maintains optimum soil temperature and soil water content, thereby providing best suitable environment for crop growth. The review in this chapter deals with various aspects of mulches in soil moisture conservation.

KEYWORDS

- **agricultural economy**
- **mulching**
- **open-hole ratio**
- **soil moisture conservation**
- **soil structure**
- **soil temperature**
- **transpiration**
- **water management**
- **water splashing**
- **water use efficiency**
- **weed management**

REFERENCES

1. Abdul-Baki, A., Spence, C. (1992). Black polyethylene mulch doubled yield of fresh-market field tomatoes. *Hort Science, 27,* 787–789.
2. Abu-Awwad, A. M. (1998). Effect of mulch and irrigation water amounts on soil evaporation and transpiration. *Journal of Agronomy and Crop Science, 181,* 55–59.
3. Abu-Awwad, A M. (1999). Irrigation water management for efficiency water use in mulched onion. *Journal of Agronomy and Crop Science, 183,* 1–7.
4. Adeoye, K. B. (1984). Influence of grass mulch on soil temperature, soil moisture and yield of maize and gero millet in a savanna zone soil. *Samaru Journal of Agricultural Research, 2,* 87–97.

5. Aliudin, T. (1986). Effect of Soil tillage and Application of Mulch on Yield of Field Grown Garlic. Buletin-Penelitian-Hortikultura, *8*, 12–15.
6. Bachmann, J., Horton, R., van der Ploeg R. R. (2001). Isothermal and non isothermal evaporation from four sandy soils of different water repellency. *Soil Sci. Soc. Am. J., 65*, 1599–1607.
7. Barker, A. V. (1990). Mulches for Herbs. The Herb Spice and Medicinal Plant Digest, *8(3)*, 1–6.
8. Behzad, Ghorbani, (2004). Evaluation of plastic mulch effects on cucumber and tomato yield at flowering and production stages. Proceedings of ICID Interregional Conference on food production and water: social and economic issues of irrigation and drainage, Moscow, Russia, 5–11 September, pp. 2.1.3.
9. Bhelia, H. S. (1988). Tomato response to trickle irrigation and black polythlene mulch. *Journal American Society of Horticultural Science, 113*, 543–546.
10. Black, T. A., Gardner, W. R., Thurtell, G. W. (1969). The prediction of evaporation, drainage, and soil water storage for a bare soil. *Soil Sci. Soc. Am. Proc., 33*, 655–660.
11. Brisson, N., Perrier, A. (1991). A semiempirical model of bare soil evaporation for crop simulation models. *Water Resour. Res., 27*, 719–727.
12. Budelman, A. (1989). The performance of selected leaf mulches in temperature reduction and moisture conservation in the upper soil stratum, Agro. Syst., *8*, 53–66.
13. Chen, Z. (1985). Polythene mulched groundnut development in Guanzhou city. *Peanut Science Technology, 3*, 34–37.
14. Chung, D. H. (1987). Effect of polyethylene film mulching, sulfur application and different levels of nitrogen and potassium on growth, flower stalk elongation, bulbing and leaf tip yellowing of garlic (*Allium sativum cv. Enizing*). Journal of Korean Society of Horticultural Science 28, 1–8 pages.
15. Clark, J. R., Moore J. N. (1991). Southern high bush blueberry response to mulch. HortTechnology, *1(1)*, 52–54.
16. Clough, G H., Locascio S. J., Olson S. M. (1990). Yield of successively cropped polyethylene-mulched vegetables as affected by irrigation method and fertilization management. Journal American Society for Horticultural Science, *115*, 884–887.
17. Cooper A. J. (1973). *Root Temperature and Plant Growth- A Review.* Commonwealth Bureau of Horticulture and Plantation Crops, East Malling, Maidstone, Kent, UK.
18. Debashis, Chakrabortya, Shantha Nagarajanb, Pramila Aggarwala, Gupta, V. K., Tomara, R. K., Garg, R. N., Sahooa, R. N., Sarkarc, A., Chopra, U. K., Sundra, K. S., Kalra, N. (2012). Effect of mulching on soil and plant water status, and the growth and yield of wheat (*Triticum aestivum* L.) in a semiarid environment. Division of Agricultural Physics, Indian Agricultural Research Institute, New Delhi 110012.
19. Dilip, K. G., Sachin S. S., Rajesh Kumar, (1990). Importance of mulch in crop production. *Indian J. Soil Cons., 18*, 20–26.
20. Friake, N. N., Bangal, G. B., Kenghe, R. N., More, G. M. (1990). Plastic tunnel and mulches for water conservation. *Agricultural Engineering Today*, 14(3–4):35–39.
21. Fritton, D. D., Kirkham, D., Shaw, R. H. (1970). Soil water evaporation, isothermal diffusion, and heat and water transfer. *Soil. Sci. Soc. Am. Proc., 34*, 183–189.
22. Gardner, H. R., Gardner, W. R. (1969). Relation of water application to evaporation and storage of soil water. *Soil. Sci. Soc. Am. Proc., 33*, 192–196.
23. Gardner, W. R. (1959). Solutions of the flow equation for the drying of soils and other porous media. *Soil. Sci. Soc. Am. Proc., 23*, 183–187.
24. Goswami, S. B., Saha, S. (2006). Effect of organic and inorganic mulches on soil-moisture conservation, weed suppression and yield of elephant-foot yam (*Amorphophallus peoniifolius*). *Indian J. Agron., 51(2)*, 154–156.

25. Ingman, M. (2012). The Role of plastic mulch as a water conservation practice for desert. Oasis Communities of Northern China Mark Ingman's thesis defense for his MS in Water Resources Policy and Management.

26. Jack, C. V., Brind W. D., Smith R. (1955). *Mulching Tech. Comm. No. 49*, Commonwealth Bulletin of Soil Science, UK.

27. McDonald, K. (2013). *Thirty-five Water Conservation Methods for Agriculture, Farming, and Gardening*. Part II.

28. Kratsch, H. (2007). Water-wise landscaping: Mulch. HG/Landscaping/2007–01pr, Utah State University.

29. Lascano, R. J., C. H. M. van Bavel, (1986). Simulation and measurement of evaporation from a bare soil. *Soil. Sci. Soc. Am. J., 50,* 1127–1133.

30. Leskovar, D. I., Ward, J. C., Sprague, R. W. (2001). Yield, quality, and water use efficiency of muskmelon are affected by irrigation and transplanting versus direct seeding. *HortScience, 36(2),* 282–289.

31. Li, M. S. (2006). Effect of drip emitter discharge on the soil moisture, heat, solutes dynamics and crop water use under drip irrigation with plastic mulch. North-west A & F University, pp. 20–60.

32. Li, S. X., Wang, Z. H., Li, S. Q., Gao, Y. J., Tian, X. H. (2013). Effect of plastic sheet mulch, wheat straw mulch, and maize growth on water loss by evaporation in dry land areas of China. *Agricultural Water Management, 116,* 39–49.

33. Li, Y., Wang, Q. J., Wang, W. Y. (2005). Distribution and movement characteristics of soil water and soil salt during evaporation from perforated plastic mulch. *Plant Nutrition and Fertilizing Science, 11(2),* 187–193.

34. Li, Y., Wang, Q. J., Wang, W. Y. (2005). Soil evaporation under perforated plastic mulch. *Chinese Journal of Applied Ecology, 16(3),* 445–449.

35. Louise, E. B., Lassoie, J. P., Fernandes, E. C. M. (1999). Agroforestry in Sustainable Agricultural Systems. CRC Press, Boca Raton, New York, Washington, D. C., p 53–54.

36. Mata, V. H., Nunez, R. E., Sanches, P. (2002). Soil temperature and soil moisture in Serrano pepper (*Capsicum annuum* L.) with fertigation and mulching. Proceeding of the 16th International Pepper Conference Tampico, Tamaulipas, Mexico, November 10–12.

37. Men, Q., Li, Y., Feng, G. (2003). Effects of plastic film mulch patterns on soil surface evaporation. Journal of Irrigation and Drainage, *22(2),* 17- 20.

38. Menezes, S. M., Navais D. E., Sontos H. L., Dos M. A. (1974). The effect of nitrogen fertilization, plant spacing and mulching on the yield of garlic cultivar Amarante. *Revista Ceres, 21,* 203–211.

39. Milly, P. C. D. (1984). A simulation analysis of thermal effects on evaporation from soil. *Water Resour. Res., 20,* 1087–1098.

40. Munley, W. G. Jr., Hipps, L. E. (1991). Estimation of regional evaporation for a tall grass prairie from measurements of properties of the atmospheric boundary layer. *Water Resour. Res., 27,* 225–230.

41. Orzolek, M. D., Murphy J., Ciardi J. (1993). The effect of colored polyethylene mulch on the yield of squash, tomato and cauliflower. Final Report to the Pennsylvania Vegetable Marketing and Research Commodity Board. The Pennsylvania State University, USA.

42. Philip, J. R. (1957). Evaporation, moisture and heat fields in the soil. *J. Meteorol., 14,* 354–366.

43. Qi, G. P. (2008). Combined mechanism of root-soil water-salt in drip irrigation under mulch on saline-alkaline land. Gansu Agriculture University, pp. 19–45.

44. Ramakrishna, A., Tam, H. M., Wani, S. P., Long, T. D. (2006). Effect of mulch on soil temperature, moisture, weed infestation and yield of groundnut in northern Vietnam. *Field Crops Res., 95,* 115–125.

45. Reynolds, W. D., Walker, G. K. (1984). Development and validation of a numerical model simulating evaporation from short cores. *Soil. Sci. Soc. Am. J.*, *48*, 960–969.
46. Robinette, G. O., Sloan, K. W. (1984). *Water Conservation Landscape Design and Management*. Van Nostrand Reinhold Company, Berkshire, England, p: 147–158.
47. Sood, B. R., Sharma V. K. (1996). Effect of intercropping and planting geometry on the yield and quality of forage maize. *Forage Research,* 24, 190–192.
48. Suleiman, A. A., Ritchie, J. T. (2003). Modeling soil water redistribution during second-stage evaporation. *Soil Sci. Soc. Am. J.*, *67,* 377–386.
49. Thurston, H. D. (1997). Mulch systems – sustainable methods for tropical agriculture. West-view Press, A Division of HarperCollins Publishers, Inc., Colorado, USA, 19–29.
50. Vickers, A. (2001). *Water Use Conservation*. Water Flow Press, Amherst, Massachusetts, USA, 215–217 & 380.
51. Yadahalli, G. S., Vidyavathi G. Y and Srinivasareddy, G. V. (2011). Mulching: One of the means to mitigate drought. *Agrobios Newsletter*, *9(10),* 36–37.
52. Yi, L., Mingan Shao, Wenyan Wang, Robert Q. Wang, (2003). Open-hole ratios: effects of perforated plastic mulches on soil water evaporation. *Horton Soil Science*, *168(11),* 751–758.
53. Zajicek, J M and Heilman J. L. (1991). Transpiration by crape myrtle cultivars surrounded by mulch, soil and turf grass surfaces. *Hort. Science*, *26(9),* 1207–1210.

CHAPTER 18

PERFORMANCE OF DRIP IRRIGATED GROUNDNUT

ANGREJ SINGH, KAMAL G. SINGH, RAMESH. P. RUDRA, and PRADEEP K. GOEL

18.1 INTRODUCTION

Groundnut (*Arachis hypogea L.*) can be grown over a wide range of climatic conditions. It has been reported that under low temperature conditions, polythene mulch applied at beginning of crop growth accelerated emergence, seedling growth, flowering and increased pod number and 100 seed mass of groundnut by increasing soil temperatures. The normal planting time of spring groundnut in Punjab is February and its harvesting coincides with the onset of the monsoon which results in significant harvest losses. The soil temperature is quite low under Punjab conditions in the months of January and February so groundnut germination is affected. The plastic mulches are known for the increase in soil temperature and for conserving higher soil moisture in the root zone. Therefore, this study explores possibility of planting groundnut in the month of January and harvesting it before the onset of the rainy

18.2 MATERIALS AND METHODS

A field experiment was conducted during winter/spring season of 2007 and 2008 in the department of Soil and Water Engineering, Punjab Agricultural University Ludhiana. The soil of the experimental plot was loamy sand, with pH 8.2 and EC of 0.14 mmhos/cm. Soil was low in organic carbon (0.36%), available N (244.8 kg ha^{-1}), high in available P (25.0 kg ha^{-1}) and medium in available K (240 kg ha^{-1}).

The experiment was laid in factorial split plot design with three factors viz. method of planting (Flat and Bed), mulches (Transparent plastic mulch (15 μ), black biodegradable mulch(15 μ), and no mulch) and two dates of sowing (23 Jan

*In this chapter, the currency is expressed in Indian Rupees (1.00 US$ = Rs. 60.93; 1.00 Rs. = 0.02 US$).

2007 and 27 Jan 2008) as D_1 and (22 Feb 2007 and 15 Feb 2008) as D_2. All the treatments were replicated thrice.

The recommended dose of nitrogen fertilizer @ 15 Kg N/ha (33 Kg urea/ha) was applied as basal dose at the time of sowing. The crop was sown as per treatment after giving heavy pre sowing irrigation. Flat sowing was done at spacing of 30×15 cm² while bed planting was done at 45×10 cm². The plot size was 4.5×2.7 m². The seed of groundnut variety SG-99 was hand dibbled at the desired spacing. The mulches were applied on the same day after sowing and were removed after one month from date of sowing. The crop was given one hand weeding 30 DAS. The irrigation was applied using drip irrigation system at IW/CPE ratio of 0.75. During both the years D_1 was harvested in 2nd week of June and D_2 was harvested in 4th week of June.

18.3   RESULTS AND DISCUSSION

The data on soil temperature revealed that morning soil temperature recorded at 7.30 a.m. was 1.2–1.5°C higher in biodegradable mulch and 2.4–3.1°C higher under plastic mulch as compared to no mulch plots during 2007 and 2008, respectively. The increase in afternoon (14.30 hrs.) temperature in biodegradable and plastic mulch film as compared to the no mulch plot was 1.3 and 2.1°C in 2007 and 1.5 and 2.0° C during 2008, respectively. The results are in line with those obtained by Huwenguang et al. [1], and Kumar and Ngachan [2].

The planting methods significantly affected the germination percentage (Table 18.1) and recorded 3.7% higher germination in bed planting as compared to flat planting method. The data on number of mature pods/plant, single/double seeded pods and shelling percent did not differ significantly at the time of harvest. The 100-kernel weight was significantly higher in flat planting than bed planting system. This may be because from the edges of bed, the pegs take more time to contact the ground as compared to flat soil surface, so effective time for seed maturity in pod was comparatively less in bed planting system. The results confirm the findings of Kumar and Ngachan [2].

TABLE 18.1   Groundnut Germination, Yield Attributes and Yield As Influenced by Planting Method, Mulch and Dates of Sowing*

Treatment	Germination (%) 30 DAS	Percentage of mature pods	No. of mature pods/plant		100 kernel weight (g)	Shelling percentage	Pod yield, per 100 kg/ha
		At harvest	Single	Double			
Bed planting	57.7	71.4	9.4	23.9	63.8	62.6	30.7

TABLE 18.1 *(Continued)*

Flat planting	54.0	69.9	9.0	24.3	64.9	63.3	33.6
CD (p=0.05)	2.5	NS	NS	NS	1.10	NS	1.63
No mulch	**38.3**	**69.7**	**8.2**	**22.7**	**61.6**	**60.5**	**30.3**
Plastic mulch	68.1	77.7	9.5	25.4	66.6	65.3	33.2
Biodegradable mulch	61.1	75.4	10.0	24.2	64.9	63.2	32.9
CD (p=0.05)	**3.0**	**6.34**	**1.02**	**1.16**	**1.35**	**1.40**	**1.99**
D_1	31.0	59.9	8.9	23.6	63.8	62.4	29.5
D_2	80.7	62.9	9.5	24.6	64.9	63.5	33.9
CD (p=0.05)	**2.3**	**NS**	**NS**	**NS**	**NS**	**NS**	**1.60**

*Pooled data 2007–2008.

The application of biodegradable mulch and plastic mulch film improved the groundnut germination. The early germination may be due to increased soil temperature and favorable soil moisture in the seeding zone. At the time of harvest the percentage of mature pods was significantly higher under plastic mulch film as compared to no mulch but at par with biodegradable mulch. The no. of mature pods/ plant, 100-kernel weight and shelling % were significantly better both under the plastic and biodegradable mulch as compared to no mulch (Table 18.1). The highest pod yield of 33.2 q/ha was obtained under plastic mulch was at par with the biodegradable mulch, but both the mulches were significantly superior to the no mulch conditions. The results confirm the findings of Subrahmaniyan and Kaliaselvan [4, 5], and Raskar and Bhoi [3].

Planting date had a significant influence on the germination percentage. The mid February sown crop recorded significantly higher germination percentage as compared to January (4th week) sown crop (Table 18.1). The reduced germination in January sown crop is mainly due to low temperature. The planting dates did not have any significant effect on the yield attributes of groundnut. However, pod yield was significantly higher in the second date of sowing. This may be because of the cumulative effect of the yield attributes and due to extremely cold winters during 2008 in which first date-sown crop remained under frost spell for more than two weeks resulting in reduced plant stand and hence poor yield.

18.4. SUMMARY

The application of biodegradable and plastic mulch increased the pod yield of groundnut to the tune of 8.6 and 9.6% over the control plot. The yield reduction due to early planting was 15%, which is considered as better option if the crop damage (quality deterioration) by rain is considered.

KEYWORDS

- biodegradable mulch
- crop damage
- germination
- groundnut
- mulch
- planting date
- planting method
- plastic mulch
- pod yield

REFERENCES

1. Huwenguang, Duan, S. Sui, Q. (1995). High-yield technology for groundnut. International Arachis Newsletter, *15*, 14–15.
2. Kumar, S., Ngachan, S. V. (2001). Performance of winter groundnut (*Arachis hypogaea L.)* with polythene mulch under rained condition of Manipur valley. *Indian J. Agron.*, *46*, 151–155.
3. Raskar, B. S., Bhoi, P. G. (2003). Response of summer groundnut (*Arachis hypogaea*) to irrigation regimes and mulching. *Indian J. Agron.*, *48*, 210–213.
4. Subrahmaniyan, K., Kalaiselvan, P. (2005). Flowering behavior and reproductive growth of polyethylene film mulched groundnut (*Arachis hypogaea*) intercropped with cotton (Gossypium hirsutum) under irrigated situation. *Indian J. Agron.*, *50*, 126–128.
5. Subrahmaniyan, K., Kalaiselvan, P., Balasubramanian, T. N. (2008). Micro climate variations in relation to different types of polyethylene film mulch on growth and yield of groundnut (*Arachis hypogaea*). *Indian J. Agron.*, *53*, 184–188.

CHAPTER 19

PERFORMANCE OF DRIP IRRIGATED POTATO

AMANPREET KAUR CHAWLA and KAMAL G. SINGH

19.1 INTRODUCTION

Agricultural crop production is witnessing a rapid transition to agricultural commodity production and in this emerging global economic order, potato is appearing as an important crop, poised to sustain and diversify food production in this millennium. India, the biggest producer after Russia and China, has 128.48 thousand hectare under potato crop with an annual output of 23,271.8 thousand metric tons. In Punjab, potato is most prominent crop in winter occupying an area of 66.5 thousand hectare with the total production of 1,382.6 thousand metric tons and with an average yield of 20,791 Kg/ha [22].

Potato (*Solanum tuberosum* L.) a rich source of starch, vitamins especially B_1 and C and minerals, is one of the rare noncereal food that can meet the nutritional requirements of the fast growing population particularly in developing countries where it has not yet been adopted as a staple food. In India, about 98% of the crop production is used for table purpose, where as in developed countries about 15% is used for processing purpose. Potato production is seasonal and quiet fluctuating causing large variations in its prices in India [22, 23].

The alternative use of potatoes as processed food can help to stabilize the potato production vis-à-vis its prices. The major bottleneck for potato processing is suitability of tuber for processing. The quality requirements for processing depends mainly upon dry matter and reducing sugar content as these determine the yield, texture and quality of processed products. A dry matter content of 18–20% and reducing sugar less than 0.25% is desirable for processing [21]. Prevailing night temperature during tuberization affects the quality parameters. If the night temperature is less than 10°C, the dry matter content is low (<18%) and reducing sugar content is more than 0.25%, where as night temperature of 10–12°C improves the quality parameters and bring these parameters near the threshold limit. Further increase in

---

*In this chapter, the currency is expressed in Indian Rupees (1.00 US$ = Rs. 60.93; 1.00 Rs. = 0.02 US$).

the night temperature above 12°C, favors the quality parameters to remain within the desirable limits. However, tuber yield is inversely affected with the increase in night temperature.

If some agro-techniques are developed which can maintain the optimum temperature for quality parameters while optimizing the tuber yield, the potato cultivation can become more remunerative. Mulches are known to modify soil hydrothermal conditions. Mulches act as covering soil surface vapor barriers or reflective materials that can help in checking evaporation by reducing the intensity with which external factors, such as radiations and wind act upon soil surface. Therefore, the use of mulches results in moisture and heat conservative and retard weed growth. Mulches also reduce soil and water losses considerably thus allowing more water intake into soil. Today different type of mulches such as polyethylene (PE), biodegradable and organic mulches are available. Due to high specific heat of water it is known to stabilize the temperature. Maintenance of soil moisture content at higher levels raises the minimum soil temperature [12]. This task can be better accomplished by drip method of irrigation, which supplies water in relatively less quantity but frequently in the root zone only.

Drip irrigation is an efficient method of water application and represents a definite advancement in irrigation technology with wider applications. This system typically applies water from a point source called emitter/dripper on daily basis under low pressure and at slow rate to the root zone to keep the soil always in wet condition or near field capacity. It offers considerable flexibility in fertigation because of frequent or nearly continuous application of plant nutrients along with irrigation water. Adoption of this irrigation technology has not only ensured better crop yields but also better quality produce.

Studies on the use of mulches along with drip irrigation for vegetable crop especially potato are very few. Therefore, the present study was planned with the following objectives:

1. to modify the hydrothermal soil environment using mulches during tuberization and drip irrigation for quality potato production.
2. to study the effect of conditions so obtained on the tuber yield and quality of drip irrigated potatoes.

19.2   REVIEW OF LITERATURE

19.2.1   EFFECTS OF MULCHING ON POTATO CROP

Covering the soil surface with any material, which may be organic or synthetic in nature, is known to conserve the soil moisture, obstruct the irradiation of heat from the soil and suppress the weed. Several investigators have observed the favorable effect of mulches on the soil and plant parameters. Awan [2] indicated that an increase in soil temperature from mulching caused significant increase in potato yields. Mulching also helped in the conservation of soil moisture.

Grewal and Singh [12] found the possibility of manipulating hydrothermal regime of the soil with different organic mulches in order to improve the quality and yield of potato. They observed that mulches lower the maximum soil temperature at a depth of 10 cm by 1.5°C during autumn and 3.5°C during spring. Midmore et al. [17] reported that mulch maintains soil moisture reserves more effectively than bare soil. It was further reported that mulches retained more heat in the soil at night when combined with agronomic practices and thus increase soil heat retention [22, 23].

Khalak and Kumaraswamy [14] reported an increase in the tuber yield and promotion of plant growth. It was observed that mulches help to improve the growth components like dry matter accumulation/plant, plant height, shoots/plant and leaf-area duration. Jain et al. [13] reported the increase in yield and the water use efficiency by using the plastic mulch on potato crop as compared to the nonmulched crop. Chandra et al. [7] reported an increase in tuber number and weight/plant, tuber yield and size owing to mulches that provide congenial environment for tuber development by maintaining soil temperature in optimum range and conserving soil moisture. It was further found that canopy temperature was lower in mulch treatment during early phase of crop growth.

Lamont et al. [16] reported that row covers had no significant effect on yield and grades of the crop. All the mulch treatments significantly increased marketable yields compared to no mulch treatments. Sahoo et al. [19] revealed that plastic mulching increases the tuber yield and plant height of potatoes and water use efficiency. The transparent and black polyethylene film mulches increase total tuber yield by 16% and 8%, respectively, and average tuber weight by 14% and 12%, respectively, compared with no mulch (21.6 t/ha and 72.2 g/tuber).

19.2.2 EFFECTS OF DRIP IRRIGATION ON POTATO

Water is known to stabilize the temperature due to its high specific heat. Drip method of irrigation can better accomplish the task of maintaining soil moisture as it can supply water frequently in relatively less quantity in the root zone only. Awari and Hiwase [3] reported an increase in growth, leaf area, tuber yield and water use efficiency with drip irrigation compared with furrow/basin irrigation. Chawla and Narda [8, 9, 10] observed increased value of root density in the top layer of trickle irrigated potato crop as compared with furrow irrigation. An increase of 40% in tuber yields with trickle irrigation than furrow irrigation was also observed in the same study. Ahire et al. [1] observed that drip irrigation system produced higher number of tubers per plant (5.58), larger size of tubers (16.08 cm), more weight of tubers per plant (145.5 g) and higher tuber yield (20.43t/ha) as compared to surface irrigation in potato crop. Drip irrigation saved 46% irrigation water over surface irrigation.

Chawla and Narda [8] showed that leaf area index, percent ground cover and dry matter accumulation were higher in the trickle irrigated than in conventionally irrigated potato crop which in-turn increased the tuber yield. Jain et al. [13] reported greater water saving by reducing the conveyance, application and deep percolation losses, in trickle irrigation as compared to surface irrigation. The increase in plant height, number of branches and yield of potato was also observed. Chawla and Narda [9] indicated that about 30% of water saving could be done using trickle irrigation in comparison to furrow irrigation on potato crop. It was further reported the increased yield of fresh tubers with trickle irrigation. Singh et al. [20] revealed that the drip irrigation produced significantly higher tuber yield of potato by improving the crop water use efficiency as compared to conventional furrow method.

19.2.3   EFFECTS OF MULCHING AND DRIP IRRIGATION ON OTHER CROPS

Easson [11] reported that mulching increased yield by 2–4 t/ha and dry matter content and advanced maize maturity, particularly in crops sown early. Palada et al. [18] revealed that the use of mulches and drip irrigation lead to the increased total number and weight of fruits of bell pepper. Also mulching resulted in more efficient use (35%) of irrigation water compared to bare plots and 33–46% increase in economic returns from bell pepper production.

Waterer [23] reported that plastic mulches improved stand establishment and fruit yields relative to nonmulches control. Clear mulch was observed superior to black plastic in some cases. Khan [15] observed increased pod yield of groundnut and higher water use efficiency were recorded under plastic mulches. Minimum depletion of moisture was also observed in plastic mulches.

Suresh and Kumar [21] studied the effects of drip irrigation and mulches on the performance of pointed gourd. It was observed that drip irrigation and mulches resulted in the greatest fruit length (9.53 cm), fruit circumference (5.35 cm), number of fruits per vine (165), fruit weight (51.0 g), yield per vine (7.90 kg), yield per hectare (19.75tons), benefit cost ratio (9.18) and net profit (Rs. 690/mm of water). Verma [22] indicated that drip irrigation and polyethylene mulch besides the 55.6% water saving has also resulted to 37% higher crop yield compared with surface irrigation only for capsicum crop.

19.3   MATERIALS AND METHODS

To investigate the effect of mulches on potato, field studies were undertaken using drip and conventional methods of irrigation. Plant growth parameters like leaf area, dry matter accumulation, yield and quality attributes of drip irrigated were compared with conventionally irrigated potato crop.

Field studies on potato crop were conducted at the Research Farm of the Department of Soil and Water Engineering, PAU Ludhiana from October 2006–January 2007 using mulches with drip and conventional irrigation system. Ludhiana is situated at 30°56' N latitude, 70°52' longitude and 247 m above mean sea level.

The summers of Ludhiana are quite hot and winters are equally cold. Soil physio-chemical properties at field site are given in Table 19.1.

TABLE 19.1   Soil Characteristics At the Experimental Site

Soil depth, (cm)	Percentage of			Texture
	Sand	**Silt**	**Clay**	
0–15	69.19	20.68	10.12	Sandy Loam
15–30	69.00	19.28	11.72	Sandy Loam

To check the level of initial N, P and K in the soil, it was considered appropriate to have the soil tested for initial fertility. The initial fertility status of soil is given in Table 19.2. The results showed that there was no need to add KCI (Muriate of Potash) to the soil because the soil had sufficient K content.

TABLE 19.2   Initial Status of Soil Fertility

Soil depth (cm)	(kg/ha)			Organic C	–pH	EC (ds/cm)
	N	**P**	**K**	**(%)**		
0–15	Medium	11	187.5	0.64	8.3	0.8
15–30	Medium	8	157.5	0.52	8.4	0.7

Drip irrigation system with inline emitters was installed at the site. A plot measuring 864 m² (having 36 subplots each measuring 24 m²) was used for drip-irrigated crop and on other side a plot measuring 288 m² (having 12 subplots each measuring 24 m²) was used for conventionally irrigated crop. Henceforth, the former would be termed as drip irrigated plot while the latter one as conventionally irrigated plot.

Prior to the installation of drip irrigation system, it was tested for design discharge and also to rectify of emitters if any. It was found that at a pressure of 1.5 kg/cm² the average discharge per emitter was 2.0 l/hr. And during the entire course of investigation, the system was operated at this pressure.

In the field experimental layout, mulches (after 50 days of planting) and irrigation (drip and conventional) were studied. Drip irrigation was having three levels, that is, low, medium and high, based on IW/CPE ratio of 0.6, 0.8 and 1.0, respectively. The 80% of area was considered for irrigation water requirement.

19.3.1 CONVENTIONAL IRRIGATION TREATMENT

At the time of sowing 100% of recommended dose of DPA and 50% of urea was applied after 50 days of planting. Irrigation was applied using (IW/CPE) = 1, where IW is irrigation water and CPE is cumulative pan evaporation. The crop was conventionally irrigated, that is, by flooding through siphon tubes, having discharges of 1 lps, which were run for 2.0 min; and six siphon tubes of discharges 1 lps (one in each row) were used to flood one plot of 24 m² during each irrigation.

19.3.1.1 LOW LEVEL DRIP IRRIGATION TREATMENT (LDI)

In this treatment, irrigation was applied by adopting IW/CPE = 0.6 and the system was run for 30 min during each irrigation for 12 plots.

19.3.1.2 MEDIUM LEVEL DRIP IRRIGATION TREATMENT (MDI)

In this treatment, irrigation was applied by adopting IW/CPE = 0.8 and the system was run for 45 min during each irrigation for 12 plots.

19.3.1.3 HIGH LEVEL DRIP IRRIGATION TREATMENT (HDI)

In this treatment, irrigation was applied by adopting IW/CPE = 1.0 and the system was run for 55 min during each irrigation for 12 plots.

A 75% of recommended dose of urea was applied in all the drip-irrigated treatments as this dose gave maximum yield when applied with drip irrigation. Whole or DAP and 50% of urea (out of 75% of recommended dose of urea) was applied at the time of sowing, while remaining 50% of urea was fertigated in four equal splits after 45 days of planting at an interval of 10 days.

19.3.2 MULCHING TREATMENTS

In all irrigation treatments, three types of mulches, that is, plastic mulch (15 micron thickness and 5 m long), biodegradable mulch (15 micron thick and 5 m long) and organic mulch (rice straw) were used to modify the hydrothermal regime of the soil. Out of total of 48 plots, 12 plots at random were left nonmulched in all above treatments. This experiment had a total of 16 treatments and with three replications.

19.3.3 CROP CULTIVATION

The experimental area measuring about 1200 m² was tilled twice with a disc harrow followed by a cultivator and planking. Farm yard manure (FYM) @ 50 t/ha based on

"*Package of Practices of PAU for Vegetables and Horticultural Crops*" was added about one month prior to the field preparation so that it got decomposed and mixed up thoroughly in the soil by the time of sowing.

Certified seed of potato (Solanum tuberosum L. variety Kufri Sutlej) was spread in the shade to break its dormancy and then treated with Emissan @ 5 g/L just two days before sowing. The experimental field was divided into two parts. One part (which was further divided into three parts each having 12 plots) was kept for drip irrigated potato and the other part (having 12 plots) was kept for conventionally irrigated potato. To get the required amount of N and P in the soil, 120 kg of DAP/ha and half of 120 kg/ha of urea were applied to the field before placement of seed. The potato seed was placed at 20 cm spacing along markings with 60 cm row-to-row spacing in each plot. The seeds were then covered with soil and irrigated by flooding to ensure uniform germination. Planting was done in the second week of October. As per the recommendations of PAU, full package of plant protection measures were adopted during the growth period of crop so as to have a disease free and weed free crop

19.3.4   FERTIGATION PLAN

Bucks and Nakayama [4] have listed factors, which act as a guide while selecting fertilizer for application through the drip system of irrigation. Any chemical added to drip system of irrigation must meet the following criteria:
1.   Should not corrode or clog any component of the system;
2.   Be safe for field use;
3.   Be water soluble; and
4.   Not react adversely with salts or other chemicals present in the irrigation water.

Urea fulfills the above-mentioned requirements and is well suited for injection into drip irrigation. It is relatively soluble in water. It dissolves in nonionic form so that it does not react with other substances in irrigation water. Thus it is not likely to cause precipitation problem and will move deeper into the soil profile. Therefore, urea was selected as a source of N-fertigation.

Fertilizer through drip irrigation can be applied by three principal methods: (i) pressure differential; (ii) venturi (vacuum) pump; and (iii) a metering pump. In the present study, venturi system was used. It was fitted on the suction side of the mono block pump used for lifting water from the storage tank and delivering it to the drip system. The quantity of urea to be applied on any fertigation day was first dissolved in a small quantity of water and then the solution was made up to 10 L. The intake pipe of the venturi was kept immersed in the fertilizer solution. The solution was sucked in through this intake pipe and delivered to the system after passing through the filter. The system was run for different durations (as mentioned earlier) for each drip irrigated treatment which was calculated on the basis of cumulative pan evaporation (CPE): 0.6, 0.8 and 1.0 times of CPE and accordingly for different irrigation treatments.

In drip irrigation, the single most important factor is water quality being pumped into the irrigation system. Water samples were analyzed for suspended solids, dissolved solids, pH, carbonates, Ca, Mg, Iron and Electrical Conductivity, etc., according to the criteria given by Bucks and Nakayama [4] for classification of irrigation water. It was found that all the parameters of irrigation water were within the safe limits.

Bucks et al. [4–6] recommended that drip lines should be flushed at least once after every six months of use. The laterals were flushed with water before use to remove the previous sediment build-up. By this method, already-clogged emitters were partially opened-up.

19.3.5   MULCHING

Mulches were applied over the whole field after 50 days of planting. Three types of mulches, that is, plastic, biodegradable and organic mulches were applied over each irrigation treatment with three replications. Some of the plots were left nonmulched in each treatment. Mulching was done in the 1st week of December. Thermometers were installed at a depth of 15 cm in 32 plots (both mulched and nonmulched), to evaluate the effects of mulches on soil temperature. The minimum temperature was recorded at 6:30 a.m. and maximum temperature was recorded at 2:30 p.m.

19.3.6   BIOMETRIC PARAMETERS

Different crop parameters like plant height, leaf area index (LAI), dry matter accumulation (DMA), yield and quality parameters were measured at different intervals during the crop season to compare different treatments and effects of mulches. Plant height, LAI and DMA were measured at an average interval of two weeks. The number of haulms per plant was measured 10 days before harvesting. Temperature was recorded daily. Yield and quality attributes were studied at the end of the season.

Plant height: Height of five plants at random from each of the irrigation treatments was measured at 30, 45, 60, 75, and 90 days after planting and then arithmetic mean was found to get an average value of plant height.

Leaf area index (LAI): Two plants from each treatment were taken at 30,45,60,75 and 90 days after planting and green leaves were removed from the shoots. The leaf area was measured using LI 3050. An electronic leaf area meter (Model LI 3000 with conveyor belt) and the average of the two readings was taken. Leaf area index was then calculated as the ratio of leaf area of a plant to ground area (row-to-row spacing × plant-to-plant spacing) commanded by that plant.

Dry matter accumulation (DMA): The plant leaves along with shoots and branches were first air dried and then oven dried at 60°C to a constant weight to record the dry matter accumulation of plants at 30, 45, 60, 75, and 90 days after planting.

Number of haulms per plant was observed after 80 days of planting, that is, 10 days before harvesting. Average data of five plants at random from each irrigation treatment under different mulches was recorded for this purpose.

Crop yield: The potato crop was dug out manually from each treatment 100 days after planting. Yield was noted for every plot separately and fresh tubers were graded in three lots according to the average size: A grade – each tuber weighing more than 50 g; B grade – weighing between 20–50 g; and C grade – weighing less than 20 g. Grade wise tubers corresponding to each treatment were weighed and added up for calculating total yield for each plot. Then average yield (100 kg/ha) was calculated for each treatment.

19.3.7 FRUIT QUALITY

Fresh potato tubers were collected from each treatment and were immediately analyzed for biochemical properties (dry matter content and reducing sugar). Reducing sugar content was estimated immediately after the harvesting of the crop using Uristix method. Potato tubers were cut into two pieces and then a strip was placed over it for 30 seconds. The color of the strip changed and was matched with the color displayed on the strip bottle and % reducing sugar was noted accordingly. Two samples from each plot were taken and the average of two values was calculated, which represented the value of reducing sugar content for the same plot.

To determine the dry matter content of potato tubers, the tubers were cut into small pieces and put into petri dishes. Each petri dish was weighed before and after adding the sample to it. This sample was then oven dried at 65°C for 24 h. The petri dishes were again after drying. The percent dry matter content was calculated by using the following formula.

$$DMC\ (\%) = (Final\ weight/Initial\ weight) \times 100 \tag{1}$$

19.3.8 PROCESSING ATTRIBUTES

To study the effects of various treatments on processing qualities of potato tubers, the B grade (medium size) tubers were processed into chips after 100 days of planting. Randomly selected five tubers per sample were sliced into chips (about 1.4 mm thickness) with hand operated rotary slicer. From these 15 slices, 3 slices per tuber were taken and washed with water. After drying with paper napkins, these were fried in the refined oil at 180°C temperature and organolaptic evaluation was done. Organolaptic observations were recorded to study the consumer preference for processed product. For this purpose, a panel of five judges was set up to evaluate various samples. A testing format was prepared based on attributes such as flavor, texture, appearance and overall acceptability. The chips were evaluated according to 1–9 Hedonic Scale Sensory Evaluation Form. The grades were recorded.

19.3.9 STATISTICAL ANALYSIS

For the purpose of statistical analysis, mulches were considered as main treatments and irrigation levels as subtreatments. Therefore, analysis was done on the basis of split plot design.

19.4 RESULTS AND DISCUSSION

19.4.1 PLANT HEIGHT

The Figs. 19.1–19.3 indicate the variation of plant height for the drip irrigated crop and conventionally irrigated crop as observed during the crop growth season. These figures indicate that during the growth phase, plant grew slowly in the beginning in all the treatments and thereafter showed a rapid increase in later stages. All the treatments showed a maximum plant height in the growth period between 60–75 days after planting followed by decline thereafter in senescence period.

Figure 19.4 shows the mean potato height of drip irrigated crop (averaged over all the treatments) and conventional irrigated crop. It clearly revealed that throughout the crop season, plant height in drip irrigation treatments was higher as compared to conventionally irrigated crop. Among all the drip irrigation treatments, HDI gave the maximum plant height throughout the season followed by MDI and LDI treatment. The difference in plant height after 60 days of planting was more pronounced, that is, drip irrigated crop had greater height as compared with control. This may be due to the application of split dose of nutrient along with the drip irrigation throughout the crop season, which helped in better growth of crop and hence led to greater height. In the later stages, plant height in both the case showed decline because of maturity.

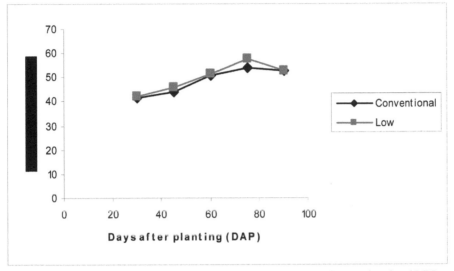

FIGURE 19.1   Variation of plant height with days after planting: Conventional and LDI.

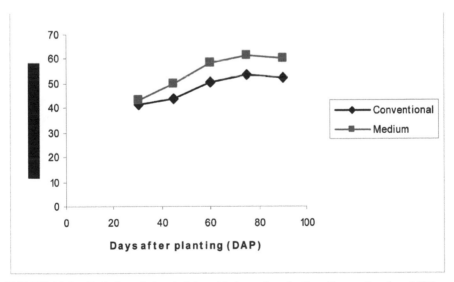

FIGURE 19.2   Variation of plant height with days after planting: Conventional vs. MDI.

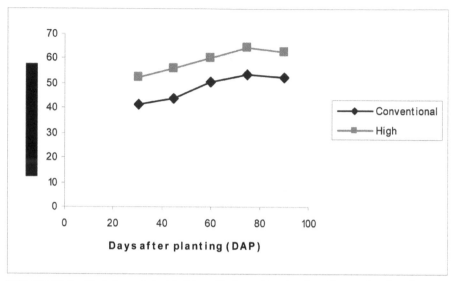

FIGURE 19.3   Variation of plant height with days after planting: Conventional vs. HDI.

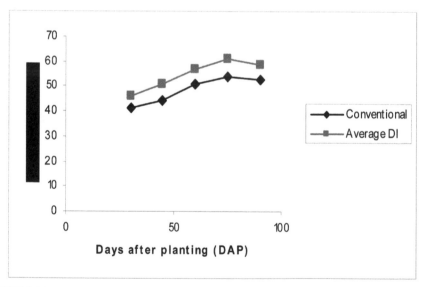

FIGURE 19.4   Variation of plant height with days after planting: Conventional vs. Average DI.

19.4.2 LEAF AREA INDEX (LAI)

The Figs. 19.5–19.7 show the values of LAI for all the drip-irrigated treatments and the conventional irrigated crop, during the growing season. These figures depict that during the growth phase, LAI increased in the beginning at a slow rate till 30 DAP, after which it showed a rapid increase between 30 and 60 DAP. It peaked at around 70 DAP in all the treatments followed by an abrupt decline thereafter in the senescence phase. The initial slow growth rate of leaf area of the plant was getting established. Once the plants got established, the growth of leaf area was resulted because of root development and proliferation that increased the uptake of nutrients. Leaf area declined during the late season owing to shedding of leaves as the potato crop reaches physiological maturity after 70 DAP.

The highest LAI of 3.6 was observed for treatment HDI at 75 DAP, that is, when irrigation was applied at IW/CPE = 1.0. The fresh tuber yield was directly correlated to LAI, since yield was the highest for the treatment corresponding to the highest value of LAI. In conventional treatment, the yield was less owing to low value of LAI of 2.7 at 75 DAP. When canopy was well developed and well spread like that in the drip plots, the leaves synthesized carbohydrates at a fast rate, which were translocated to the tubers and thus contributed towards the increased yield.

Figure 19.8 compares LAI for the drip-irrigated crop (averaged over all the treatments) with that of conventionally irrigated crop. It clearly shows that LAI of drip irrigated potato crop remained higher after 40 DAP than the conventionally irrigated crop, indicating thereby that the number of leaves (though not counted) as well as the canopy spread, had consistently been much higher in drip crop as compared to the conventional crop. This factor later on contributed towards higher yield of crop from the drip irrigated crop.

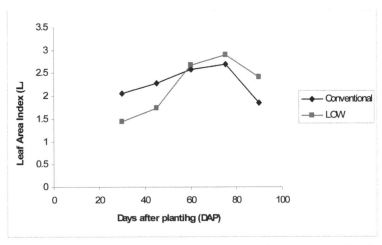

FIGURE 19.5   Variation of leaf area index with days after planting: Conventional vs. LDI.

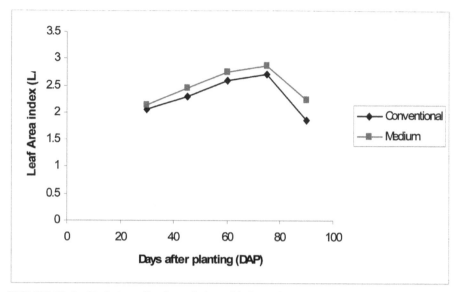

FIGURE 19.6   Variation of leaf area index with days after planting: Conventional vs. MDI.

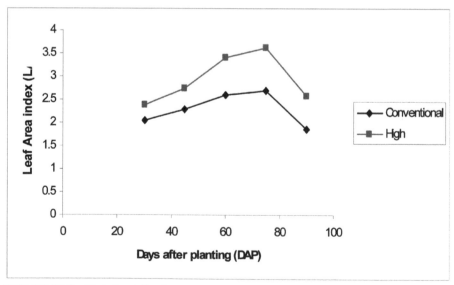

FIGURE 19.7   Variation of leaf area index with days after planting: Conventional vs. HDI.

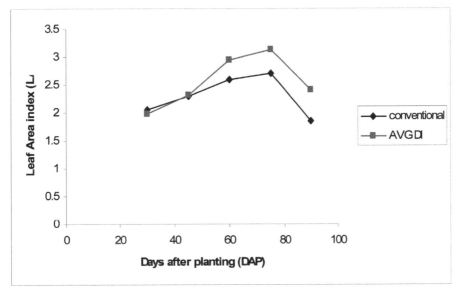

FIGURE 19.8 Variation of leaf area index with days after planting: Conventional vs. Average DI.

19.4.3 DRY MATTER ACCUMULATION (DMA)

The Figs. 19.9–19.11 show the DMA of the above ground parts under different treatments, during the advancement of crop season. These figures show that the DMA of the crop increased with the crop season because of increasing uptake of nutrients and spreading of canopy. The maximum DMA of 28.9 g/plant was observed in the treatment HDI, which was closely followed by the treatment MDI (24.2 g/plant) on 75 DAP. At 45, 60 and 75 DAP, HDI showed higher value of DMA than other treatments.

The dry matter accumulation averaged for all the drip treatments was compared with conventional treatment in Fig. 19.12. The DMA in drip-irrigated crop was always higher than conventional crop throughout the season. The better performance of drip irrigated treatments over the conventional irrigated treatments may be due to the split dose of nutrient which might have increased the fertilizer use efficiency along with higher water use efficiency in drip irrigated crop due to better management of moisture and nutrients. Average value of DMA for drip-irrigated treatments was about 1.3 times the conventional irrigated crop at the end of the season. The DMA decreased slightly in all the treatments at around 90 DAP because of maturity and senescence period. DMA and LAI behaved in an identical fashion for the drip irrigated and the conventional treatments.

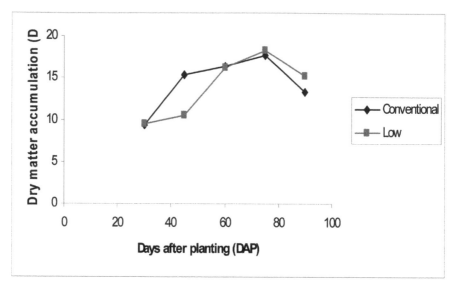

FIGURE 19.9   Variation of Dry matter accumulation with days after planting: Conventional vs. LDI

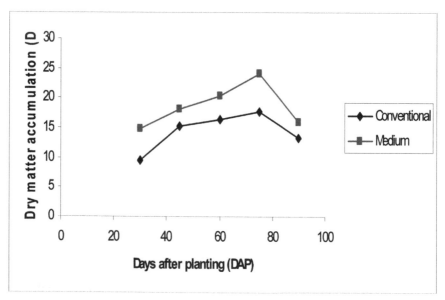

FIGURE 19.10   Variation of dry matter accumulation with days after planting: Conventional vs. MDI.

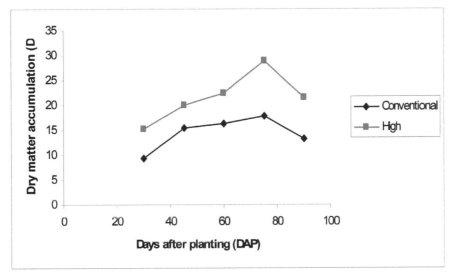

FIGURE 19.11   Variation of dry matter accumulation with days after planting: Conventional vs. HDI.

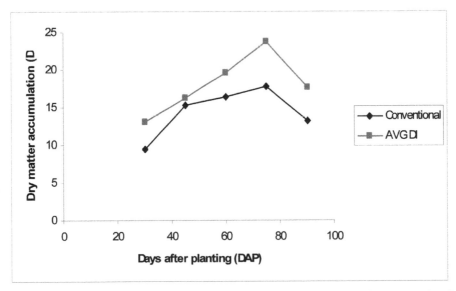

FIGURE 19.12   Variation of dry matter accumulation with days after planting: Conventional vs. Average DI.

19.4.4   NUMBER OF HAULMS PER PLANT

The average data recorded for the number of haulms per plant is given in Table 19.3. The data revealed that the number of haulms per plant was more in case of drip-irrigated crop as compared to conventional irrigated crop. Also, the mulches increased the number of haulms. Plastic and biodegradable mulches gave comparable number of haulms while organic mulch gave less number. It was observed that all the drip irrigation treatments gave higher number of haulms per plant as compared to conventional treatment.

TABLE 19.3   Average Number of Haulms Per Plant For Potato Crop

Treatments	Plastic mulch	Biodegradable mulch	Organic mulch	No mulch
LDI	4.5	4.3	4.1	4.8
MDI	5.1	4.0	4.2	4.3
HDI	5.2	5.5	4.5	4.4
Conventional	3.7	3.7	3.1	4.0

TABLE 19.4   Statistical Analysis of Number of Haulms Per Plant

Mulch, M	Irrigation treatment, I				
	Conventional	Medium	High	Low	Mean, I
		Drip irrigation, DI			
Plastic	3.7	5.1	5.2	4.5	**4.6**
Biodegradable	3.7	4.0	5.5	4.3	**4.4**
Organic	3.1	4.2	4.5	4.1	**3.9**
No mulch	4.0	4.3	4.4	4.8	**4.4**
Mean, M	**3.6**	**4.4**	**4.9**	**4.4**	

I = 0.041 at CD (5%); M = 0.392 at CD (5%); IxM = NS; I = Irrigation, and M = Mulch.

Statistical analysis for different irrigation treatments and different mulch treatments given in Table 19.4 revealed that there was a significance effect of irrigation and mulches on number of haulms per plant while the interaction of irrigation level and mulching was found to be non significant.

19.4.5   EFFECTS OF MULCHING ON SOIL TEMPERATURE

The minimum and maximum soil temperature data were recorded for the whole crop season. The Figs. 19.13–19.16 represent the average data recorded before the application of mulches and the data of temperature after the application of mulches (after 50 days after planting) for different mulches including no mulch. The diurnal variation in soil temperature revealed that the soil temperature was the lowest in early morning hours, that is, 6:30 a.m., thereafter it increased up to 2:30 p.m. and reached the maximum at 2:30 p.m. A gradual decrease was observed from 2:30 p.m. onwards.

In the beginning when the mulches were not applied and also the canopies were not fully developed, there was no significant difference in the soil temperature among various treatments. But at later stages, when the mulches were applied and also the crop was fully developed (canopy cover fully developed, that is, canopy cover fully developed), it affected the soil temperature in different treatments. It might be due to the application of mulches to maintain soil temperature and also radiation reaching the ground surface was affected by different canopy structures. The minimum temperature data showed that the mulches increased the minimum soil temperature by 2 to 3°C and thus helped to maintain required soil moisture within the soil.

The graphs between minimum soil temperature and days after mulches were applied (three days average) for different type of mulches including control under different treatments are shown in Figs. 19.13–19.16. The figures revealed that in all the irrigation treatments, the plastic mulch increased the minimum temperature to an optimum level that is required for better growth of crop, which further helped in increasing the yield of crop, the highest yield was obtained from the treatment under plastic mulch. A possible reason for this may be the more favorable soil moisture and more favorable condition, which produce vigorous growth of crop.

Biodegradable mulch was found to be helpful in maintaining the soil temperature and hence the soil moisture followed by organic mulch while the control was having the lowest minimum temperature. The mulches thus improved the quality and quantity of the yield. The reason might be that the mulches maintained the soil moisture and thus improved the growth. The pooled analysis revealed that mulches increased the yield over no mulch. Singh et al. [20] also reported 26% increase in yield of potato owing to mulch.

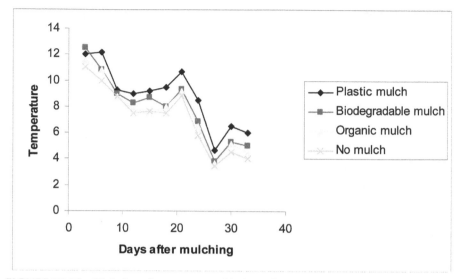

FIGURE 19.13   Variation of temperature with days after mulching for LDI: Four mulching treatments.

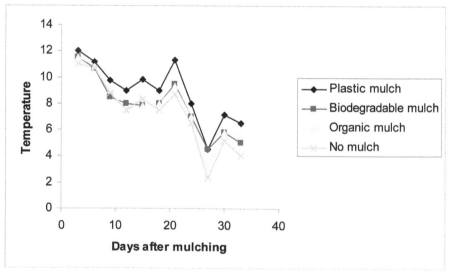

FIGURE 19.14   Variation of temperature with days after mulching for MDI: Four mulching treatments.

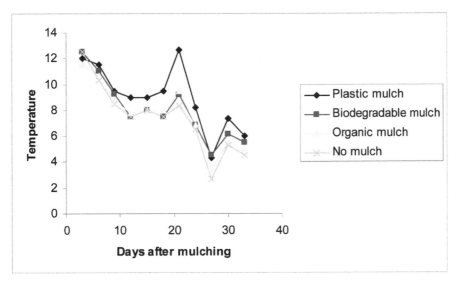

FIGURE 19.15   Variation of temperature with days after mulching for HDI: Four mulching treatments.

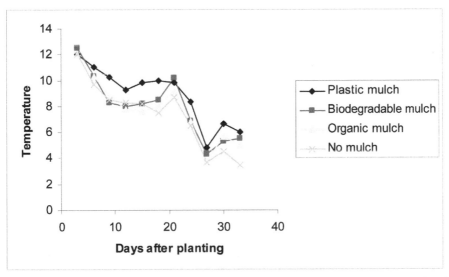

FIGURE 19.16   Variation of temperature with days after mulching for conventional irrigation: Four mulching treatments.

19.4.6 CROP YIELD

Average fresh tuber yield is given in the Table 19.5. The table shows that higher yield was recorded in the drip-irrigated plots under mulches. The highest yield of 331.5 per 100 kg/ha was obtained from the HDI drip irrigated plot under plastic mulch followed by 309.3 per 100 kg/ha from the same irrigation treatment under biodegradable mulch, while a yield of 188.8 per 100 kg/ha was obtained from the nonmulched conventionally irrigated treatment.

TABLE 19.5   Average Yield (100 kg/ha) of Potato

Treatment	Plastic mulch	Biodegradable mulch	Organic mulch	No mulch
LDI	269.4	252.5	278.3	156.7
MDI	298.1	274.4	299.4	257.4
HDI	331.5	309.3	296.3	262.4
Conventional	227.8	248.1	240.3	188.8

On the overall, the yield obtained from the drip-irrigated crop was higher as compared to conventional plots. Table 19.6 presents the statistical analysis of yield data. It revealed that the interaction of mulches and drip irrigation had a significant effect on yield of the crop. The yield obtained from the drip irrigated plots under mulches was higher than the yield obtained from the conventionally irrigated non-mulched crop.

TABLE 19.6   Statistical Analysis of Yield (100 kg/ha)

Mulch, M	Conventional	Medium	High	Low	Mean
Plastic	227.8	298.1	331.5	269.4	**281.7**
Biodegradable	248.1	274.4	309.3	252.5	**271.1**
Organic	240.3	299.4	296.3	278.3	**278.6**
No mulch	188.8	257.4	262.4	156.7	**216.3**
Mean	**226.3**	**282.3**	**299.3**	**299.9**	

CD (5%), I = 13.76,
CD (5%), M = 13.28,
CD (5%), IxM = 26.56; I = Irrigation and M = Mulch.

The maximum yield obtained in drip irrigation under mulching may be due to the availability of uniform moisture in the soil wetted by drippers, along with better moisture conservation, proper utilization of nutrients and less weed growth around the potato plant. Similar results were obtained by Jain et al. [13].

19.4.6.1 EFFECTS OF MULCHING

The average yield data for potato crop under different irrigation treatments with various mulches are given in Table 19.7. The Table 19.7 shows that the yield recorded from drip irrigation under mulched conditions was higher as compared to mulched conditions under conventional irrigation. The highest average yield was 299.7 per 100 kg/ha in the treatment under plastic mulch with drip irrigation, thus showing 32.9% increase followed by organic and biodegradable mulches with an increase of 29.2% and 23.6%, respectively, over no mulch drip irrigated crop.

Table 19.7 shows that drip irrigated crop under all types of mulches gave comparable yield while the nonmulched crop under drip irrigation gave much less yield, that is, 225.5 per 100 kg/ha. The overall mean yield (based on average of all the mulched drip irrigated treatments) was 289.9 per 100 kg/ha which was 28.6% higher than the nonmulched drip irrigated crop. It can be concluded that the average mulched drip irrigated crop yield (289.9 per 100 kg/ha) was 21.4% higher than the conventionally irrigated mulched crop yield (238.7 per 100 kg/ha) and 53.5% higher than conventionally irrigated nonmulched crop yield (188.8 per 100 kg/ha).

TABLE 19.7   Average Yield (100 kg/ha) of Potato Under Mulching

Treatment	Fresh yield (100 kg/ha)		% increase over no mulch	
	Average for all DI	Conventional	Average for all DI	Conventional
Plastic mulch	299.7	227.8	32.9	20.7
Biodegradable mulch	278.7	248.1	23.6	31.4
Organic mulch	291.3	240.3	29.2	27.3
Average mulch	289.9	238.7	28.6	26.4
No mulch	225,5	188.8		

Statistical analysis showed that there was a significant difference between the yields obtained from the mulched and nonmulched treatments. The differences within the three mulch treatments were not so significant and out of the three mulches plastic mulch was found to be the best to obtain the higher yield. This indicates that the total tuber yield of fresh potatoes were higher under mulching compared to control treatments.

19.4.6.2   EFFECTS OF DRIP IRRIGATION ON POTATO YIELD

Table 19.8 includes average yield of fresh tubers of potato for all irrigation and mulching treatments. The table indicates that the highest yield was 312.3 per 100 kg/ha in treatment HDI under mulches against of 238.7 per 100 kg/ha in conventional irrigated mulched plots, registering an increase of 30.8%. The overall mean yield (based on average of all the drip irrigated mulched treatments) was 289.9 per 100 kg/ha, which was 21.4% higher than conventional irrigated mulched plots. Even the lowest average yield of 266.7 per 100 kg/ha (LDI) in drip irrigated mulched treatment was higher than the control. It can be concluded that the drip irrigated nonmulched crop yield (225.5 per 100 kg/ha) was 19.4% higher than conventional irrigated nonmulched yield (188.8 per 100 kg/ha). Therefore, by applying 25% less fertilizer than the recommended dose of nitrogen, increased yield can be obtained by using drip irrigation under mulching.

TABLE 19.8   Average Yield (q/ha) of Potato on Irrigation Basis

Treatment	Fresh tuber yield (100 kg/ha)		% increase over conventional	
	Average of all mulches	No mulch	Average of all mulches	No mulch
LDI	266.7	156.7	11.7	-
MDI	290.6	257.4	21.7	36.3
HDI	312.3	262.4	30.8	38.9
Average DI	289.9	225.5	21.4	19.4
Conventional	238.7	188.8		

It can be concluded that there was significant difference among various irrigation treatments. Tuber yield obtained from the drip-irrigated treatment was significantly different from conventional treatment. Out of the three drip irrigation treatments, HDI was the best followed by MDI, indicating higher tuber yield of fresh tubers against conventional grown potato.

19.4.7 GRADING OF POTATO

The grade wise tuber yield data for potato crop is given in Table 19.9. It may be seen that the treatment HDI had the highest percentage of grade-A tubers and was best in terms of total yield followed by MDI treatment with 47.6 grade-A tubers. It was also observed that all drip-irrigated treatments gave higher percentage of grade-A tubers but conventional treatment gave much less percentage of grade-A tubers. Also it was found that HDI was the best in terms of LAI and DMA. Therefore, it may be concluded that the yield and plant growth parameters like LAI and DMA were directly correlated.

TABLE 19.9   Average Yield (100 kg/ha) of Potato

Treatment	Grade wise produce*(100 kg/ha)		
	A	**B**	**C**
LDI	104.9 (43.7%)	96.6 (40.3%)	38.4 (15.99%)
MDI	134.5 (47.9%)	110.9 (39.3%)	36.9 (13.2%)
HDI	152.6(51.1%)	108.4 (36.3%)	38.9 (12.9%)
Average DI	130.7 (47.6%)	105.3 (38.6%)	38.1 (13.9%)
Conventional	85.5 (38.7%)	95.5 (42.2%)	42.6 (20.0%)

*Grade A >50 g; Grade B 20–25 g; and Grade C < 20 g.

All the drip-irrigated treatments gave higher percentage of grade-A tubers as compared to grade-B and grade-C tubers. Comparing the grade wise average percentage of drip irrigated potato and conventionally irrigated potato, it was observed that drip irrigated crop gave higher percentage of grade-A tubers, and conventionally irrigated crop gave higher percentage of grade-B and grade-C tubers. Therefore, it can be concluded that by using less amount of water and fertilizer in potato crop, increased yield with better quality of crop can be obtained with drip irrigation.

Pie diagram in Fig. 19.17 shows the percentage distribution of grade-wise produces of potato tubers for drip irrigated and conventional irrigated plots. The diagram illustrates that the drip irrigated crop had more percentage of grade-A tubers than conventional irrigated potatoes. On an average, drip irrigated crop had 47.6% of grade-A size, 38.6% of grade-B size and only 13.9% of grade-C size tubers as compared to 38.7%, 42.2% and 20.0%, respectively, in conventional treatment.

This shows that conventionally irrigated crop had higher proportion of medium sized potato tubers, where as drip irrigated potatoes comprised of higher proportion of grade-A potatoes, which resulted in higher yield. Thus drip irrigation in addition to increase in total yield of potato crop, also improved the fruit quality in terms of size. It also saves money in terms of applied fertilizer, because the yield and grade-A sized tubers in 75% of recommended dose were higher than 100% of N application. The maximum yield in HDI treatment can be attributed to the efficient utilization of applied fertilizer under drip irrigation as compared to conventional irrigation.

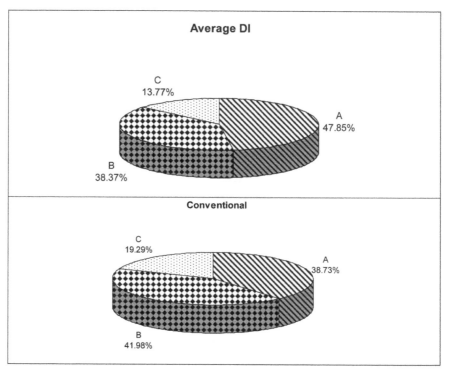

FIGURE 19.17   Percentage distribution of grade wise produce in potatoes: top – average of all drip irrigation treatments; and bottom – conventional irrigation.

19.4.8   FRUIT QUALITY

19.4.8.1   DRY MATTER CONTENT

Dry matter content is one of the most important quality factors in potato intended for processing by frying. The average data for dry matter content (%) of potato crop is given in Table 19.10. It can be observed that the optimum dry matter content of 15–21% was observed in all drip irrigated treatments, while lower dry matter content was observed in the conventional treatment. The HDI and MDI gave better values of dry matter content than LDI under all mulch treatments. While the values obtained in LDI were higher (21.1% and 21.5% under plastic and organic mulch, respectively) than the optimum ranges (18.20%), the values obtained from conventional irrigated crop under biodegradable and organic mulch were much less (16.3% and 16.4%, respectively) than the optimum range because of which poor quality of chips were obtained. The average lowest value of dry matter content 14.9% was obtained with no mulch, conventionally irrigated treatment.

TABLE 19.10   Average Dry Matter Content % For Potato Crop

Treatments	Plastic mulch	Biodegradable mulch	Organic mulch	No mulch
LDI	21.1	19.9	21.5	21.0
MDI	19.0	19.1	18.9	15.3
HDI	20.1	18.7	18.6	16.3
Conventional	15.3	16.3	16.4	14.9

TABLE 19.11   Statistical Analysis of Dry Matter Content

	Conventional	Medium	High	Low	Mean
Plastic	15.3	19.0	20.1	21.1	**18.9**
Biodegradable	16.3	19.1	18.7	19.9	**18.5**
Organic	16.4	18.9	18.6	21.5	**18.9**
No mulch	14.9	15.3	16.3	21.0	**16.9**
Mean	**15.7**	**17.7**	**18.5**	**21.3**	

CD (5%), I = 2.11, CD (5%), M = NS, CD (5%), IxM = NS; I = Irrigation and M = Mulch.

Statistical analysis of dry matter content (Table 19.11) showed that the irrigation played significant role in improving the dry matter content of crop while the mulches had no significant effect on dry matter content. Also, the interaction of irrigation and mulches had nonsignificant effect on dry matter content.

19.4.8.2   REDUCING SUGAR CONTENT

The average data on reducing sugars is presented in Table 19.12, which revealed that the plastic mulch reduced the reducing sugar to a lower level than the biodegradable and organic mulch. Under LDI, MDI, HDI and conventional treatments, the average reducing sugar content were lower 0.067, 0.058, 0.067 and 0.075 in plastic mulch; followed by 0.075, 0.083, 0.067 and 0.083 in biodegradable mulch; and 0.142, 0.100, 0.116 and 0.092 in organic mulch; and no mulch had higher reducing sugar values of 0.208, 0.192, 0.183 and 0.125, respectively. Good quality of chips was obtained from the potato with lower values of reducing sugar content. Statistical analysis of reducing sugar content is presented in Table 19.13. The data revealed that mulches played significant role in affecting reducing sugar content in tubers while different irrigation treatments had no effect on the reducing sugar content. The effect of interaction of mulches in irrigation was no significant for reducing sugar content.

TABLE 19.12   Average Content of Reducing Sugar For Potato Crop

Treatments	Plastic mulch	Biodegradable mulch	Organic mulch	No mulch
LDI	0.067	0.075	0.142	0.208
MDI	0.058	0.083	0.100	0.192
HDI	0.067	0.067	0.116	0.183
Conventional	0.075	0.083	0.092	0.125

TABLE 19.13   Statistical Analysis of Reducing Sugar Content

	Conventional	Medium	High	Low	Mean
Plastic	0.075	0.058	0.075	0.067	**0.069**
Biodegradable	0.083	0.083	0.067	0.075	**0.077**
Organic	0.092	0.100	0.117	0.142	**0.113**
No mulch	0.125	0.191	0.183	0.208	**0.177**
Mean	**0.093**	**0.108**	**0.110**	**0.123**	

CD (5%), I = NS, CD (5%), M = 0.023, CD (5%), IxM = NS; I = Irrigation and M = Mulch.

19.4.9 QUALITY OF PROCESSING

To check the suitability of potato for processing, it was considered appropriate to test the quality of produce under mulched and nonmulched treatments. For this, chips for tubers obtained under each treatment were fried and organolaptic evaluation was recorded. According to these observations, chips obtained from all the mulched treatments were found better in all respects, that is, color, texture, taste and acceptability than the nonmulched treatment. The overall acceptability of all mulched treatments rated was 7–8, while nonmulched bagged lower rating of 6–7 in organolaptic evaluations. Thus the mulches improved the processing quality of potato.

19.4.10 WATER SAVING UNDER DRIP IRRIGATION

Drip irrigation saves water, ensures better yield and better quality of produce. Most of the research studies in this system are directed towards on these aspects.

In this study, the total quantity of water used by the crop comprised of two components: water used by the crop consumptively and preemergence irrigation that was provided to the entire field after sowing to ensure uniform germination. Fertilizer was applied along with the irrigation and the crop was harvested before the attack of frost. Also, the rainwater was excluded, as it was common to both the drip and conventional methods.

19.4.10.1 CONSUMPTIVE USE

Quantity of water applied under drip irrigation was based on IW/CPE ratio of 0.6, 0.8 and 1.0 for different irrigation treatments (classified as low, medium and high), respectively. The total cumulative pan evaporation for the entire crop season was 16.3 cm of water.

The quantity of water used consumptively by the crop under LDI treatment = $16.3 \times 0.6 = 9.78$ cm

Pre emergence irrigation = 3 cm
Therefore, total water used $(Q_L) = 9.78 + 3 = 12.78$ cm
Similarly, water used under MDI $(Q_M) = 16.04$ cm
And water used under HDI $(Q_H) = 19.30$ cm

19.4.10.2 QUANTITY OF WATER USED IN FURROW IRRIGATION

During each irrigation, water was applied with the help of siphon tubes of 1 lps discharge (1 in each row and 6 rows in one plot of 24 m^2). Irrigation was applied for two minutes in each plot (12 plots) and 7 such irrigations were applied. Therefore, total quantity of water delivered under conventional treatment:

$Q_F = (1 \times 60 \text{ lpm}) \times (6 \text{ rows}) \times (2.0 \text{ min.}) \times (12 \text{plots}) \times (7 \text{ irrigations})$

$= 60480 \text{ L} = 60.48 \text{ m}^3$ or $60.48 \times 100/288 = 21.0 \text{ cm}$

Total water used including 3 cm preemergence irrigation $= 21+3 = 24$ cm

The percentage saving in irrigation water using drip irrigation can be worked out as under:

Water saving in HDI treatment $= [(Q_F - Q_H)/ Q_F] \times 100 = [(24-19.3)/24] \times 100 = 19.58\%$

Similarly, water saving were 33.3% in LDI and 46.6% in MDI, respectively. Also the results are in tune with the findings of Suresh et al. [21], who reported saving of 40–60% by drip irrigation.

19.4.10.3 WATER USE EFFICIENCY (WUE)

Quantity of water used along with yield and WUE for various treatments are given in Table 19.14.

TABLE 19.14 Yield, Quantity of Water Used, and Water Use Efficiency of Potato For Various Treatments

Treatments	Yield (100 kg/ha)	Water used (cm)	WUE 100 kg/(ha-cm)
LDI	239.2	12.8	18.7
MDI	282.3	16.0	17.6
HDI	299.9	19.3	15.5
Average DI	273.8	16.0	17.1
Conventional	226.3	24.0	9.4

WUE of all the drip-irrigated treatments was higher than the furrow irrigation. It was found that average WUE of drip-irrigated treatments was about 2.0 times than that of conventional treatment (9.4 per 100 kg/ha-cm). These results are in agreement with Chawla [8] and Singh [20], who reported more than double water use efficiency of drip treatment as compared to furrow treatment in potato.

19.5 SUMMARY

Field studies to observe the effects of mulches and drip irrigation on potato (*Solanum tuberosum* L.) were undertaken using different type of mulches (plastic, biodegradable and organic) and irrigation levels (LDI, MDI and HDI based on 0.6, 0.8 and 1.0 IW/CPE ratios, respectively). Performance of drip irrigated mulched crop was compared with drip irrigated nonmulched and conventionally irrigated mulched and nonmulched crop. In drip-irrigated plots, there were three (LDI, MDI and HDI) irrigation treatments. In all the three treatments, 75% of recommended dose of urea was applied. Out of 75%, 50% of urea was applied at the time of sowing, serving as a base dose and remaining 50% was applied in four equal splits starting 45 days of planting at an interval of 10 days, while 100% of DAP was applied at the time of sowing. In all the drip irrigation treatments, plastic mulch, biodegradable mulch and organic mulch were used after 50 days of planting, that is, at the time of tuberization.

Various plant performance parameters, that is, plant height, leaf area index, dry matter accumulation were measured at 15 days interval during the crop growth season. Number of haulms per plant were counted 10 days before harvesting, while yield and quality attributes, that is, dry matter content and reducing sugar content were determined only at the end of season.

Daily soil temperature data was recorded to study the effects of different mulches on crop growth. The usual plant protection measures were undertaken to have a disease-free and healthy crop. The crop performance parameters were also statistically analyzed. The following important conclusions were drawn from the present study.

1. Plant height throughout the growth season was more in drip-irrigated crop than the conventionally irrigated crop. The maximum plant height of 64.4 cm was observed in HDI treatment at 75 DAP as compared to 53.8 cm for conventional treatment.

2. The maximum leaf area index of 3.6 was observed in HDI treatment at 75 DAP as compared to 2.7 for the conventionally irrigated crop. This fact accounts for higher dry matter accumulation in the same treatment. There was a marginal decrease in leaf area index of all treatments around 90 DAP which may be attributed to physiological maturity and senescence.

3. Highest DMA of 28.9 g/plant at 75 DAP was observed in HDI treatment as compared to 17.7 g/plant at 75 DAP in case of conventional treatment. Average value of DMA for drip-irrigated crop was about 1.3 times the conventional irrigated crop at the end of the season.

4. Plastic mulch increased the minimum soil temperature by 2–3°C and brought it to an optimum range (10–12°C), that is, required for better crop growth, which further helps in improving quality of crop.

5. Plastic mulch with drip irrigation resulted in 32.9% increase in yield over nonmulched drip irrigated crop, followed by organic and biodegradable mulch with 29.9% and 23.6%, respectively. The average yield of all three mulch treatments under drip irrigation was 28.6% higher than the no mulch drip irrigation treatment.

6. The highest yield of 312.3 per 100 kg/ha was obtained under HDI treatment (averaged over all mulches) against a yield of 238.7 per 100 kg/ha realized for the conventionally irrigated mulched potato, registering an increase of 30.8%.

7. The average (of all the treatments) mulched drip irrigated crop yield (289.9 per 100 kg/ha) was 53.5% higher than the conventionally irrigated nonmulched crop (188.8 per 100 kg/ha).

8. Drip irrigated potato with mulch had 47.6% of A grade, 38.6% of B grade and 13.9% of C grade tubers as compared to conventionally grown mulched crop which had respectively 38.7, 42.2 and 20.0% of A, B and C grade tubers, respectively.

9. Mulches improve the processing quality of potatoes. The optimum range (18.20%) of dry matter content for best processing quality of potato was obtained from all the drip irrigated treatments compared to conventional treatment that gave lower dry matter content value. The average lowest dry matter content 14.97% was obtained from no mulch conventionally irrigated treatment.

10. The dry matter content values obtained in LDI were higher (21.1% and 21.5% under plastic and organic mulch, respectively) than the optimum range (18.20%). The values obtained from conventional irrigated crop under biodegradable and organic mulch were much less (16.3% and 16.4%, respectively) than the optimum range and was found that poor quality of chips were obtained.

11. Plastic mulches lowered the reducing sugar content to an optimum range (<0.25). It was 0.067, 0.058, 0.067 and 0.075 in LDI, MDI, HDI and conventional treatments, respectively, followed by biodegradable and organic mulches in which it was 0.075, 0.083, 0.067 and 0.083 and 0.142, 0.100, 0.116 and 0.092, respectively. Thus mulching improves the processing quality of potato resulting with better – quality chips.

KEYWORDS

- biodegradable mulch
- conventional irrigation
- cumulative pan evaporation, CPE
- days after planting, DAP
- drip irrigation
- dry matter accumulation, DMA
- dry matter content
- furrow irrigation
- haulms per plant
- high drip irrigation, HDI
- irrigation water, IW
- leaf area index, LAI
- low drip irrigation, LDI
- medium drip irrigation, MDI
- organic mulch

- pan evaporation
- plastic mulch
- polyethylene film
- potato processing
- reducing sugar content
- soil temperature
- water use efficiency, WUE

REFERENCES

1. Ahire, N. R., Bhoi, P. G., Solanke, A. V. (2000). Effect of row spacing and planting system on growth and yield of potato under surface and drip irrigation. *Journal of Indian Potato Association, 27,* 59–60.
2. Awan, Abdul Bari, (1964). Influence of mulch on soil moisture, soil temperature and yield of potatoes. *American Potato Journal, 4,* 337–339.
3. Awari, H., W., Hiwase, S. S. (1994). Effect of irrigation systems on growth and yield of potato. *Annals of Plant Physiology, 8,* 185–187.
4. Bucks, D. A., Nakayama, F. S. (1980). Injection of fertilizer and other chemicals for drip irrigation. *Proc. Agri. Turf. Irrig. Cong.,* Houston, TX, USA.
5. Bucks, D., A, Nakayama, F. S., Gilbert, R. G. (1979). Trickle irrigation water quality and preventive maintenance. *Agri. Water Mgmt., 2,* 149–162.
6. Bucks, D. A., Nakayama, F. S., Warrick, A. W. (eds), (1982). Principles, Practices and Potentialities of Trickle Irrigation. In: *Advances in Irrigation* by Daniel Hillel, Vol. I: 220–290.
7. Chandra, S., Singh, R. D., Bhatnagar, V. K., Bishit, J. K. (2002). Effect of mulch and irrigation on tuber size, canopy temperature, water use and yield of potato (*Solanum tuberosum*). *Indian Journal of Agronomy, 47,* 443–448.
8. Chawla, J. K., Narda, N. K. (1999). Root growth characteristics and tuber yield of trickle fertigated potatoes. *Journal of Agricultural Engineering, 36,* 65–78.
9. Chawla, J., K, and Narda, N. K. (2000). Growth parameters of trickle fertigated potato (*Solanum tuberosum*). *Indian Journal of Agricultural Sciences, 70,* 747–752.
10. Chawla, J. K., Narda, N. K. (2001). Economy in water and fertilizer use in trickle fertigated potato. *Irrigation and Drainage, 50,* 129–137.
11. Easson, D. L. (2000). Annual Report Agricultural Research Institute of Northern Ireland. No. 1999–2000, 41–49.
12. Grewal, S. S., Singh, N. T. (1974). Effect of organic mulches on the hydrothermal regime of soil and growth of potato crop in Northern India. *Plant and Soil, 40,* 33–47.
13. Jain, V. K., Shukla K. N., Singh, P. K. (2001). Response of potato under drip irrigation and plastic mulching. In: *Micro Irrigation* by Eds. Singh, H. P., Kaushish, S. P., Kumar, A., Murthy, T. S., Proceedings National Congress on Micro Irrigation by IARI, New Delhi, 413–417.
14. Khalak, A., Kumaraswamy, A. S. (1992). Effect of irrigation schedule and mulch on growth attributes and dry-matter accumulation in potato (*Solanum tuberosum*). *Indian Journal of Agronomy, 37,* 510–513.
15. Khan, A. R. (2002). Mulching effects on soil physical properties and peanut production. *Italian Journal of Agronomy, 6,* 113–118.

16. Lamont, W. J. Jr., Orzolek, M. D., Dye, B. (2002). Production of drip irrigated potatoes as affected by plastic mulches and row covers. *Journal of vegetable crop production, 8,* 39–47.
17. Midmore, D. J., Berrios, D., Roca, J. (1986). Potato (*Solanum spp.*) in hot tropics, II: Soil temperature and moisture modification by mulch in contrasting environments. *Field Crop Research, 15,* 97–108.
18. Palada, M. C., Davis, A. M., Crossman, S. M. A. (2002). Comparison of organic and synthetic mulch for bell pepper production at three levels of drip irrigation. *Proceedings of the 113th Annual Meeting Hort. Soc.,* Lake Buena Vista, Florida, USA, 23–25 July.
19. Sahoo, R. C., Mohapatra, B. K., Khanda, C. M., Lenka, D. (2002). Water use efficiency and yield of potato as influenced by plastic mulching. *Crop Res., 24,* 338–342.
20. Singh, N. T., Sood, M. C., Sharma, R. C. (2002). Effect of irrigation levels, cultural practices and nitrogen application on potato production under drip and sprinkler methods of irrigation. *Potato Global Research and Development,* II:923–925.
21. Suresh, R., Kumar, A. (2006). Effect of drip irrigation and mulches on pointed gourd in calcareous soil in North Bihar. *Indian Journal of Soil Conservation, 34,* 83–85.
22. Verma, S. C. (1991). *Potato Processing in India.* Technical Bulletin 34, Central Potato Research Institute (CPRI), Shimla, India, 34 pages.
23. Waterer, D. (2000). Influence of soil mulched and method of crop establishment on growth and yields of pumpkins. *Canadian Journal of Plant Science, 80,* 385–388.

PART IV
CROP SEQUENCE AND ECONOMICS

CHAPTER 20

EVALUATION OF DIFFERENT CROP SEQUENCES USING DRIP IRRIGATION SYSTEM

A. K. SAINI and KAMAL G. SINGH

20.1 INTRODUCTION

Water is one of the vital inputs for agricultural production. The availability of water is decreasing every year due to increased demand of water for agriculture, industrial and domestic's activities. Due to the inefficient use of water in the agriculture sector, water table all over the Punjab has declined very rapidly during the last two decades. That means availability of water in crop production is becoming a constraint implying that irrigation water must be used efficiently by adopting advanced irrigation methods. Drip irrigation is one of the fastest expanding technologies in modern irrigated agriculture with great potential to achieve savings in water and fertilizer.

A research study revealed that when the same quantity of water was used through drip and furrow irrigation for potato crop, drip irrigation resulted in higher number of tubers/plant, yield/plant and number of marketable tubers/plant [1]. Shelke et al. [3] reported that the water saving under drip irrigation was 24% and increase in yield by 13.4% over surface irrigation method for banana. Sivakumar et al. [4] reported that for the sunflower crop drip irrigation at 0.5 Epan was more economical with a shortest payback period of 0.49 year under paired row planting method. Chawla and Narda [2] reported water and fertilizer savings of 30% and 70% respectively with comparable yield levels under trickle fertigated crop as compared to furrow-irrigated crop of potatoes. These studies revealed that drip irrigation system caused a substantial saving of irrigation water and increase in yield.

This research study evaluates the yearly performance of drip irrigation system for different crop sequences.

---

*In this chapter, the currency is expressed in Indian Rupees (1.00 US$ = Rs. 60.93; 1.00 Rs. = 0.02 US$)

## 20.2	MATERIALS AND METHODS

A field study was conducted at the research farm of Department of Soil and Water Engineering at Punjab Agricultural University (PAU), Ludhiana for two consecutive years (2002–2004). Different crop sequences consisted of only vegetable crops, only field crops, and a combination of vegetable and field crops, as follows:

1. Cauliflower – Hybrid chili
   (October–February) (March–October)
2. Sunflower – Maize – Potato
   (January–May) (June–September) (October–December)
3. Sunflower – Cotton
   (January–May End) (May End –December)

A built-in type of drip irrigation system, was used to irrigate the crop, with a dripper spacing of 30 cm. Three levels of drip irrigation scheduling were selected on the basis of IW/CPE ratios (ratio of irrigation water applied to cumulative pan evaporation): 0.5 (low), 0.75 (medium) and 1.0 (high). The 10 mm depth of irrigation water was applied during irrigation. The interval between the two irrigations varied depending upon the IW/CPE ratio for different treatments of irrigation. Irrigation to furrow irrigated plots were based on the recommendations in package practices by the university (PAU, Ludhiana). If the effective rainfall was less than 10 mm, it was considered that irrigation was equal to the effective rainfall; and if effective rainfall was more than 10 mm, it was considered as irrigation equal to 10 mm.

The experiment was laid out in randomized block design with 4 treatments for each crop and each treatment was replicated thrice. The size of each plot was 5.10 × 2.70 m² for hybrid chili and cauliflower; 5.10 × 3.6 m² for sunflower and potato crop; 5.1 × 3.0 m² for maize and 5.1 × 4.05 m² for cotton. The soil at the experimental site was sandy loam, having pH 8.5 and available N, P, K of 51.6, 80.6 and 148.9 kg/ha, respectively. For the 0–30 am soil depth, the water content (volume basis) was 0.1739 cm³/cm³ at – 0.03 MPa (field capacity), and 0.0523 cm³/cm³ at – 1.5 MPa (permanent wilting point). The soil bulk density was 1.58 g/cm³. Recommended dose of fertilizer was applied according to the package of practices for cultivation of vegetables and field crops of PAU. To reduce the number of lateral lines, the recommended single row planting method was altered to paired row planting method for all crops and yield of all the crops were obtained on per plot basis, as shown below:

### 20.2.1	CROP RECOMMENDED SPACING PAIRED ROW PLANTING

Hybrid chili and Cauliflower 45×30 cm, 60:30×30 cm
   Sunflower and Potato 60×30 cm, 90:30×30 cm
   Maize 50×22 cm, 75:25×22 cm
   Cotton 67.5×15 cm, 90:45×15 cm

Paired row planting method was used for all the crop sequences, because this will decrease the cost of lateral line to half. The treatments are listed below:

T_1 = Paired row planting, drip irrigated at low level of irrigation, IW/CPE = 0.5

T_2 = Paired row planting, drip irrigated at medium level of irrigation IW/CPE=0.75

T_3 = Paired row planting, drip irrigated at high level of irrigation, IW/CPE=1.0

T_4 = Paired row planting, irrigated as per recommended practices by PAU.

20.3 RESULTS AND DISCUSSION

The average yield, irrigation water applied and water use efficiency for all the three sequences are presented in Tables 20.1–20.6. Table 20.1 revealed a nonsignificant difference for average fruit length among the irrigation treatments for chili crop. Statistically, there was significant increase in yield within drip irrigation treatments as well as when compared to furrow irrigation method. Drip irrigation at low level of irrigation gave 16% more yield as compared to drip irrigation at high level of irrigation. This may be due to the fact that chili crop is sensitive to excess moisture condition. Similarly, drip irrigation at low level of irrigation gave 29% more yield and saved 63% of irrigation water as compared to furrow irrigation method. Water use efficiency was maximum at low level of drip irrigation and was 252% higher compared to furrow irrigation method.

Table 20.2 depicts the variation in head diameter, head weight and yield in different irrigation treatments for cauliflower crop. Statistically, head diameter of cauliflower was significantly higher in T_1 as compared to other treatments. Similarly, head weight was also significantly higher in drip irrigation treatments. Head weight at low level of drip irrigation was 35.4% more as compared to furrow irrigation method. The increase in the yield with drip irrigation was due to the higher number of heads and head weights. Drip irrigation at low level of irrigation gave 38% higher yield and saved 40% of irrigation water as compared to furrow irrigation. The water use efficiency with drip irrigation at low level of irrigation was 130% more as compared to conventional furrow irrigation method.

TABLE 1   Average Fruit Length, Yield, Irrigation Water Applied and WUE of Chili Under Drip and Furrow Irrigation

Treatments	Average fruit length (cm)	Yield (t/ha)	Water depth applied (cm)	IWUE (kg/ha-cm)
T_1	4.47	21.91	31.9	687
T_2	4.38	20.11	42.8	470
T_3	4.36	18.85	53.8	350
T_4	4.26	16.96	87.0	195
CD at 5%	NS	0.746		
C.V.	2.71	0.192		

TABLE 20.2   Head Diameter, Head Weight, Yield, Irrigation Water Applied and IWUE of Cauliflower Under Drip and Furrow Irrigation

Treatments	Head diameter (cm)	Head weight (gm)	Yield (t/ha)	Water depth applied (cm)	IWUE (kg/ha-cm)
T_1	22.7	272.5	23.55	12.6	1869
T_2	21.3	265.2	23.50	14.9	1577
T_3	21.3	250.8	23.03	17.2	1339
T_4	20.4	201.2	17.07	21.0	813
CD at 5%	0.41	4.59	2.01		
C.V.	0.96	0.93	0.462		

TABLE 20.3   Head Diameter, Yield, Irrigation Water Applied and IWUE of Sunflower Under Drip and Furrow Irrigation

Treatments	Head diameter (cm)	Yield (t/ha)	Water depth applied (cm)	IWUE (kg/ha-cm)
T_1	20.2	2.83	28.4	100
T_2	20.2	2.83	36.6	77
T_3	19.1	2.64	44.9	59
T_4	18.8	2.53	60.0	42
CD at 5%	NS	0.118		
C.V.	5.77	0.218		

TABLE 20.4   Yield, 1000 Grain Weight, Mean Cob Length, Irrigation Water Applied and IWUE of Maize Under Drip and Furrow Irrigation

Treatments	Mean cob length (cm)	1000 grain weight (gm)	Yield (t/ha)	Water depth applied (cm)	IWUE (kg/ha-cm)
T_1	13.3	203.7	3.11	17.3	180
T_2	13.2	201.8	3.10	21.9	142
T_3	13.5	197.2	2.87	26.5	108
T_4	13.1	182.8	2.60	36.5	71
CD (5%)	0.19	NS	NS		
C.V.	0.72	5.91	0.763		

TABLE 20.5   Plant Height, Dry Matter, Yield, Irrigation Water Applied and IWUE of Potato Under Drip and Furrow Irrigation

Treatments	Final plant height (cm)	Dry matter (gm)	Yield (t/ha)	Water depth applied (cm)	IWUE (kg/ha-cm)
T_1	48.3	25.3	25.32	13.4	1890
T_2	47.8	25.7	25.13	14.8	1698
T_3	48.6	26.5	25.54	16.2	1577
T_4	48.0	24.2	24.96	21.0	1188
CD at 5%			NS		
C.V.			0.458		

TABLE 20.6   Plant Height, Dry Matter, Yield, Irrigation Water Applied and IWUE of Cotton Under Drip and Furrow Irrigation

Treatment	Final plant height (cm)	Dry matter (gm/plant)	Yield (t/ha)	Water depth applied (cm)	IWUE (kg/ha-cm)
T_1	94.2	83.6	1.75	15.0	116.6
T_2	96.5	76.6	1.84	19.8	92.7
T_3	97.0	93.1	1.88	24.0	78.5
T_4	92.0	86.6	1.84	31.0	58.3
CD at 5%			NS		
C.V.			6.39		

TABLE 20.7   Yield and Depth of Water Applied of All Three Crop Sequences

Crop	Yield (t/ha)			Depth of water applied (cm)		
	Drip	Furrow	Yield increase, %	Drip	Furrow	Water saving, %
Cauliflower	23.55	17.07	37.96	12.6	21.0	40.33
Cotton	1.75	1.837	–4.84	15.0	31.0	51.61
Hybrid chili	21.91	16.96	29.2	31.9	87.0	63.33
Maize	3.11	2.60	19.62	17.3	36.5	52.60
Potato	25.32	24.96	1.44	13.4	21.0	36.19
Sunflower	2.83	2.53	11.86	28.4	60.0	52.67

Table 20.3 indicates statistically nonsignificant differences in head diameter among all different irrigation treatments for sunflower crop. Drip irrigation at low level of irrigation gave statistically more yield and saved 53% water as compared to furrow irrigation. The water use efficiency at low level of drip irrigation was 138% higher as compared to conventional furrow irrigation method.

Table 20.4 revealed statistically nonsignificant differences in 1000-grain weight and yield among all irrigation for maize crop. Drip irrigation at low level of irrigation saved 52% water as compared to furrow irrigation method. Water use efficiency with drip irrigation at low level of irrigation was 153% more as compared to furrow irrigation method.

Table 20.5 shows statistically significant differences in yield for all irrigation treatments for potato crop. Drip irrigation at low level of irrigation saved 38% of water and 3.6 q/ha more yield as compared to conventional furrow irrigation method.

Table 20.6 indicates nonsignificant differences in cotton yield and final plant height between drip and conventional furrow irrigated treatments. Drip irrigation at low level of irrigation saved substantial quantity of water as compared to conventional furrow irrigation. Hence, with the major objective of saving water without affecting yield, drip irrigation at low level of irrigation can be adopted.

Table 20.7 shows drip-irrigated and furrow irrigated treatments having maximum yield and amount of water applied for all the crops. Data revealed that cauliflower-hybrid chili (vegetable crop sequence) was best because there was significant increase in yield as well as substantial amount of water saving. Whereas in other crop sequences, there was substantial amount of water saving without affecting the yield. Thus drip irrigation is also beneficial for field crops in water scarcity areas

20.4   SUMMARY

A field study on drip irrigation was conducted for different crop sequences consisting of only vegetable crop, only field crops and combination of vegetable and field crops. The study revealed that cauliflower-hybrid chili (vegetable crop sequence) with drip irrigation at low level of irrigation (IW/CPE=0.50) gave more yield and saving of water as compared to other crop sequences. The other crop sequences which comprised of vegetable and field crops showed nonsignificant increase in yield with drip irrigation as compared to furrow irrigation, but resulted in substantial water saving without affecting the crop yield.

KEYWORDS

- banana
- cauliflower
- chili
- crop sequences
- drip irrigation
- field crops
- furrow irrigation
- irrigation
- paired row planting
- plant height
- potato
- sunflower
- vegetable
- water saving
- water use efficiency

REFERENCES

1. Boujelben, A. M., Barek, K., Chartzoulakis, K. S. (1997). Potato crop response to drip irrigation system. *Acta Horti.*, *449(I)*, 241–243.
2. Chawla, J. K., Narda, N. K. (2001). Economy in fertilizer use in trickle fertigated potato. *J. Irrigation and Drainage*, *50*, 129–137.
3. Shelke, D. K., Vaishnava, V. G., Oza, S. R., Jadhav, G. S. (1999). Management of irrigation in banana through drip irrigation systems. *J. Maharashtra Agric. Univ.*, *23(3)*, 317–318.
4. Sivakumar, H. K., Ramachandrappa, B. K., Nanjappa, H. V. (2001). Economics of drip irrigation in Sunflower. *Karnataka J. Agric. Science*, *14(4)*, 924–927.

CHAPTER 21

ECONOMICS OF DRIP IRRIGATED CROP SEQUENCES

A. K. SAINI, KAMAL G. SINGH, and M. SIAG

21.1 INTRODUCTION

Water is one of the primary inputs in agricultural production. Efficient and proper utilization of this scarce resource is crucial for increasing agricultural production. This can be achieved by adopting methods having higher water application and distribution efficiencies. The modern methods of irrigation (sprinkler irrigation; drip irrigation and microsprinkler irrigation) are becoming increasingly popular. Application efficiency varies from 90 to 95% with these modern irrigation methods as compared to 50–60% in conventional irrigation systems.

With drip irrigation, there are substantial water savings as well as increase in yield. Grimes et al. [5] reported substantially higher tomato yield in drip-irrigated plots over furrow-irrigated plots. Jadhav et al. [6] reported that there was a saving of 31.5% of water by drip method and benefit-cost ratio for tomato was 5.15 compared to 2.96 for flood method. Sivakumar et al. [7] reported that for sunflower drip irrigation at 0.5 Epan was more economical with a shortest payback period of 0.49 year under paired row method planting. Chawla and Narda [4] indicated water and fertilizer savings to the extent of 30% and 70%, respectively with comparable yield under trickle fertigated potatoes as compared to furrow-irrigated crop.

The United States Department of Commerce [1], in its comprehensive market assessment of drip irrigation systems in India, has reported that increase in yield under drip irrigation system ranged as high as 100% in bananas, 40–50% in sugarcane, pomegranate, tomato and chilly and around 25% in grapes, cotton and groundnut. In these crops, the irrigation water saving was from 40 to 70% compared to conventional methods. The pay back period ranged from 1 to 4 years only for different crops.

The drip irrigation is well suited for high value crops and was found economically viable for crops having wider row-to-row and plant-to-plant spacing. In case of closely spaced crops, installation of drip irrigation system is quite expensive. But

---

*In this chapter, the currency is expressed in Indian Rupees (1.00 US$ = Rs. 60.93; 1.00 Rs. = 0.02 US$).

the use of drip irrigation in vegetables not only saves water use but also increases yield with good quality produce. Since the use of drip irrigation system requires a high initial investment, it is not a viable option to use the system only once a year for taking a single crop. Therefore the present study was conducted to identify different field and vegetable crops, which may be grown with drip irrigation method using agronomically feasible and economically viable crop sequences so that this system can be used effectively and efficiently throughout the year. This study presents the economical analysis of different crop sequences.

21.2   MATERIALS AND METHODS

The experiments were conducted at Punjab Agricultural University – Ludhiana with the following crop sequences consisting of field and vegetable crops:
1. Cauliflower – Hybrid chili
    (October–Feb.) (March–October)
2. Sunflower – Maize – Potato
    (January–May) (June–September) (October–December)
3. Sunflower – Cotton
    (January–May End) (May End–December)

The first crop sequence comprised of vegetable crops only, whereas the second sequence consisted of both vegetable (potato) and field crops (sunflower and maize), and the third crop sequence included only field crops. Within the same sequence, the crops having almost the same row-to-row spacing were taken so that the same lateral network in drip irrigation can be used throughout the year.

Treatments for different crops: In case of drip irrigation, paired row planting on raised beds was adopted so that two rows were irrigated with a single lateral drip line, in order to optimize the cost of drip irrigation. In case of surface irrigation, paired row planting was done because it use less quantity of water as compared to single row planting method without any adverse effects on crop yield [2]. All the treatments were replicated thrice and experiment was conducted for two years (2002–2003 and 2003–2004). The recommended package of practices for the different crops by Punjab Agricultural University, Ludhiana was followed. The treatments were:

T_1 = Drip irrigated, paired row planting, IW/CPE is 0.50
T_2 = Drip irrigated, paired row planting, IW/CPE is 0.75
T_3 = Drip irrigated, paired row planting, IW/CPE is 1.00
T_4 = Surface irrigated, paired row planting, IW/CPE is 1.00

Cost of laterals varied from crop to crop depending upon the row-to-row spacing. The subsidy of Rs. 25,000/- per hectare was deducted from the cost of drip irrigation system. Cost of cultivation and selling price of produce was taken as per statistics provided by Department of Agricultural Economics, PAU, Ludhiana

[3]. An additional cost of operation and maintenance of drip irrigation system @ Rs.500/- per month per hectare was added to the cost of cultivation for drip system. Crop yield was taken from experimental data from two years of study.

Depending upon the number of crops per sequence, the annual cost of drip irrigation system was divided equally among these crops. The duration of the individual crop was not considered for distributing the annual cost of drip irrigation system; because the net returns from the complete crop sequence were compared by taking the total annual cost of the drip system.

21.3 RESULTS AND DISCUSSION

Each crop was sown twice for two consecutive years hence two sets of data were generated. Paired row planting method was used for both drip and surface irrigation treatments. For individual crops, the trends observed were similar during both the years. The yields and water savings (with respect to conventional surface irrigation) achieved for all the crops are shown in Table 21.1.

TABLE 21.1 Yield and Water Saving For Different Crops Under Various Treatments

Treatments	Yield (100 kg/ha)					
	Cauliflower	Chilly	Cotton	Sunflower	Maize	Potato
T_1	235.5	219.1	17.48	28.3	31.1	253.2
	(40.0)	(63.3)	(51.6)	(52.6)	(52.6)	(36.2)
T_2	235.0	201.1	18.36	38.3	31.0	251.3
	(29.0)	(50.8)	(36.1)	(39.0)	(40.0)	(29.5)
T_3	230.3	188.5	18.84	26.4	28.7	255.4
	(18.1)	(38.2)	(22.6)	(25.1)	(27.4)	(22.8)
T_4	170.7	169.6	18.37	25.3	26.0	249.6

Values in parenthesis are water savings in percent.

Table 21.1 concludes that:
1. There is a significant saving of water in T_1, T_2 and T_3 as compared to conventional irrigation treatment T_4. Saving of water is maximum in T_1 treatment where water application was restricted to 50% of the accumulated pan evaporation.
2. There is a significant increase in yield under drip irrigation system for cauliflower and hybrid chili crops as compared to conventional irrigation where-

as for other crops, yield difference under drip and conventional irrigation methods are relatively same.

21.3.1 ECONOMIC ANALYSIS

Economic viability of the drip irrigation system was performed for two different situations.

21.3.1.1 CASE I – UNLIMITED WATER

Tables 21.2–21.4 show the calculations of net returns for the three different crop sequences for conventional and drip irrigation systems. All cost and incomes are for one-hectare area. Table 21.5 gives a summary of the net income from different crops individually and from the crop sequences.

TABLE 21.2 Economic Analysis of Drip Irrigation Versus Conventional Irrigation For Cauliflower – Hybrid Chili Crop Sequence

	Parameters	Cost (Rs.)	
		Drip	**Conventional**
		Cauliflower	
1	Main line, submain, fertilizer tank, control valve, filters, pumping unit, etc.		Nil
		52,470	
		20	
	a) Fixed cost	1312	
	b) Life (years)	1312	
	c) Depreciation per crop	2624	
	d) Interest (10% per year) per crop		
	e) Total (c+d)		
2	Laterals with inbuilt drippers		Nil
	a) Fixed cost	98,200	
	b) Life (years)	10	
	c) Depreciation per crop	4910	
	d) Interest (10% per year) per crop	2455	
	e) Total (c+d)	7365	
3	Cost of cultivation	56,557	53,557
4	Seasonal total cost {1(e)+2(e) + 3}	66,546	53,557

5	Yield (100 kg)	235.5	170.7
6	Selling price (Rs./100 kg)	400	400
7	Income from produce (6 × 7)	94,200	68,280
8	Net seasonal income (8 − 4)	27,654	14,723
		Hybrid Chili	
1	Main line, submain, fertilizer tank, control valve, filters, pumping unit, etc.	2624	Nil
2	Laterals with inbuilt drippers	7365	Nil
3	Cost of cultivation	41,665	38,665
4	Seasonal total cost {1+2+ 3}	51,654	38,665
5	Yield of produce (100 kg)	219.1	169.6
6	Selling price (Rs./100 kg)	350	350
7	Income from produce (6 × 7)	76,685	59,360

Parameters		Cost (Rs.)	
		Drip	Conventional
8	Net seasonal income (8–4)	25,031	20,695
		Cauliflower – Hybrid Chili	
	Total cost	118,200	92,222
	Total income	170,885	127,640
	Net income	52,685	35,418
	Benefit–Cost ratio	1.45:1	1.38:1

TABLE 21.3 Economical Analysis of Drip Irrigation Verses Conventional Irrigation For Sunflower-Maize-Potato Crop Sequence

Different Components Taken		Cost (Rs.)	
		Drip	Conventional
		Sunflower	
1	Main line, submain, fertilizer tank, control valve, filters, pumping unit, etc.		Nil
		52,470	
	a) Fixed cost	20	
	b) Life (years)	875	
	c) Depreciation per crop	875	
	d) Interest (10% per year) per crop	1750	
	e) Total (c+d)		

TABLE 21.2 *(Continued)*

2	Laterals with inbuilt drippers		Nil
	a) Fixed cost	85,000	
	b) Life (years)	10	
	c) Depreciation per crop	2834	
	d) Interest (10% per year) per crop	1417	
	e) Total (c+d)	4251	
3	Cost of cultivation	14,710	12,710
4	Seasonal total cost {1(e)+2(e) + 3}	20,711	12,710
5	Yield of produce (100 kg)	28.3	25.3
6	Selling price (Rs./100 kg)	1200	1200
7	Income from produce (6 × 7)	33,960	30,360
8	Net seasonal income (8–4)	13,249	17,650

	Different Components Taken	Cost (Rs.)	
		Drip	**Conventional**
		Maize	
1	Main line, submain, fertilizer tank, control valve, filters, pumping unit, etc.	1750	Nil
2	Laterals with inbuilt drippers	4251	Nil
3	Cost of cultivation	17,325	15325
4	Seasonal total cost {1+2+ 3}	23,326	15325
5	Yield of produce (100 kg)	31.1	26.0
6	Selling price (Rs./100 kg)	600	600
7	Income from produce (6 × 7) (+ by products @ Rs. 3000/ha)	21,660	18600
8	Net seasonal income (8–4)	–1666	3275
		Potato	
1	Main line, submain, fertilizer tank, control valve, filters, pumping unit, etc.	1750	Nil
2	Laterals with inbuilt drippers	4251	Nil
3	Cost of cultivation	44,103	42103
4	Seasonal total cost {1+2+ 3}	50,104	42103
5	Yield of produce (100 kg)	255.4	249.6
6	Selling price (Rs./100 kg)	250	250

TABLE 21.3 *(Continued)*

7	Income from produce (6 x 7)	63,850	62400
8	Net seasonal income (8 – 4)	13,746	20297
		Sunflower-Maize-Potato	
	Total cost	94,141	70,138
	Total income	11,9470	11,1360
	Net income	25,329	41,222
	Benefit – cost ratio	1.26:1	1.58:1

TABLE 21.4 Economical Analysis of Drip Irrigation Verses Conventional Irrigation For Cotton-Sunflower Crop Sequence

	Different Components Taken	Cost (Rs.)	
		Drip	**Conventional**
		Cotton	
1	Main line, submain, fertilizer tank, control valve, filters, pumping unit, etc.		Nil
		52,470	
	a) Fixed cost	20	
	b) Life (years)	1312	
	c) Depreciation per crop	1312	
	d) Interest (10% per year) per crop	2624	
	e) Total (c+d)		
2	Laterals with inbuilt drippers		Nil
	a) Fixed cost	67,400	
	b) Life (years)	10	
	c) Depreciation per crop	3370	
	d) Interest (10% per year) per crop	1685	
	e) Total (c+d)	5055	
3	Cost of cultivation	19,485	16,485
4	Seasonal total cost {1(e)+2(e) + 3}	27,164	16,485
5	Yield of produce (100 kg)	18.84	18.37
6	Selling price (Rs./100 kg)	1800	1800
7	Income from produce (6 × 7)	33,912	33,066
8	Net seasonal income (8–4)	6748	16,581

TABLE 21.4   *(Continued)*

	Sunflower		
1	Main line, submain, fertilizer tank, control valve, filters, pumping unit, etc.	2624	Nil
2	Laterals with inbuilt drippers	5055	Nil
3	Cost of cultivation	15,710	12,710
4	Seasonal total cost {1+2+ 3}	23,389	12,710
5	Yield of produce (100 kg)	28.3	25.3
6	Selling price (Rs./100 kg)	1200	1200
7	Income from produce (6 × 7)	33,960	30,360
8	Net seasonal income (8–4)	10,571	17,650

	Cotton-Sunflower		
	Total cost	50,553	29,195
	Total income	67,872	63,426
	Net income	17,319	34,231
	Benefit – cost ratio	1.34:1	2.17:1

Table 21.5 shows net returns per hectare for three crop sequences (both under drip and conventional irrigation) outlined for economical analysis. The analysis clearly shows an increase in income under drip irrigation for the vegetable crop sequence of cauliflower and hybrid chili. The extra cost due to drip irrigation is exceeded by additional returns due to higher production. However, in the second and third crop sequences, which consisted mainly of field crops, the net return under drip irrigation was less than that under conventional surface irrigation. This was mainly due to insignificant increase in yield under drip irrigation and hence the additional cost of installation of drip system reduced the net returns as compared to conventional furrow irrigation. Therefore, it may be concluded from the present study that adoption of drip irrigation system was economically viable for vegetable crop sequence but not for crop sequences with field crops.

TABLE 21.5 Comparison of Net Profit For Different Crop Sequences Under Unlimited Water Supply

Crops included in different sequences	Net income (Rs.)		Conventional
	Drip		
Cauliflower-Hybrid Chili			
Cauliflower	27,654	14,723	
Hybrid chili	25,031	20,695	
Net income from sequence	**52,685**	**35,418**	
Sunflower-Maize-Potato			
Sunflower	13,249	17,650	
Maize	−1666	3275	
Potato	13,746	20,297	
Net income from sequence	**25,329**	**41,222**	
Sunflower-Cotton			
Sunflower	10,571	17,650	
Cotton	6748	16,581	
Net income from sequence	**17,319**	**34,231**	

21.3.1.2 CASE II – LIMITED WATER

The total water saving for T_1 treatment under different crop sequences and additional area that can be irrigated has been presented in Table 21.6.

TABLE 21.6 Total Water Requirement For the Different Crop Sequence

Crop sequence	Water requirement (m³/ hectare)		Saving over the conventional system (m³/ha)	Additional area irrigated with drip (hectare)
	Drip	Conventional		
Sunflowe –cotton	2840+1500 =4340	6000+3100 =9100	4760	1.100
Cauliflower- hybrid chili	1260+3190 =4450	2100+8700 =10800	6350	1.420
Sunflower-maize- potato	2840+1730+1340 =5910	6000+3650+ 2100 =11750	5840	1.000

Tables 21.7–21.9 show the calculations of net returns for different crop sequences for both conventional and drip irrigation system taking into consideration the additional area that can be irrigated by saving in water by drip irrigation method. Again all cost and incomes are for one hectare.

TABLE 21.7   Net Return From Cauliflower–Hybrid Chili Crop Sequence Under Drip and Conventional Furrow Irrigation Method With Water as a Limiting Factor

	Different Components taken	Cost (Rs.)	
		Drip	Conventional
		Cauliflower	
1	Seasonal total cost	66,546	53,557
2	Yield of produce (100 kg)	235.5	170.7
3	Selling price (Rs./100 kg)	400	400
4	Income from produce (2 × 3)	94,200	68,280
5	Net seasonal income (5 – 1)	27,654	14,723
	Different Components taken	**Cost (Rs.)**	
		Drip	Conventional
		Hybrid chili	
1	Seasonal total cost	51,654	38,665
2	Yield of produce (100 kg)	219.1	169.6
3	Selling price (Rs./100 kg)	350	350
4	Income from produce (2 × 3)	76,685	59,360
5	Net seasonal income (5 – 1)	25,031	20,695
		Cauliflower – Hybrid chili	
	Net income/ha	52,685	35,418
	Additional area irrigated (ha)	1.42	-
	Additional income	74,813	
	Total income	**127,498**	**35,418**

TABLE 8 Economic Analysis of Drip Versus Conventional Irrigation For Sunflower-Maize-Potato Crop Sequence With Water as a Limiting Factor

	Different Components taken	Cost (Rs.)	
		Drip	**Conventional**
	Sunflower		
1	Seasonal total cost	20,711	12,710
2	Yield (100 g)	28.3	25.3
3	Selling price (Rs./100 kg)	1200	1200
4	Income from produce (6 × 7)	33,960	30,360
5	Net seasonal income (8 – 4)	13,249	17,650
	Maize		
1	Seasonal total cost	23,326	15325
2	Yield (100 kg)	31.1	26.0
3	Selling price (Rs./100 kg)	600	600
4	Income from produce (6 × 7) (+ by products @ Rs. 3000/ha)	21,660	18,600
5	Net seasonal income (8 – 4)	–1666	3275
	Potato		
1	Seasonal total cost	50,104	42,103
2	Yield (100 kg)	253.2	249.6
3	Selling price (Rs./100 kg)	250	250
4	Income from produce (6 × 7)	63,300	62,400
5	Net seasonal income (8 – 4)	13,196	20,297
	Sunflower-maize-potato		
	Net income/ha	24779	44,282
	Additional area irrigated, ha	1.005	-
	Additional income	24902	-
	Total income, Rs.	**49,681**	**44,282**

TABLE 21.9 Economical Analysis of Drip Irrigation Verses Conventional Irrigation For Cotton-Sunflower Crop Sequence

	Different Components Taken	Cost (Rs.)	
		Drip	Conventional
	Cotton		
1	Seasonal total cost	27,164	16,485
2	Yield (100 kg)	17.48	18.37
3	Selling price (Rs./100 kg)	1800	1800
4	Income from produce (6 x 7)	31,464	33,066
5	Net seasonal income (8 – 4)	4300	16,581
	Sunflower		
1	Seasonal total cost	23,389	12,710
2	Yield (100 kg)	28.3	25.3
3	Selling price (Rs./100 kg)	1200	1200
4	Income from produce (6 × 7)	33,960	30,360
5	Net seasonal income (8 – 4)	10,571	17,650
	Cotton-sunflower		
	Net income/ha	14,871	34,231
	Additional area irrigated, ha	1.120	-
	Additional income, Rs.	16,656	
	Total income, Rs.	**31,527**	**34,231**

Table 21.10 gives a comparison of net income from different crop sequences under both drip and conventional irrigation system. The table shows that under water scarcity situation, adoption of drip irrigation system gives very high returns with vegetable crop sequence as compared to that obtained with conventional irrigation system. Adoption of drip irrigation under water scarcity conditions also becomes viable for field crop sequences, as net incomes under drip irrigation are comparable with conventional irrigation.

TABLE 21.10   Net Income For Different Crop Sequences Under Limited Water

Crop included in different sequences	Net income (Rs.)	
	Drip	Conventional
Cauliflower-hybrid chili	127,498	35,418
Sunflower-maize-potato	49,681	44,282
	31,527	34,231
Sunflower-cotton		

21.4   CONCLUSIONS

1. Utilizing the drip irrigation system round the year for different crops, reduces its annual cost and hence can help in making the system viable.
2. Under unrestricted water supply, drip irrigation is viable for the crop sequence of cauliflower-hybrid chili but not for crop sequences with field crops.
3. Under water scarce conditions, adoption of drip irrigation system is also viable for crop sequence with field crops.

21.5   SUMMARY

A drip irrigation system was evaluated for vegetable crop sequence, vegetable and field crop sequence and only field crop sequence. Economic analysis of these crop sequences was performed for two different situations.

Under unlimited water supplies: Out of three crop sequences, vegetable crop sequence of cauliflower-hybrid chili gave maximum net return with drip irrigation method as compared to other crop sequences with drip and conventional irrigation method. The net return for this crop sequence with drip irrigation method was 1.49 times higher as compared to conventional irrigation method.

Under Limited water supplies: Drip irrigation method gave substantially higher net returns for vegetable crop sequence, because water saved with drip irrigation method can be used to irrigate the additional area. The net return with drip was 3.6 times higher as compared to conventional irrigation method. Other crop sequences with drip irrigation method also gave nearly equivalent net return as compared to conventional method of irrigation.

KEYWORDS

- cauliflower
- conventional irrigation
- cotton
- crop sequence
- drip irrigation
- fertigation
- maize
- micro irrigation
- micro sprinkler
- net profit
- net return
- plastics
- potato
- soil water
- sprinkler irrigation
- sunflower
- vegetable crop
- water application
- water distribution

REFERENCES

1. Anonymous, (1999). India – Drip and Micro Irrigation Systems – Market Assessment. STAT. *US Department of Commerce, 202,* 482.
2. Anonymous, (2000). Annual progress report on: Application of plastics in Agriculture. Department of Soil and Water Eng., P. A. U., Ludhiana – India.
3. Anonymous, (2004). Unpublished report of Department Economics, P. A. U., Ludhiana.
4. Chawla, J. K., Narda, N. K. (2001). Economy in fertilizer use in trickle fertigated potato. *Irrigation and Drainage, 50,* 129–137.
5. Grimes, D. W., Miller, R. J., Sehweeers, V. H., Smith, R. B., Wiley, P. L. (1972). Soil strength modification of root development and soil water extraction. *California Agricultural Journal, 26,* 12–14.
6. Jadhav, S. S., Gutal, G. B., Chougale, A. A. (1990). Cost economics of drip irrigation system for tomato crop. Proceedings of the 11th International Congress on the Use of Plastics in Agriculture, March, New Delhi, India. pp. B.171–B.176.
7. Sivakumar, H. K., Ramachandrappa, B. K., Nanjappa, H. V. (2001). Economics of drip irrigation in Sunflower. *Karnataka Journal of Agricultural Sciences, 14(4),* 924–927.

CHAPTER 22

ECONOMICS OF DRIP IRRIGATED CLOSELY SPACED CROPS

KAMAL G. SINGH, H. A. W. S. GUNATHILAKE,
RAMESH P. RUDRA, and PRADEEP K. GOEL

22.1 INTRODUCTION

Cost of a drip irrigation system is comprised of: the capital investment; cost due to water pump, head control components, main line, laterals, emitters; installation cost; the operating cost of pumping energy, pump maintenance, lateral maintenance; and system operating labor cost, etc. Comparatively, the capital cost is predominant in any drip irrigation system. Therefore, in designing the system, the priority is given to plot an optimal geometrical layout of the system (lateral spacing and emitter spacing) without affecting the number of plants (plant density).

Efficient application of water through drip irrigation system depends on the understanding of hydraulics of moisture advance in horizontal and vertical directions within the soil profile. The flow phenomenon in a drip irrigation system is a function of application rate and duration of irrigation, both of which are of paramount importance in the design of a drip irrigation system. Several studies have been conducted to investigate the soil wetting and water front advance within the soil profile under point source of water application. It is reported that the rate of horizontal water movement in the soil and metric potential at different distances from the dripper are functions of soil type and drip discharge rate. It was also reported that increase in drip discharge rate resulted in an increase in the horizontal wetted area and a decrease in the depth of wetted soil [1, 2, 5]. Goel et al. [3] studied the effects of drip discharge rate on the soil moisture distribution patterns under bare soil conditions using 12, 8, 4 and 2 lph discharge rates of drippers operated for 25, 37.5, 75 and 150 min, respectively, with the aim to provide equal quantity of water. They found that in the beginning the water at higher application rate saturated the soil near the dripper and infiltration was slower; whereas the water penetrated deeper with lower

---

*In this chapter, the currency is expressed in Indian Rupees (1.00 US$ = Rs. 60.93; 1.00 Rs. = 0.02 US$).

application rate because of availability of more time for infiltration. Sharma et al. [7] studied two dimensional moisture movement flow under soil tank model and reported the shape of moisture movement to be semielliptical. Similar works have also been reported by Brandt et al. [1], Kaul [4] and Sivanappan et al. [8]. For closely spaced field crops, it is important to determine the optimal spacing of laterals and drippers.

The present study attempts at developing rational criteria for determining the optimal spacing of laterals and drippers of a drip system so that the drip irrigation technology is an economically viable for closely spaced field crops.

22.2   DEVELOPMENT OF DESIGN CRITERIA FOR THE ESTIMATION OF THE CAPITAL COST

The least cost drip system will be the one where lateral spacing is kept corresponding to the operation irrigation time that results into maximum horizontal advance of soil moisture but this system will not be the most efficient as it will result into avoidable deep percolation losses. To avoid the percolation losses, the maximum permissible vertical advance should not be allowed to exceed the root zone depth of the crop. Therefore, the operation time of the system must be adjusted so that the vertical moisture advance remains within the crop root zone. Horizontal advance corresponding to the operation duration of the system that results into a vertical advance equal to the root zone depth therefore be considered as the maximum allowable lateral spacing [6].

22.2.1   CASE STUDY

A field of 80 m length and 40 m width, having sandy loam soils was considered for designing the drip irrigation system using the developed design criteria for irrigating okra crop. The crop root zone depth was 40 cm. Drippers of 2 and 4 lph were considered for determining the optimal spacing of drippers and laterals for most economic drip system. Kaul and Michael [4] developed the relationships between horizontal and vertical advance of soil moisture front with operation time of the drip system for sandy loam soils. The Eqs. (1) and (2)present the relationships between horizontal moisture advance and operation time of the system for dripper discharges of 2 and 4 lph, respectively:

$$Ha^2 + 2.03 \times 10^{-3} T^2 - 3.06 T - 250 = 0 \tag{1}$$

$$Ha^2 + 3.33 \times 10^{-3} T^2 - 4.99 T - 500 = 0 \tag{2}$$

$$Va = 4.037 T^{0.444} \tag{3}$$

$$Va = 5.164\ T^{0.424} \tag{4}$$

where: Ha = horizontal advance, cm; and T = operation time, min.; and Va = vertical advance, cm. The relationships between vertical moisture front advance with operation time of the system for dripper discharges of 2 and 4 lph are expressed in Eqs. (3) and (4).

Horizontal and vertical advance of moisture fronts were estimated using Eqs. (1)–(4) at different times of operation of the drip system (30, 60, 90, 120, 150, 180, 210, 300, 480, 720 and 780 min.) under the dripper discharge of 2 and 4 lph. Since the horizontal moisture front moves equally on either side of the dripper, the horizontal moisture spread was taken equal to twice the horizontal advance (estimated from the above relationships), for the purpose of determining the lateral spacing. Based on the lateral spacing and the field, length number of laterals required in the field was calculated. Twenty percent overlap was taken in determining the number of drippers. The number of drippers was therefore calculated taking the dripper spacing equal to 80% that of the lateral spacing.

22.3 RESULTS AND DISCUSSION

Maximum lateral spacing and maximum irrigable root zone depth under dripper discharge of 2 and 4 lph and for different operating durations of a drip system from 30 min to 210 min are presented in Table 22.1. Increasing operating time from 30 min to 210 min resulted into larger allowable lateral spacing for 2 and 4 lph from 37.0 to 56.7 and 50.8 to 74.9 cm, respectively (Table 22.1). Higher dripper discharge attained the maximum allowable lateral spacing in lesser duration of operation of the system. However, it may be noted in Table 1 that the vertical advance of moisture did not stabilize and kept increasing with increasing dripper discharge and increasing operation time. Vertical advance of moisture reached the root zone depth of 40 cm in operation time of 180 min and 120 min under the dripper discharges of 2 and 4 lph (Table 22.1), respectively.

TABLE 22.1 Maximum Lateral Spacing and Vertical Depth Under the Given Application Rates

Operation time (min)	Maximum lateral spacing under dripper discharge (cm)		Maximum root zone depth (vertical advance) of moisture (cm)	
	2.0 lph	4.0 lph	2.0 lph	4.0 lph
30	37.0	50.8	18.3	21.6
60	42.0	56	24.9	29.0
90	45	60.7	29.8	34.5

120	48.5	64.8	33.8	**38.9**
150	51.5	68.5	37.3	42.8
180	54.2	71.8	**40.5**	46.2
210	56.7	74.9	43.4	49.4
300	60.9	82.4	50.7	58.0
480	70.7	92.2	62.5	70.8
720	74.8	97.2	74.8	84.0
780	74.8	97.2	77.5	86.9

Based on the values, the maximum allowable lateral spacing (Table 22.1), number of laterals required in the field (80 m × 40 m) under different discharge and operation time were determined and are presented in Table 22.2. As mentioned earlier, the total number of drippers required in the field was estimated based on the assumption of 20% overlap. Total number of drippers required under different dripper discharge and operation time is also presented in Table 22.2.

Total length of laterals and the number of the drippers may decrease with increasing dripper discharge and operation time. Minimum length of the laterals (4240 m) and number of drippers (7102) were observed in case of operation time of 210 min. It may be noted in Table 2 that operation time more than 720 min did not result in any reduction in the required length of lateral and number of drippers. However, further increase in operation time resulted in the deep percolation losses (Table 22.1).

Optimal operation durations of drip system for 2 and 4 lph were 180 and 120 min based on the crop root zone (Table 22.1). The length of laterals and number of drippers required at optimal operation durations for 2 and 4 lph are presented in Table 22.3. It is evident in Table 22.3 that drip system with 4 lph drippers will be the most economical compared to drippers of 2 lph.

TABLE 22.2　Total Length Laterals and Total Number of Drippers For a Field of 80 m × 40 m Under Different Operating Times and Dripper Discharges

Discharge (lph)	Operation time (min.)	No. of laterals	No. of emitters/ lateral	Total length of laterals (m)	Total number of emitters
2	30	108	270	8640	29,160
2	60	95	238	7600	22,610
2	90	89	222	7120	19,758
2	120	82	206	6560	16,892
2	150	78	194	6240	15,132
2	**180**	**74**	**185**	5920	**13,690**

2	210	71	176	5680	12,496
2	300	66	164	5280	10,824
2	480	57	141	4560	8037
2	720	53	134	4240	7102
2	780	53	134	4240	7102
4	30	79	196	6320	15,484
4	60	71	179	5680	12,709
4	90	66	165	5280	10,890
4	**120**	**62**	**154**	**4960**	**9548**
4	150	58	146	4640	8468
4	180	56	139	4480	7784
4	210	53	134	4240	7102
4	300	49	121	3920	5929
4	480	43	108	3440	4644
4	720	41	103	3280	4223
4	780	41	103	3280	4223

TABLE 22.3   Overall Capital Cost (Rs.) of the Drip System

Dripper discharge (lph)	Operation time (min)	Length of laterals (m)	Number of drippers	Cost of installation (IR)	Total cost (IR)
2-On-line	180	5920	13,690	2000	83,141
2-In-line	180	5920	-	-	69,760
4-On-line	120	4960	9548	1700	64,109
4-In-line	120	4960	-	-	58,480

*Cost of laterals (Online) poly tube IR = Rs. 7.00/m

Cost of laterals (Inline) IR = Rs. 11.75/m

Cost of emitters IR = Rs. 2.90/each on November 9, 2008.

Therefore, in capital cost-wise the most economical drip system for the given field is 4Lph dripper system; installed at 64.8 cm lateral spacing; operated for 120 min (02 h) duration.

22.4   SUMMARY

The present study attempts at developing economic criteria for developing the drip system for closely spaced field crops. Water front advance under a point source of water application depends on soil type, dripper discharge and the operation time of the system. Larger operation time results into larger lateral spacing but may simultaneously result into deep percolation losses because of consequential larger vertical advances. Horizontal advance corresponding to the operation duration of the system that results into a vertical advance equal to the root-zone depth should therefore be considered as the maximum allowable lateral spacing. The procedure of determining the optimal spacing of laterals and drippers has been discussed for irrigating okra crop in sandy loam soils. Optimal duration of operation of drip irrigation system for 2 and 4 lph drippers were 180 and 120 min based on the horizontal and vertical advance of soil moisture. Drip irrigation system with 4 lph dripper discharge, laterals spaced at 64.8 cm apart was most economical system for irrigating a crop of 40 cm root zone depth in sandy loam soils.

KEYWORDS

- capital investment installation
- closely spaced crop
- drip irrigation
- duration of irrigation
- emitter spacing
- head control
- horizontal advance
- infiltration
- installation cost
- labor cost
- lateral maintenance
- lateral spacing
- main line
- micro irrigation
- moisture distribution pattern
- moisture front advance
- okra
- operating cost
- optimal geometrical layout

- **optimal spacing**
- **pump maintenance**
- **pumping energy**
- **root-zone depth**
- **sandy loam**
- **soil moisture**
- **trickle irrigation**
- **trickle source**
- **vertical advance**
- **water pump**
- **wetted area**

REFERENCES

1. Brandt, A., Bresler, E. N., Ben-Asher, J., Goldberg, D. (1971). Infiltration from a trickle source-1. *Soil Sci. Soc. Amer. Proc., 35,* 683–689.
2. Bresler, E., Heller, J., Diner, N., Ben-Asher, J. (1971). Infiltration from a trickle source-II. *Soil Sci. Soc. Amer. Proc.,* 35.
3. Goel, A. K., Kumar, R. (1993). Effect of drip discharge rate on soil moisture distribution pattern. *J. Water Management, 1(1),* 50–51.
4. Kaul, R. K., Michael, A. M. (1986). Moisture front advance under a point source of water application. *J. Agric. Eng.* XIIX(2).
5. Keller, J., Karmeli, D. (1974). *Trickle Irrigation Design.* page 135. Rain Bird Sprinkler Manufacturing Corporation, CA.
6. Patel, N., Rajput, T. B. S. (2001). Minimization of cost of drip system for field crops. In: Singh, H., P, Kaushish, S. P., Kumar, A., Murthy, T. S., Samuel, J. C. (eds), *Micro Irrigation.* pp. 569–573.
7. Sharma, K. N., Paul, J. C., Nayak, S. C., Mohanty, D. (1997). Water movement in soil from drip source. *Journal of Soil and Water Conservation, 4,* 134–142.
8. Sivanappan, R. K. (1987). Status and prospects of drip irrigation in India. *Proceedings of National Seminar on Use of Plastics in Agriculture,* held at New Delhi, during February, 13–28.

CHAPTER 23

ECONOMICS OF DRIP IRRIGATED CAULIFLOWER-CHILI SEQUENCE

KAMAL G. SINGH, G. MAHAJAN, and MUKESH SIAG

23.1 INTRODUCTION

Viewed from the perspective of water stress due to lowering of ground water table, the purpose of irrigation is to keep water status at a level that maximize yield within the constraints of irrigation supply and weather. In this connection, drip irrigation is a relatively new technology of irrigation especially in water scarcity areas. The system has proved its superiority over other conventional methods of irrigation, especially in vegetable crops owing to precise application of water in the root zone. The major drawback of the drip irrigation system is its high initial investment; however, cost can be recovered in a short span if proper water management and design principles are followed. Among the various components of the drip irrigation system, the cost of the lateral is a major factor, which influences the total cost of installation. Any effort made to reduce the length of lateral per unit area will result in reduction of system cost. Many scientists reported that drip irrigation in chill-cauliflower is very economical by reducing the cost and water use by 50%, when these crops were planted in paired row patterns [1, 2]. Further, it has been observed that it is not a viable option to use the drip irrigation system only once a year for taking a single crop. Therefore, the present study was conducted to study the yield response and economic viability of drip irrigation system in cauliflower-hybrid chili sequence.

23.2 MATERIALS AND METHODS

23.2.1 EXPERIMENTAL SETUP

The field experiment was conducted at the research farm of Department of Soil and Water Engineering, PAU, Ludhiana for cauliflower-hybrid chili sequence. Cauliflower (cultivar PG 26) and hybrid chili (cultivar, CH-1) were used in this study. The

---

*In this chapter, the currency is expressed in Indian Rupees (1.00 US$ = Rs. 60.93; 1.00 Rs. = 0.02 US$).

experimental set up included 5 treatments comprising of: three levels of drip irrigation (1.0 Epan, 0.75 Epan and 0.5 Epan) and two check basin methods of irrigation
when the crops were sown in either paired row or single row in randomized block
design with three replications. In normal sowing method, both the crops were sown
in sequence at a spacing of 45 cm between row-to-row and 30 cm between plant-to-
plant. However, in paired sowing, the row-to-row spacing between paired rows was
30 cm and row spacing between pair rows was 60 cm but plant-to-plant spacing was
kept same as 30 cm. Therefore, in paired sowing total as well as number of rows and
plants were same. As there was one lateral for two rows of each pair, the number of
lateral, cost and water needs were reduced to 50%.

In check-basin method (surface flooding), the irrigations were provided on the
basis of 1.0 cumulative pan evaporation (Epan). In both the years, cauliflower cultivar PG 26 was sown in the first week of October. Whereas, hybrid chili (CH-1) was
sown in the first week of March in both the years. Both the crops were raised with
recommendations in package of practices of PAU. The recommended fertilizers in
cauliflower were 125 **kg** N and 62.5 **kg** P_2O_5/ha. The recommended fertilizers in
hybrid chili were 75 **kg** N and 25 **kg** P_2O_5/ha. All phosphorus dose was basal applied
(before transplanting of each crop) in all the treatments. The drip system consisted
of polyethylene laterals of 12 mm in diameter that were laid parallel (each lateral
served 2 rows of crop). The laterals were provided with online emitters of 3 L/hour
capacity at 0.3 m apart.

23.2.2   BENEFIT COST ANALYSIS

To calculate the net return from different crop sequence, the cost of different inputs
included: expenses incurred on preparation of field, plowing, seed, sowing, cost of
fertilizers, manure and their application, weeding, crop protection measure and cost
of irrigation water, harvesting and selling prices; and these were based on the data
from the Department of Agricultural Economics, PAU, Ludhiana. Market prices
were taken to compute cost of drip irrigation system. Cost of laterals varied from
crop to crop depending upon row-to-row spacing.

The subsidy of Rs. 25,000 per hectare was deducted from the cost of irrigation
system. An additional cost of operation and maintenance of drip irrigation system @
Rs. 500 per month was added to the cost of cultivation for drip system. The annual
cost of drip irrigation system was divided equally between the two crops. The seasonal cost of drip irrigation included: depreciation, prevailing bank rate of interest
(8%/year/crop). The useful life was 20 years for main line, sub main, fertilizer tank,
valve, filter and pumping unit; and while for inbuilt drippers, it was considered 10
years. Economic analysis was done for two different situations, that is, for water
hunger area and for unlimited water supply. Under unlimited water supply situation,
comparisons were made for the same area (1 ha) under drip and conventional system
of irrigation. The maximum yield obtained under drip and conventional method of

irrigation was taken to find out the net returns per hectare and then the net income under both systems of irrigation for different crop sequences was compared and analyzed. For water hunger area, it was assumed that the water saved per hectare using drip irrigation system for cauliflower-hybrid chili sequence can be used to bring additional area under irrigation using the same sequence; and the additional income generated was added to calculate the net return under drip irrigation system. This income was compared with the per hectare income generated from conventional surface irrigation method. Whereas maximum yield of different crops under conventional irrigation was taken; and in case of drip irrigation the yield from best treatment was considered. The best treatment was the one that gave maximum water saving and an additional area can be brought under the crops with the water saving.

23.3 RESULTS AND DISCUSSION

23.3.1 EFFECTS OF IRRIGATION LEVELS ON CAULIFLOWER-CHILI SEQUENCE

The Table 23.1 indicates that drip irrigation at 0.5 Epan caused significantly higher cauliflower (198.2 100 kg) and hybrid chili (219.1 100 kg) yield as compared to check basin method of irrigation when both the crops were sown in either paired rows or in normal sowing method. Normal sown crop under check basin method of irrigation produced 26.1% and 25.8% lesser cauliflower and hybrid chili yield, respectively than the paired sown crop, although the same quantity of water was applied. In cauliflower, drip irrigation at all the levels of irrigation (0.50 Epan, 0.75 Epan and 1.0 Epan) proved significantly superior over check basin method of irrigation, when the crop was sown either normally or in paired rows.

TABLE 23.1 Effects of Different Treatments on Water Use Efficiency, Cauliflower and Hybrid Chili Yield in Cauliflower-Hybrid Chili Sequence

Treatments	Cauliflower yield (100 kg/ha)				Hybrid Chili yield (100 kg/ha)			
	Year I	Year II	Mean	Mean WUE 100 kg/ (ha-cm)	Year I	Year II	Mean	Mean WUE 100 kg/ (ha-cm)
D 0.5 Epan PR	235.0	161.48	198.2	15.7	176.6	261.5	219.1	6.89
	(13.32)	(11.90)	(12.6)		(29.3)	(34.4)	(31.8)	
D 0.75 Epan PR	235.0	154.44	194.2	13.0	159.2	243.0	201.1	4.71
	(14.98)	(14.80)	(14.9)		(38.9)	(46.6)	(42.7)	

TABLE 23.1   *(Continued)*

D 1.0 Epan PR	230.3	145.40	187.8	10.9	157.4	219.8	188.5	3.51
	(16.64)	(17.80)	(17.2)		(48.6)	(58.9)	(53.7)	
C.B. 1.0 Epan PR	170.7	86.74	128.7	6.27	143.6	195.5	169.6	1.75
	(20.0)	(21.0)	(20.5)		(94.0)	(100.0)	(97.0)	
C.B. 1.0 Epan NS	102.3	88.0	95.1	4.63	141.7	110.0	125.9	1.30
	(20.0)	(21.0)	(20.5)		(94.0)	(100.0)	(97.0)	
LSD (0.05)	**40.4**	**30.6**	-	-	**14.2**	**36.2**	-	-

D = Drip irrigation, C.B. = Check basin, PR = Paired rows, NS = Normal sowing.

Values in parentheses indicate irrigation water applied to crop in cm

Drip irrigated cauliflower crop resulted in statistically same yield at all the levels of irrigation with maximum water use efficiency (15.7 per 100 kg/ha-cm) at 0.5 Epan. In hybrid chili crop, drip irrigated crop at 0.5 and 0.75 Epan gave significantly more yield over check basin method of irrigation, when the crop was sown either normally or in paired rows. The highest chili yield obtained, when the crop was drip irrigated at 0.5 Epan with maximum water use efficiency (6.89 per 100 kg/ha-cm). It was observed that drip irrigated chili crop resulted in statistically same yield at all the levels of irrigation (0.50 Epan, 0.75 Epan and 1.0 Epan). Further, the study indicates that as the water supply increased through drip irrigation in both the crops, the water use efficiency was decreased.

23.3.2   BENEFIT COST ANALYSIS

Table 23.2 presents the economic analysis of drip versus check basin method of irrigation for cauliflower-hybrid chili sequence. For computing economic benefits in both the drip irrigated crops, the treatment which gave maximum yield and water saving were selected. As in both the crops, drip irrigation at 0.5 Epan caused highest yield with maximum water use efficiency. Therefore, yields under these treatments were selected for computing the economics in both the crops. It was observed that the net seasonal income was higher in drip-irrigated crops. In cauliflower, due to lesser yield in check basin method, the net return was negative, while it was profitable in drip-irrigated crop and the net profit was Rs. 12,734/ha. In hybrid chili, net profit was observed in both the methods of irrigation, but further it was more in drip-irrigated crop (Rs. 25,031/ha) than check basin method of irrigation (Rs. 20,695/ha). Further analysis showed that cauliflower-hybrid chili sequence gave net return of Rs. 37,765/ha in drip irrigation method as compared to Rs. 18,618/ha in check basin method of irrigation.

The net return was also computed in view of water hunger area. In water hunger area, it was assumed that due to limited water supply, 1.65 ha more area (Table 23.3) can be irrigated due to drip irrigation under the cauliflower-hybrid chili sequence, which otherwise remain barren (un cultivated). The additional income from the more area (1.65 ha) was added to calculate the net return. It was seen in Table 3 that in water hunger area, the net return boosted to Rs. 100,077 from Rs. 37,765 in drip irrigation.

TABLE 23.2 Economic Analysis of Drip Versus Check Basin Method of Irrigation For Cauliflower-Hybrid Chili Sequence

Parameters	Drip	Check basin
Cauliflower		
1. Seasonal total cost (Rs./ha)	66,546	53,557
2. Yield of produce (100 kg/ha)	198.2	128.7
3. Selling price (Rs./100 kg)	400	400
4. Income from produce (2x3) (Rs./ha)	79,280	51,480
5. Net seasonal income (4–1) (Rs./ha)	12,734	–2,077
Parameters	Drip	Check basin
Hybrid chili		
1. Seasonal total cost (Rs./ha)	51,654	38,665
2. Yield of produce (q/ha)	219.1	169.6
3. Selling price (Rs./q)	350	350
4. Income from produce (2x3) (Rs./ha)	76,685	59,360
5. Net seasonal income (4–1) (Rs./ha)	25,031	20,695
Cauliflower-hybrid chili		
1. Net income/ha (Rs.)	37,765	18,618
2. Additional area irrigated (ha)	1.65	-
3. Additional income	62,312	-
4. Total Net income	**100,077**	**18,618**

TABLE 23.3   Total Water Requirement for the Cauliflower-Chili Sequence in Different Irrigation Methods

a. Water requirement of cauliflower-hybrid chili sequence in drip irrigation method (m³/ha)	1260 + 3180 = 4440
b. Water requirement of cauliflower-hybrid chili sequence in check basin method of irrigation (m³/ha)	2050 + 9700 = 11,750
c. Saving of water due to drip irrigation over the check basin method of irrigation (m³/ha) = b – c	7310
d. Additional area that can be irrigated with drip irrigation (ha)	1.65

It can be concluded that drip irrigation is very useful method of irrigation in cauliflower-hybrid chili sequence in terms of higher yield and water saving. In water hunger area, its use can bring a revolution in vegetable sequence and can improve the economic conditions of the farmers of that area.

23.4   SUMMARY

A field experiment was conducted at the research farm of Department of Soil and Water Engineering, PAU, Ludhiana to study the response of cauliflower-hybrid chili sequence to drip irrigation and also to study the economic feasibility of drip irrigation system. The results revealed that in hybrid chili crop, drip irrigation at lowest level of irrigation (0.5 Epan) caused highest yield (219 per 100 kg/ha) with highest water use efficiency and proved significantly better than all other levels of drip irrigation including check basin method of irrigation. In cauliflower crop, the yield was statistically same at all the levels of drip irrigation, but significantly more than check basin method of irrigation when the crop was sown normal or in paired row pattern.

Under unlimited water supply, the economic analysis indicated that drip irrigation in cauliflower-hybrid chili sequence gave a net return of Rs. 37,565/ha as compared to Rs. 18,618/ha in check basin method of irrigation. However, under limited water supply or in water scarcity areas, drip irrigation boosted the net return to the tune of Rs. 100,077 due to increase in yield resulted from additional area covered under irrigation.

KEYWORDS

- benefit cost ratio
- cauliflower
- check basin irrigation
- chili
- crop sequence
- drip irrigation
- economics
- Epan
- limited water supply
- liquid fertilizer
- nitrogen
- okra
- paired row pattern
- water requirement
- water saving
- water scarcity

REFERENCES

1. Singh, V. Y., Joshi, N. L., Singh, D. V., Saxena, A. K. (1999). Response of chili to water and nitrogen under drip and check basin method of irrigation. *Ann. Arid Zone*, *38*, 9–13.
2. Tumbare, A. D., Shinde B. N., Bhoite, S. V. (1999). Effects of liquid fertilizers through drip irrigation on growth and yield of okra. *Indian J. Agron.*, *44*, 176–178.

APPENDICES

(Modified and reprinted with permission from: Megh R. Goyal, 2012. Appendices. Pages 317–332. In: *Management of Drip/Trickle or Micro Irrigation* edited by Megh R. Goyal. New Jersey, USA: Apple Academic Press Inc.)

APPENDIX A

CONVERSION SI AND NON-SI UNITS

To convert the Column 1 in the Column 2	Column 1	Column 2	To convert the Column 2 in the Column 1
	Unit	Unit	
Multiply by	SI	Non-SI	Multiply by

LINEAR

0.621 —— kilometer, km (10^3 m)	miles, mi ———————	1.609
1.094 —— meter, m	yard, yd ———————	0.914
3.28 —— meter, m	feet, ft ———————	0.304
3.94×10^{-2} — millimeter, mm (10^{-3})	inch, in ———————	25.4

SQUARES

2.47 ——hectare, he	acre ———————	0.405
2.47 —— square kilometer, km²	acre ———————	4.05×10^{-3}
0.386 —— square kilometer, km²	square mile, mi² ———	2.590
2.47×10^{-4} — square meter, m²	acre ———————	4.05×10^{-3}
10.76 —— square meter, m²	square feet, ft² ———	9.29×10^{-2}
1.55×10^{-3} — mm²	square inch, in² ———	645

CUBICS

9.73×10^{-3} — cubic meter, m³	inch-acre ———————	102.8
35.3 —— cubic meter, m³	cubic-feet, ft³ ———	2.83×10^{-2}
6.10×10^4 — cubic meter, m³	cubic inch, in³ ———	1.64×10^{-5}

2.84 × 10⁻² —liter, L (10⁻³ m³)   bushel, bu ——————— 35.24
1.057 ——— liter, L   liquid quarts, qt ———— 0.946
3.53 × 10⁻² —liter, L   cubic feet, ft³ ————— 28.3
0.265 ——— liter, L   gallon —————— 3.78
33.78 ——— liter, L   fluid ounce, oz ———— 2.96 × 10⁻²
2.11 ——liter, L   fluid dot, dt ———— 0.473

2.84×10^{-2} —liter, L (10^{-3} m³)	bushel, bu ———	35.24
1.057 ——— liter, L	liquid quarts, qt ———	0.946
3.53×10^{-2} —liter, L	cubic feet, ft³ ———	28.3
0.265 ——— liter, L	gallon ———	3.78
33.78 ——— liter, L	fluid ounce, oz ———	2.96×10^{-2}
2.11 ——liter, L	fluid dot, dt ———	0.473

WEIGHT

2.20×10^{-3} —gram, g (10^{-3} kg)	pound, ——— 454	
3.52×10^{-2} —gram, g (10^{-3} kg)	ounce, oz ———	28.4
2.205 ——— kilogram, kg	pound, lb ——— 0.454	
10^{-2} ——— kilogram, kg	quintal (metric), q ———	100
1.10×10^{-3} — kilogram, kg	ton (2000 lbs), ton ———	907
1.102 ——— mega gram, mg	ton (US), ton ———	0.907
1.102 ——— metric ton, t	ton (US), ton ———	0.907

YIELD AND RATE

0.893 ——kilogram per hectare	pound per acre ———	1.12
7.77×10^{-2} — kilogram per cubic meter	pound per fanega ———	12.87
1.49×10^{-2} — kilogram per hectare	pound per acre, 60 lb ——	67.19
1.59×10^{-2} — kilogram per hectare	pound per acre, 56 lb ——	62.71
1.86×10^{-2} — kilogram per hectare	pound per acre, 48 lb ——	53.75
0.107 ——liter per hectare	galloon per acre ———	9.35
893 ———ton per hectare	pound per acre ———	1.12×10^{-3}
893 ———mega gram per hectare	pound per acre ———	1.12×10^{-3}
0.446——ton per hectare	ton (2000 lb) per acre ——	2.24
2.24 ———meter per second	mile per hour ———	0.447

SPECIFIC SURFACE

10 ———	square meter per kilogram	square centimeter per gram ———	0.1
10^{3} ———	square meter per kilogram	square millimeter per gram ———	10^{-3}

PRESSURE

9.90 ———	megapascal, MPa	atmosphere ———	0.101
10 ———	megapascal	bar ———	0.1

1.0 ————	megagram per cubic meter	gram per cubic centimeter ————	1.00
2.09×10^{-2} —	pascal, Pa	pound per square feet ——	47.9
1.45×10^{-4} —	pascal, Pa	pound per square inch ——	6.90×10^3

To convert the Column 1 in the Column 2	Column 1	Column 2	To convert the Column 2 in the Column 1
	Unit	*Unit*	
Multiply by	SI	Non-SI	Multiply by

TEMPERATURE

1.00 (K-273)—	Kelvin, K	centigrade, °C ————	1.00 (C+273)
(1.8 C + 32)—	centigrade, °C	Fahrenheit, °F ————	(F–32)/1.8

ENERGY

9.52×10^{-4} —	Joule J	BTU ————————	1.05×10^3
0.239 ————	Joule, J	calories, cal ————	4.19
0.735 ————	Joule, J	feet-pound ————	1.36
2.387×10^5 —	Joule per square meter	calories per square centimeter —	4.19×10^4
10^5 ————	Newton, N	dynes ————————	10^{-5}

WATER REQUIREMENTS

9.73×10^{-3} —	cubic meter	inch acre ————————	102.8
9.81×10^{-3} —	cubic meter per hour	cubic feet per second ——	101.9
4.40 ————	cubic meter per hour	galloon (US) per minute —	0.227
8.11 ————	hectare-meter	acre-feet ————————	0.123
97.28 ——	hectare-meter	acre-inch ————————	1.03×10^{-2}
8.1×10^{-2} —	hectare centimeter	acre-feet ————————	12.33

CONCENTRATION

1 ————	centimol per kilogram	milliequivalents per 100 grams ————	1
0.1 ————	gram per kilogram	percents ————	10
1 ————	milligram per kilogram	parts per million ————	1

NUTRIENTS FOR PLANTS

2.29 —— P	P_2O_5 ————	0.437
1.20 —— K	K_2O ————	0.830
1.39 —— Ca	CaO ————	0.715
1.66 —— Mg	MgO ————	0.602

NUTRIENT EQUIVALENTS

Column A	Column B	Conversion A to B	Equivalent B to A
N	NH_3	1.216	0.822
	NO_3	4.429	0.226
	KNO_3	7.221	0.1385
	$Ca(NO_3)_2$	5.861	0.171
	$(NH_4)_2SO_4$	4.721	0.212
	NH_4NO_3	5.718	0.175
	$(NH_4)_2HPO_4$	4.718	0.212
P	P_2O_5	2.292	0.436
	PO_4	3.066	0.326
	KH_2PO_4	4.394	0.228
	$(NH_4)_2HPO_4$	4.255	0.235
	H_3PO_4	3.164	0.316
K	K_2O	1.205	0.83
	KNO_3	2.586	0.387
	KH_2PO_4	3.481	0.287
	KCl	1.907	0.524
	K_2SO_4	2.229	0.449
Ca	CaO	1.399	0.715
	$Ca(NO_3)_2$	4.094	0.244
	$CaCl_2 \times 6H_2O$	5.467	0.183
	$CaSO_4 \times 2H_2O$	4.296	0.233
Mg	MgO	1.658	0.603
	$MgSO_4 \times 7H_2O$	1.014	0.0986

S	H_2SO_4	3.059	0.327
	$(NH_4)_2SO_4$	4.124	0.2425
	K_2SO_4	5.437	0.184
	$MgSO_4 \times 7H_2O$	7.689	0.13
	$CaSO_4 \times 2H_2O$	5.371	0.186

APPENDIX B

PIPE AND CONDUIT FLOW

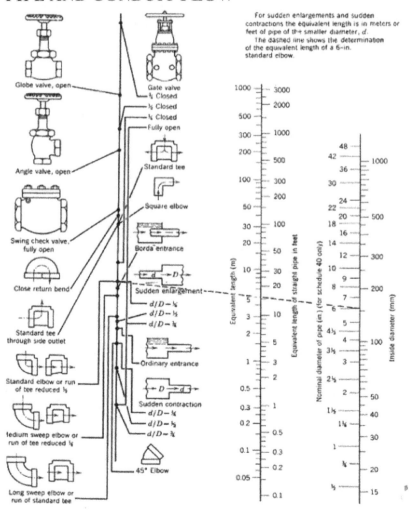

For sudden enlargements and sudden contractions the equivalent length is in meters or feet of pipe of the smaller diameter, d. The dashed line shows the determination of the equivalent length of a 6-in. standard elbow.

APPENDIX C

PERCENTAGE OF DAILY SUNSHINE HOURS: FOR NORTH AND SOUTH HEMISPHERES

Latitude	Jan	Feb	Mar	Apr	May	Jun	Jul	Aug	Sep	Oct	Nov	Dec
					NORTH							
0	8.50	7.66	8.49	8.21	8.50	8.22	8.50	8.49	8.21	8.50	8.22	8.50
5	8.32	7.57	8.47	3.29	8.65	8.41	8.67	8.60	8.23	8.42	8.07	8.30
10	8.13	7.47	8.45	8.37	8.81	8.60	8.86	8.71	8.25	8.34	7.91	8.10
15	7.94	7.36	8.43	8.44	8.98	8.80	9.05	8.83	8.28	8.20	7.75	7.88
20	7.74	7.25	8.41	8.52	9.15	9.00	9.25	8.96	8.30	8.18	7.58	7.66
25	7.53	7.14	8.39	8.61	9.33	9.23	9.45	9.09	8.32	8.09	7.40	7.52
30	7.30	7.03	8.38	8.71	9.53	9.49	9.67	9.22	8.33	7.99	7.19	7.15
32	7.20	6.97	8.37	8.76	9.62	9.59	9.77	9.27	8.34	7.95	7.11	7.05
34	7.10	6.91	8.36	8.80	9.72	9.70	9.88	9.33	8.36	7.90	7.02	6.92
36	6.99	6.85	8.35	8.85	9.82	9.82	9.99	9.40	8.37	7.85	6.92	6.79
38	6.87	6.79	8.34	8.90	9.92	9.95	10.1	9.47	3.38	7.80	6.82	6.66
40	6.76	6.72	8.33	8.95	10.0	10.1	10.2	9.54	8.39	7.75	6.72	7.52
42	6.63	6.65	8.31	9.00	10.1	10.2	10.4	9.62	8.40	7.69	6.62	6.37
44	6.49	6.58	8.30	9.06	10.3	10.4	10.5	9.70	8.41	7.63	6.49	6.21
46	6.34	6.50	8.29	9.12	10.4	10.5	10.6	9.79	8.42	7.57	6.36	6.04
48	6.17	6.41	8.27	9.18	10.5	10.7	10.8	9.89	8.44	7.51	6.23	5.86
50	5.98	6.30	8.24	9.24	10.7	10.9	11.0	10.0	8.35	7.45	6.10	5.64
52	5.77	6.19	8.21	9.29	10.9	11.1	11.2	10.1	8.49	7.39	5.93	5.43
54	5.55	6.08	8.18	9.36	11.0	11.4	11.4	10.3	8.51	7.20	5.74	5.18
56	5.30	5.95	8.15	9.45	11.2	11.7	11.6	10.4	8.53	7.21	5.54	4.89
58	5.01	5.81	8.12	9.55	11.5	12.0	12.0	10.6	8.55	7.10	4.31	4.56
60	4.67	5.65	8.08	9.65	11.7	12.4	12.3	10.7	8.57	6.98	5.04	4.22
					SOUTH							
0	8.50	7.66	8.49	8.21	8.50	8.22	8.50	8.49	8.21	8.50	8.22	8.50
5	8.68	7.76	8.51	8.15	8.34	8.05	8.33	8.38	8.19	8.56	8.37	8.68
10	8.86	7.87	8.53	8.09	8.18	7.86	8.14	8.27	8.17	8.62	8.53	8.88
15	9.05	7.98	8.55	8.02	8.02	7.65	7.95	8.15	8.15	8.68	8.70	9.10
20	9.24	8.09	8.57	7.94	7.85	7.43	7.76	8.03	8.13	8.76	8.87	9.33
25	9.46	8.21	8.60	7.74	7.66	7.20	7.54	7.90	8.11	8.86	9.04	9.58
30	9.70	8.33	8.62	7.73	7.45	6.96	7.31	7.76	8.07	8.97	9.24	9.85

32	9.81	8.39	8.63	7.69	7.36	6.85	7.21	7.70	8.06	9.01	9.33	9.96
34	9.92	8.45	8.64	7.64	7.27	6.74	7.10	7.63	8.05	9.06	9.42	10.1
36	10.0	8.51	8.65	7.59	7.18	6.62	6.99	7.56	8.04	9.11	9.35	10.2
38	10.2	8.57	8.66	7.54	7.08	6.50	6.87	7.49	8.03	9.16	9.61	10.3
40	10.3	8.63	8.67	7.49	6.97	6.37	6.76	7.41	8.02	9.21	9.71	10.5
42	10.4	8.70	8.68	7.44	6.85	6.23	6.64	7.33	8.01	9.26	9.8	10.6
44	10.5	8.78	8.69	7.38	6.73	6.08	6.51	7.25	7.99	9.31	9.94	10.8
46	10.7	8.86	8.90	7.32	6.61	5.92	6.37	7.16	7.96	9.37	10.1	11.0

APPENDIX D

3PSYCHROMETRIC CONSTANT (Γ) FOR DIFFERENT ALTITUDES (Z)

$\gamma = 10^{-3} [(C_p.P) \div (\varepsilon.\lambda)] = (0.00163) \times [P \div \lambda]$

γ, psychometric constant [kPa C^{-1}]
c_p, specific heat of moist air = 1.013 [kJ kg^{-1} $^{\circ}$C^{-1}]
P, atmospheric pressure [kPa].

ε, ratio molecular weight of water vapor/dry air = 0.622
λ, latent heat of vaporization [MJ kg^{-1}]
= 2.45 MJ kg^{-1} at 20 $^{\circ}$C.

Z (m)	γ kPa/°C	z (m)	γ kPa/°C	z (m)	γ kPa/°C	z (m)	γ kPa/°C
0	0.067	1000	0.060	2000	0.053	3000	0.047
100	0.067	1100	0.059	2100	0.052	3100	0.046
200	0.066	1200	0.058	2200	0.052	3200	0.046
300	0.065	1300	0.058	2300	0.051	3300	0.045
400	0.064	1400	0.057	2400	0.051	3400	0.045
500	0.064	1500	0.056	2500	0.050	3500	0.044
600	0.063	1600	0.056	2600	0.049	3600	0.043
700	0.062	1700	0.055	2700	0.049	3700	0.043
800	0.061	1800	0.054	2800	0.048	3800	0.042
900	0.061	1900	0.054	2900	0.047	3900	0.042
1000	0.060	2000	0.053	3000	0.047	4000	0.041

APPENDIX E

SATURATION VAPOR PRESSURE [E$_s$] FOR DIFFERENT TEMPERATURES (T)

Vapor pressure function = e_s = [0.6108]*exp{[17.27*T]/[T + 237.3]}							
T °C	e_s kPa	T °C	e_s kPa	T °C	e_s kPa	T °C	e_s kPa
1.0	0.657	13.0	1.498	25.0	3.168	37.0	6.275
1.5	0.681	13.5	1.547	25.5	3.263	37.5	6.448
2.0	0.706	14.0	1.599	26.0	3.361	38.0	6.625
2.5	0.731	14.5	1.651	26.5	3.462	38.5	6.806
3.0	0.758	15.0	1.705	27.0	3.565	39.0	6.991
3.5	0.785	15.5	1.761	27.5	3.671	39.5	7.181
4.0	0.813	16.0	1.818	28.0	3.780	40.0	7.376
4.5	0.842	16.5	1.877	28.5	3.891	40.5	7.574
5.0	0.872	17.0	1.938	29.0	4.006	41.0	7.778
5.5	0.903	17.5	2.000	29.5	4.123	41.5	7.986
6.0	0.935	18.0	2.064	30.0	4.243	42.0	8.199
6.5	0.968	18.5	2.130	30.5	4.366	42.5	8.417
7.0	1.002	19.0	2.197	31.0	4.493	43.0	8.640
7.5	1.037	19.5	2.267	31.5	4.622	43.5	8.867
8.0	1.073	20.0	2.338	32.0	4.755	44.0	9.101
8.5	1.110	20.5	2.412	32.5	4.891	44.5	9.339
9.0	1.148	21.0	2.487	33.0	5.030	45.0	9.582
9.5	1.187	21.5	2.564	33.5	5.173	45.5	9.832
10.0	1.228	22.0	2.644	34.0	5.319	46.0	10.086
10.5	1.270	22.5	2.726	34.5	5.469	46.5	10.347
11.0	1.313	23.0	2.809	35.0	5.623	47.0	10.613
11.5	1.357	23.5	2.896	35.5	5.780	47.5	10.885
12.0	1.403	24.0	2.984	36.0	5.941	48.0	11.163
12.5	1.449	24.5	3.075	36.5	6.106	48.5	11.447

APPENDIX F

SLOPE OF VAPOR PRESSURE CURVE (Δ) FOR DIFFERENT TEMPERATURES (T)

$$\Delta = [4098. \, e°(T)] \div [T + 237.3]^2$$
$$= 2504\{\exp[(17.27T) \div (T + 237.2)]\} \div [T + 237.3]^2$$

T °C	Δ kPa/°C	T °C	Δ kPa/°C	T °C	Δ kPa/°C	T °C	Δ kPa/°C
1.0	0.047	13.0	0.098	25.0	0.189	37.0	0.342
1.5	0.049	13.5	0.101	25.5	0.194	37.5	0.350
2.0	0.050	14.0	0.104	26.0	0.199	38.0	0.358
2.5	0.052	14.5	0.107	26.5	0.204	38.5	0.367
3.0	0.054	15.0	0.110	27.0	0.209	39.0	0.375
3.5	0.055	15.5	0.113	27.5	0.215	39.5	0.384
4.0	0.057	16.0	0.116	28.0	0.220	40.0	0.393
4.5	0.059	16.5	0.119	28.5	0.226	40.5	0.402
5.0	0.061	17.0	0.123	29.0	0.231	41.0	0.412
5.5	0.063	17.5	0.126	29.5	0.237	41.5	0.421
6.0	0.065	18.0	0.130	30.0	0.243	42.0	0.431
6.5	0.067	18.5	0.133	30.5	0.249	42.5	0.441
7.0	0.069	19.0	0.137	31.0	0.256	43.0	0.451
7.5	0.071	19.5	0.141	31.5	0.262	43.5	0.461
8.0	0.073	20.0	0.145	32.0	0.269	44.0	0.471
8.5	0.075	20.5	0.149	32.5	0.275	44.5	0.482
9.0	0.078	21.0	0.153	33.0	0.282	45.0	0.493
9.5	0.080	21.5	0.157	33.5	0.289	45.5	0.504
10.0	0.082	22.0	0.161	34.0	0.296	46.0	0.515
10.5	0.085	22.5	0.165	34.5	0.303	46.5	0.526
11.0	0.087	23.0	0.170	35.0	0.311	47.0	0.538
11.5	0.090	23.5	0.174	35.5	0.318	47.5	0.550
12.0	0.092	24.0	0.179	36.0	0.326	48.0	0.562
12.5	0.095	24.5	0.184	36.5	0.334	48.5	0.574

APPENDIX G

NUMBER OF THE DAY IN THE YEAR (JULIAN DAY)

Day	Jan	Feb	Mar	Apr	May	Jun	Jul	Aug	Sep	Oct	Nov	Dec
1	1	32	60	91	121	152	182	213	244	274	305	335
2	2	33	61	92	122	153	183	214	245	275	306	336
3	3	34	62	93	123	154	184	215	246	276	307	337
4	4	35	63	94	124	155	185	216	247	277	308	338
5	5	36	64	95	125	156	186	217	248	278	309	339
6	6	37	65	96	126	157	187	218	249	279	310	340
7	7	38	66	97	127	158	188	219	250	280	311	341
8	8	39	67	98	128	159	189	220	251	281	312	342
9	9	40	68	99	129	160	190	221	252	282	313	343
10	10	41	69	100	130	161	191	222	253	283	314	344
11	11	42	70	101	131	162	192	223	254	284	315	345
12	12	43	71	102	132	163	193	224	255	285	316	346
13	13	44	72	103	133	164	194	225	256	286	317	347
14	14	45	73	104	134	165	195	226	257	287	318	348
15	15	46	74	105	135	166	196	227	258	288	319	349
16	16	47	75	106	136	167	197	228	259	289	320	350
17	17	48	76	107	137	168	198	229	260	290	321	351
18	18	49	77	108	138	169	199	230	261	291	322	352
19	19	50	78	109	139	170	200	231	262	292	323	353
20	20	51	79	110	140	171	201	232	263	293	324	354
21	21	52	80	111	141	172	202	233	264	294	325	355
22	22	53	81	112	142	173	203	234	265	295	326	356
23	23	54	82	113	143	174	204	235	266	296	327	357
24	24	55	83	114	144	175	205	236	267	297	328	358
25	25	56	84	115	145	176	206	237	268	298	329	359
26	26	57	85	116	146	177	207	238	269	299	330	360
27	27	58	86	117	147	178	208	239	270	300	331	361
28	28	59	87	118	148	179	209	240	271	301	332	362
29	29	(60)	88	119	149	180	210	241	272	302	333	363
30	30	—	89	120	150	181	211	242	273	303	334	364
31	31	—	90	—	151	—	212	243	—	304	—	365

APPENDIX H

STEFAN-BOLTZMANN LAW AT DIFFERENT TEMPERATURES (T):

$[\sigma*(T_K)^4] = [4.903 \times 10^{-9}]$, MJ K^{-4} m^{-2} day^{-1}
where: $T_K = \{T[°C] + 273.16\}$

T	$\sigma*(TK)4$	T	$\sigma*(TK)4$	T	$\sigma*(TK)4$
Units					
°C	MJ m–2 d–1	°C	MJ m–2 d–1	°C	MJ m–2 d–1
1.0	27.70	17.0	34.75	33.0	43.08
1.5	27.90	17.5	34.99	33.5	43.36
2.0	28.11	18.0	35.24	34.0	43.64
2.5	28.31	18.5	35.48	34.5	43.93
3.0	28.52	19.0	35.72	35.0	44.21
3.5	28.72	19.5	35.97	35.5	44.50
4.0	28.93	20.0	36.21	36.0	44.79
4.5	29.14	20.5	36.46	36.5	45.08
5.0	29.35	21.0	36.71	37.0	45.37
5.5	29.56	21.5	36.96	37.5	45.67
6.0	29.78	22.0	37.21	38.0	45.96
6.5	29.99	22.5	37.47	38.5	46.26
7.0	30.21	23.0	37.72	39.0	46.56
7.5	30.42	23.5	37.98	39.5	46.85
8.0	30.64	24.0	38.23	40.0	47.15
8.5	30.86	24.5	38.49	40.5	47.46
9.0	31.08	25.0	38.75	41.0	47.76
9.5	31.30	25.5	39.01	41.5	48.06
10.0	31.52	26.0	39.27	42.0	48.37
10.5	31.74	26.5	39.53	42.5	48.68
11.0	31.97	27.0	39.80	43.0	48.99
11.5	32.19	27.5	40.06	43.5	49.30
12.0	32.42	28.0	40.33	44.0	49.61
12.5	32.65	28.5	40.60	44.5	49.92
13.0	32.88	29.0	40.87	45.0	50.24
13.5	33.11	29.5	41.14	45.5	50.56
14.0	33.34	30.0	41.41	46.0	50.87

14.5	33.57	30.5	41.69	46.5	51.19
15.0	33.81	31.0	41.96	47.0	51.51
15.5	34.04	31.5	42.24	47.5	51.84
16.0	34.28	32.0	42.52	48.0	52.16
16.5	34,52	32.5	42.80	48.5	52.49

APPENDIX I

THERMODYNAMIC PROPERTIES OF AIR AND WATER

1. Latent Heat of Vaporization (λ)

$$\lambda = [2.501 - (2.361 \times 10^{-3})\,T]$$

where: λ = latent heat of vaporization [MJ kg^{-1}]; and T = air temperature [°C].

The value of the latent heat varies only slightly over normal temperature ranges. A single value may be taken (for ambient temperature = 20 °C): λ = 2.45 MJ kg^{-1}.

2. Atmospheric Pressure (P)

$$P = P_o \left[\{T_{Ko} - \alpha(Z - Z_o)\} \div \{T_{Ko}\} \right]^{(g/(\alpha R))}$$

where: P, atmospheric pressure at elevation z [kPa]

P_o, atmospheric pressure at sea level = 101.3 [kPa]

z, elevation [m]

z_o, elevation at reference level [m]

g, gravitational acceleration = 9.807 [m s^{-2}]

R, specific gas constant == 287 [J kg^{-1} K^{-1}]

α, constant lapse rate for moist air = 0.0065 [K m^{-1}]

T_{Ko}, reference temperature [K] at elevation z_o = 273.16 + T

T, means air temperature for the time period of calculation [°C]

When assuming P_o = 101.3 [kPa] at z_o = 0, and T_{Ko} = 293 [K] for T = 20 [°C], above equation reduces to:

$$P = 101.3[(293 - 0.0065Z)\,(293)]^{5.26}$$

3. Atmospheric Density (ρ)

$$\rho = [1000P] \div [T_{Kv}\,R] = [3.486P] \div [T_{Kv}], \text{ and } T_{Kv} = T_K[1 - 0.378(e_a)/P]^{-1}$$

where: ρ, atmospheric density [kg m^{-3}]

R, specific gas constant = 287 [J kg$^{-1\,K-1}$]

$T_{Kv,}$ virtual temperature [K]

$T_{K,}$ absolute temperature [K]: $T_K = 273.16 + T$ [°C]

$e_{a,}$ actual vapor pressure [kPa]

T, mean daily temperature for 24-hour calculation time steps.

For average conditions (e_a in the range 1–5 kPa and P between 80–100 kPa), T_{Kv} can be substituted by: $T_{Kv} \approx 1.01$ (T + 273)

4. Saturation Vapor Pressure function (e_s)

$$e_s = [0.6108]*\exp\{[17.27*T]/[T + 237.3]\}$$

where: e_s, saturation vapor pressure function [kPa]

T, air temperature [°C]

5. Slope Vapor Pressure Curve (Δ)

$$\Delta = [4098.\ e°(T)] \div [T + 237.3]^2$$
$$= 2504\{\exp[(17.27T) \div (T + 237.2)]\} \div [T + 237.3]^2$$

where: Δ, slope vapor pressure curve [kPa C⁻¹]

T, air temperature [°C]

e°(T), saturation vapor pressure at temperature T [kPa]

In 24-hour calculations, Δ is calculated using mean daily air temperature. In hourly calculations T refers to the hourly mean, T_{hr}.

6. Psychrometric Constant (γ)

$$\gamma = 10^{-3} [(C_p.P) \div (\varepsilon.\lambda)] = (0.00163) \times [P \div \lambda]$$

where: γ, psychometric constant [kPa C⁻¹]

c_p, specific heat of moist air = 1.013 [kJ kg⁻¹ °C⁻¹]

P, atmospheric pressure [kPa]: equations 2 or 4

ε, ratio molecular weight of water vapor/dry air = 0.622

λ, latent heat of vaporization [MJ kg⁻¹]

7. Dew Point Temperature (T_{dew})

When data is not available, T_{dew} can be computed from e_a by:

$$T_{dew} = [\{116.91 + 237.3Log_e(e_a)\} \div \{16.78 - Log_e(e_a)\}]$$

where: T_{dew}, dew point temperature [°C]

e_a, actual vapor pressure [kPa]

For the case of measurements with the Assmann psychrometer, T_{dew} can be calculated from:

$$T_{dew} = (112 + 0.9T_{wet})[e_a \div (e° T_{wet})]^{0.125} - [112 - 0.1T_{wet}]$$

8. Short Wave Radiation on a Clear-Sky Day (R_{so})

The calculation of R_{so} is required for computing net long wave radiation and for checking calibration of pyranometers and integrity of R_{so} data. A good approximation for R_{so} for daily and hourly periods is:

$$R_{so} = (0.75 + 2 \times 10^{-5} z) R_a$$

where: z, station elevation [m]

R_a, extraterrestrial radiation [MJ m^{-2} d^{-1}]

Equation is valid for station elevations less than 6000 m having low air turbidity. The equation was developed by linearizing Beer's radiation extinction law as a function of station elevation and assuming that the average angle of the sun above the horizon is about 50°.

For areas of high turbidity caused by pollution or airborne dust or for regions where the sun angle is significantly less than 50° so that the path length of radiation through the atmosphere is increased, an adoption of Beer's law can be employed where P is used to represent atmospheric mass:

$$\mathbf{R_{so}} = (\mathbf{R_a}) \exp[(-0.0018P) \div (\mathbf{K_t} \sin(\Phi))]$$

where: K_t, turbidity coefficient, $0 < K_t \le 1.0$ where $K_t = 1.0$ for clean air and $K_t = 1.0$ for extremely turbid, dusty or polluted air.

P, atmospheric pressure [kPa]

Φ, angle of the sun above the horizon [rad]

R_a, extraterrestrial radiation [MJ m^{-2} d^{-1}]

For hourly or shorter periods, Φ is calculated as:

$\sin \Phi = \sin \varphi \sin \delta + \cos \varphi \cos \delta \cos \omega$

where: φ, latitude [rad]

δ, solar declination [rad] (Eq. (24) in Chapter 3)

ω, solar time angle at midpoint of hourly or shorter period [rad]

For 24-hour periods, the mean daily sun angle, weighted according to R_a, can be approximated as:

$$\mathbf{sin(\Phi_{24})} = \sin[0.85 + 0.3 \; \varphi \sin\{(2\pi J/365) - 1.39\} - 0.42 \; \varphi^2]$$

where: Φ_{24}, average Φ during the daylight period, weighted according to R_a [rad]

φ, latitude [rad]

J, day in the year

The Φ_{24} variable is used to represent the average sun angle during daylight hours and has been weighted to represent integrated 24-hour transmission effects on 24-hour R_{so} by the atmosphere. Φ_{24} should be limited to ≥ 0. In some situations, the estimation for R_{so} can be improved by modifying to consider the effects of water vapor on short wave absorption, so that: $R_{so} = (K_B + K_D) R_a$ where:

$K_B = 0.98\exp[\{(-0.00146P) \div (K_t \sin \Phi)\} - 0.091\{w/\sin \Phi\}^{0.25}]$

where: K_B, the clearness index for direct beam radiation

K_D, the corresponding index for diffuse beam radiation

$K_D = 0.35–0.33\ K_B$ for $K_B \geq 0.15$

$K_D = 0.18 + 0.82\ K_B$ for $K_B < 0.15$

R_a, extraterrestrial radiation [MJ m^{-2} d^{-1}]

K_t, turbidity coefficient, $0 < K_t \leq 1.0$ where $K_t = 1.0$ for clean air and $K_t = 1.0$ for extremely turbid, dusty or polluted air.

P, atmospheric pressure [kPa]

Φ, angle of the sun above the horizon [rad]

W, perceptible water in the atmosphere [mm] $= 0.14\ e_a\ P + 2.1$

e_a, actual vapor pressure [kPa]

P, atmospheric pressure [kPa]

APPENDIX J

PSYCHROMETRIC CHART AT SEA LEVEL.

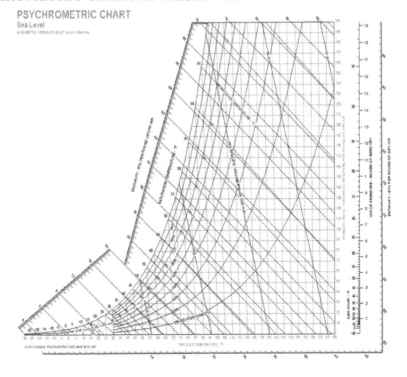

INDEX

Milton Keynes UK
Ingram Content Group UK Ltd.
UKHW031139141024
449569UK00024B/1214